SERVICES, HEATING AND EQUIPMENT FOR HOME ECONOMISTS

ELLIS HORWOOD SERIES IN FOOD SCIENCE AND TECHNOLOGY

This publishing programme provides an organised coverage on food science for professional technologists, research workers, and students in universities and polytechnics.

Series Editors:
Professor U. D. Morton, Head of Food Science, Queen Elizabeth College, University of London;
Dr. R. Scott, formerly Reading University

FUNDAMENTALS OF FOOD CHEMISTRY
W. Heimann, Director, Institute for Food Chemistry, University of Karlsruhe, Germany

HYGIENIC DESIGN AND OPERATION OF FOOD PLANT
Edited by R. Jowitt, National College of Food Technology, University of Reading

HANDBOOK OF DOMESTIC ENGINEERING
D. Kirk, Sheffield City Polytechnic and A. Milson, The Queen's College, Glasgow

PRINCIPLES OF DESIGN AND OPERATION OF CATERING EQUIPMENT
A. Milson, The Queen's College, Glasgow and D. Kirk, Sheffield City Polytechnic

TECHNOLOGY OF BISCUITS, CRACKERS AND COOKIES
J. R. Manley, Consultant, Peterborough

SENSORY QUALITY IN FOOD AND BEVERAGES
A. A. Williams and R. K. Aitken, Long Aston Research Station, Bristol

PHYSICAL PROPERTIES OF FOODS AND FOOD PROCESSING SYSTEMS
M. J. Lewis, University of Reading

FLAVOUR OF DISTILLED BEVERAGES
J. R. Piggot, University of Strathclyde

ENERGY MANAGEMENT IN FOOD SERVICE
N. Unklesbay and K. Unklesbay, University of Missouri, Columbia, Missouri, USA

THE PSYCHOBIOLOGY OF HUMAN FOOD SELECTION
L. M. Barker, Baylor University, Waco, Texas, USA

ADVANCED SUGAR CHEMISTRY
R. S. Shallenberger, Cornell University, Geneva, New York, NY, USA

SERVICES, HEATING AND EQUIPMENT FOR HOME ECONOMISTS

D. KIRK, B.Sc., A.I.F.S.T., M.Phil.
Principal Lecturer
Department of Hotel and Catering Studies and Home Economics
Sheffield City Polytechnic

and

A. MILSON, M.Sc.Tech., B.Sc., C.Eng., F.I.Gas.E., A.I.F.S.T., F.H.C.I.M.A.
Vice-Principal
The Queen's College, Glasgow

ELLIS HORWOOD LIMITED
Publishers · Chichester

First published in 1982 by

ELLIS HORWOOD LIMITED
Market Cross House, Cooper Street, Chichester, West Sussex, PO19 1EB, England

The publisher's colophon is reproduced from James Gillison's drawing of the ancient Market Cross, Chichester.

Distributors:

Australia, New Zealand, South-east Asia:
Jacaranda-Wiley Ltd., Jacaranda Press,
JOHN WILEY & SONS INC.,
G.P.O. Box 859, Brisbane, Queensland 40001, Australia

Canada:
JOHN WILEY & SONS CANADA LIMITED
22 Worcester Road, Rexdale, Ontario, Canada.

Europe, Africa:
JOHN WILEY & SONS LIMITED
Baffins Lane, Chichester, West Sussex, England.

British Library Cataloguing in Publication Data
Kirk, David
Services, heating and equipment for home economists. –
(Ellis Horwood series in food science and technology)
1. Dwellings. – Energy conservation
2. Dwellings. – Energy consumption
I. Title II. Milson, Anthony
644 TJ165.5.D86

ISBN 0-85312-134-6 (Ellis Horwood Limited – Library Edn.)
ISBN 0-85321-516-3 (Ellis Horwood Limited – Student Edn.)

Typeset in Press Roman by Ellis Horwood Ltd.
Printed in Great Britain by R. J. Acford, Chichester.

Table of Contents

Chapter 4 – Solid Fuels

Chapter 5 – Electricity

PART II – ENVIRONMENTAL CONTROL OF THE HOME

Chapter 7 – Comfort

Chapter 8 – The design of home-heating systems

Chapter 9 – Space heating equipment

Chapter 10 – Water heating

PART V – SAFETY IN THE HOME

Chapter 20 – Safety

Authors' Preface

The authors have been teaching courses on fuels and energy, domestic appliances and home heating for a number of years to students of home economics. It soon became clear that these subjects should be treated in an integrated way rather than, for example, from the points of view of individual types of fuels and the types of appliances used with that fuel. No single textbook is available which brings together all these aspects. This book aims to contribute, by its integrated approach, to the efficient running of the home and in making decisions about the choice of appliances and equipment in relation to the types of fuels and services available.

The widespread nature of the topics which must be covered to achieve an integrated approach has meant that the treatment is broad rather than deep. Nevertheless, the perspective provided will give the home economics graduate the basic knowledge necessary to communicate with experts in the various fields relating to the topics treated in the book and, at the same time, relate specialist information to the needs of the home and the family. We have included a comprehensive bibliography and list of references at the end of each chapter to help the reader find access to the more specialised material related to specific topics. The books and journals referred to should be comprehensible to the reader of this book.

The book is divided into five parts as follows:

Part I (Chapters 1 to 6): Services in the home – this gives an overview of energy sources and supplies to the home, together with a chapter on water supplies and liquid waste disposal.

Part II (Chapters 7 to 12): Environmental control of the home – this covers heating (space and water), ventilation and air conditioning together with artificial lighting.

Part III (Chapters 13 to 17): Kitchen equipment – for cooking and refrigeration together with motor driven kitchen appliances and concluding with a chapter on kitchen design.

Part IV (Chapters 18 and 19): Cleansing equipment – both laundry equipment and cleaning and polishing appliances.

Part V (Chapter 20): Safety.

We would like to thank a number of our friends and colleagues for help and advice whilst writing this book. In particular, the comments of Barbara Hanforth and Jonathan Bareham of Sheffield City Polytechnic and George Wilson of Manchester Polytechnic have been particularly valuable. We are also grateful to our long-suffering wives and families for providing encouragement and support throughout its preparation.

Acknowledgements

We acknowledge permissions to use material from a number of sources:

British Petroleum for Fig. 3.2 from *The storage and handling of domestic fuels,* 1973.

British Gas for Fig. 9.13 from *A students guide to gas catering equipment*.

Electrical Association for Women for Figs. 5.4, 9.2, 13.12, 15.7 and 18.9 from *The EAW electrical handbook*, English University Press, 1971.

Humidifier Advisory Service for Fig. 11.5 from their publication *Humidification for health, comfort and increased productivity*.

Institution of Gas Engineers for Figs. 1.3, 1.4 and 10.7 from *Future energy resources* by G. Madeley, Gas Engineering and Management, 1978, 18, p. 262–274.

Longmans Group Ltd. for Figs. 6.1, 6.4, 6.5, 6.6, 10.2, 10.4, 10.10, 12.8 and 12.17 from *Building services and equipment*, volumes 1 and 2, by F. Hall, 1976 (Longmans Technician Series) and for Figs. 9.3, 10.9 and 15.4 from *Heating, ventilating and air-conditioning,* by F. Hall, 1980 (Construction Press).

McGraw-Hill Book Company Inc. for Figs. 2.4 and 13.14 from *Refrigeration, air-conditioning, range and oven servicing,* by R. Scharff, 1976.

Thorn–EMI Lighting Ltd for Fig. 12.12 from the Thorn Lighting Catalogue.

Women's Gas Federation and Young Homemakers for Figs. 2.9, 2.10, 2.11, 9.4 and 9.8 from *Gas and its domestic uses,* by E. M. Plant, 1971.

Women's Solid Fuel Council for Fig. 13.4 from *Solid fuel in the home,* by K. V. Bradbury, 1973.

Part I

SERVICES IN THE HOME

CHAPTER 1

Present and future energy supplies

1.1 INTRODUCTION

The topics of energy and fuel supplies are frequently making the news headlines. Predictions of acute shortages in the future are followed by the announcements of the discovery of new oil and gas wells and of the development of new types of nuclear energy. Any predictions of the future situation are of a speculative nature and are likely to contain many errors because of unseen events. A detailed study of the predictions of future fuel and energy supplies is beyond the scope of this book. The objectives of this chapter are to provide a basic understanding of what is meant by the terms fuel and energy, how these relate to the domestic market, and to give some idea of the ways in which the domestic fuel market may endure in the near future.

It is obvious that fuel supplies are of considerable importance to the industrialised countries of the world. Fuel is required to provide the energy for industrial processes – boilers, furnaces, electrical generation, and energy for transport, an integral part of industrial production systems. But how significant a part of the total consumption is the use of energy in the home? In the UK approximately one quarter of energy consumed by final users in 1972 was attributed to home use (Watt Committee 1977). A similar figure for the USA in 1969 was 21% (Cook 1976). These figures show that the consumption of energy in the home forms a significant part of total energy consumption, and it follows that factors affecting world energy supplies will have an impact on the use of energy in the home and, conversely, factors affecting use in the home, such as higher standards of insulation, will make a significant contribution to the conservation of national and world energy supplies.

1.2 SOURCES OF ENERGY

So far the terms **energy** and **fuel** have been used without defining their exact meaning. Before considering any further the domestic use of fuels and energy it is necessary to define what is meant by these terms.

In order to move a chest of drawers from one place in a room to another we need to perform an amount of work, and to do this it is necessary to supply energy. In this case the energy is supplied by human effort. Anything which is capable of performing work possesses energy: this energy can exist in a number of different forms, e.g. mechanical, atomic, electrical, electromagnetic, thermal, and chemical energy.

1.2.1 Mechanical energy

This can be in a number of forms: resulting from the movement of a material (e.g. wind or flowing water), or because a body of material is at a high level (e.g. a lake or reservoir of water at a height above sea level), or because of the tension in a spring. The four forms of mechanical energy are known as:

(1) kinetic energy (e.g. energy in a moving body)
(2) potential energy (e.g. energy in a body because of its height)
(3) strain energy (e.g. energy in a spring)
(4) pressure energy (e.g. energy in a pressurised gas or liquid).

1.2.2 Nuclear energy

This is contained in the nucleus of an atom. Energy can be liberated from an atom either by splitting up parts of the atom (fission) or by combining atoms together (fusion).

1.2.3 Electrical energy

This is associated with the distribution of electrical charges. It is closely related to electromagnetic energy, which is the form of energy found in radiation, such as the radiation from the sun, light rays, and radio waves.

1.2.4 Heat or thermal energy

This is related to the vibration of molecules. Heat energy is commonly associated with an increase in temperature, but it also causes changes of state, from solid to liquid and from liquid to gas.

1.2.5 Chemical energy

This is *stored* in a molecule because of the intramolecular forces which hold molecules together. During chemical reactions there is often an absorption or release of energy. Of particular interest here is the reaction between carbon-rich chemicals and oxygen which results in the release of large amounts of energy, usually in the form of heat. Where the chemical is used specifically for the generation of heat it is known as a **fuel**, and the reaction is known as **combustion**.

Normally the term fuel refers to a material which can be burned to provide heat. There are exceptions to this, however, particularly in colloquial speech, when electricity is often classified as a fuel.

1.2.6 Combustion

Using the more restricted definition of a fuel as a material which can be burned to provide heat, fuels are invariably carbon-rich chemicals which originate from living organisms. The most abundant source of fuels are the **fossil fuels** such as coal, petroleum, and natural gas (see Chapters 2, 3 and 4). These are composed of hydrocarbons (compounds of carbon and hydrogen), and when burned in an atmosphere containing an excess of oxygen they are converted into carbon dioxide and water. As an example the combustion of methane (CH_4), the major component of natural gas, can be represented by the equation:

$$CH_4 + 2O_2 \rightarrow CO_2 + 2H_2O + \text{heat} \ .$$

The combustion of fuels is considered in more detail in the next three chapters.

1.3 UNITS OF ENERGY

Energy can be converted from one form to another. When comparing amounts of energy in different forms, for example the amount of energy used by a gas fire and that used by an electric fire, it is convenient to be able to express these amounts of energy in the same units. Energy is the ability to do work and, because of this, it has the same units as work. In the international (SI) system of units, which will be used throughout this book, the unit of both work and energy is the **joule** (J). In the case of thermal energy, a temperature rise rather than an ability to do work is used as the basis of energy measurement. To measure this in joules it is necessary to know the heat capacity of the material whose temperature is increasing, together with its mass:

$$\underset{\text{(Joules)}}{\text{thermal energy}} = \underset{\text{(kg)}}{\text{mass of material}} \times \underset{\text{(J/kg °C)}}{\text{heat capacity}} \times \underset{\text{(°C)}}{\text{temperature rise}} \ .$$

For many energy applications, the joule is a rather small unit, and for this reason multiples of the basic unit are used:

$$1 \text{ kJ} \equiv 1000 \text{ J} \qquad (\text{k} = \text{kilo})$$

$$1 \text{ MJ} \equiv 1000 \text{ kJ} \equiv 1\,000\,000 \text{ J} \quad (\text{M} = \text{Mega}) \ .$$

The term **power** is used to describe the *rate* at which energy is being used; it is therefore derived from the unit of energy divided by the unit of time. In the SI system the unit of power is the **watt** (W), which is equivalent to a consumption of 1 joule every second (1 J/s). High levels of power are expressed in thousands (kW) or millions (MW) of watts.

1.4 ENERGY CONVERSIONS

When discussing combustion we saw that chemical energy could be converted into thermal energy. It is very common for energy to change from one form to another. Some of these conversions take place naturally, such as the conversion of energy (electromagnetic) in the sun's radiation into heat when it is absorbed by a substance. Many are associated with living organisms, e.g. the conversion of sunlight (electromagnetic energy) into chemical energy via photosynthesis in the leaf of a plant, or the conversion of chemical energy into mechanical energy in the muscle of an animal. Other conversions are man-made, such as the conversion of chemical energy into electrical energy in a power station. All of these conversion processes obey the law of the conservation of energy, that is the total amount of energy must stay the same – *energy can neither be created nor destroyed*.

When considering energy conversion, the **efficiency** of the process is often measured. This is the ratio of the useful energy obtained from the process to the total energy input into the process. A gas-fired water heater, for example, may be described as having an efficiency of 80%. This means that $\frac{4}{5}$ of the energy (chemical) in the gas is converted into heat (resulting in a rise of temperature of the water) and that $\frac{1}{5}$ of the enegy is lost from the boiler, in the form of heat in the hot combustion gases (which go out of the flue) and in other heat losses. Considering the overall process from the point of view of the conservation of energy it is clear that the chemical energy contained in the unburned gas must be the same as the heat in the water together with the heat lost from the heater and the heat in the combustion products. The total amount of energy cannot increase or decrease but must remain constant. This being so, why then is there an energy crisis? The answer to this question is complex, but it may be simplified by stating that some forms of energy are more readily available than others, and the natural trend of energy conversion processes is towards the least efficiently reuseable form of energy – heat.

Many energy conversions have high efficiencies, e.g. the chemical energy in gas is all converted into thermal energy in a gas flame. When thermal energy is converted into work, however, the efficiency is very low, and for theoretical reasons, cannot be increased above a certain value. The energy which is wasted in the conversion process is always in the form of heat. This heat is usually at a temperature nearer to the temperature of the surroundings than to that of the original source of heat. This heat can be recovered, but only by supplying additional energy to the process. Heat becomes of less use to man when its temperature falls to near that of the surroundings, because its energy becomes less easily available. It is often referred to as **low-quality** energy.

The conversion of one form of energy to another always results in a loss of energy. The nearest we can get to an energy conversion without loss (i.e. 100% efficient) would be the use of an electrical immersion heater converting electrical energy into thermal energy. However, to consider one stage in the total energy conversion process can be misleading. Fig. 1.1 shows two alternative processes for the supply of hot water from natural gas.

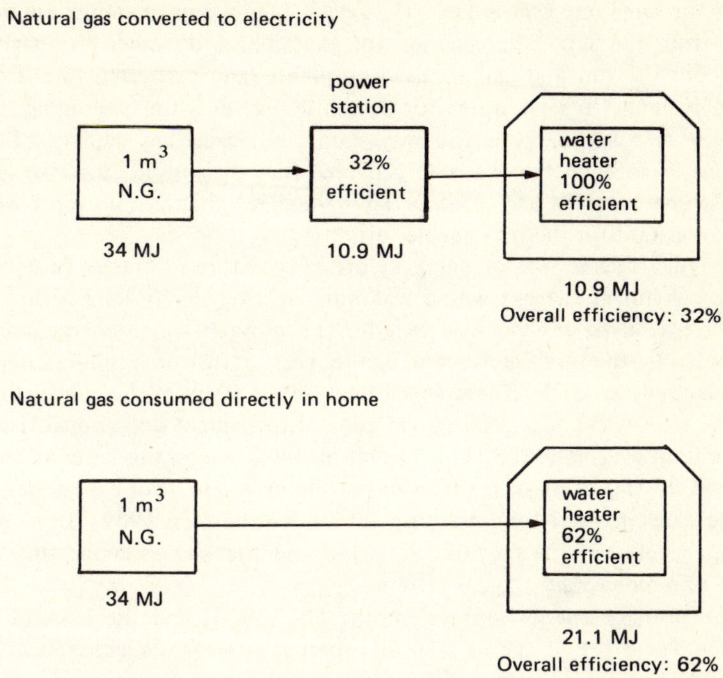

Fig. 1.1 – Energy conversion processes.

The reason for the low conversion efficiency from chemical energy (in this example from the gas) to electricity is discussed in more detail in Chapter 5.

1.5 PRIMARY AND SECONDARY FUELS

Fuels can be classified into one of two types: **primary** and **secondary** fuels. A primary fuel is used directly without an energy conversion process (although it may have undergone some preparation), e.g. coal and natural gas. A secondary fuel is a fuel which during its processing undergoes an energy conversion process. All forms of electricity, for example, are considered to be secondary fuel. When electricity is generated from coal, the chemical energy of the coal is converted first into thermal energy, then into mechanical energy, and finally into electrical energy (electricity).

1.6 RENEWABLE AND NON-RENEWABLE SOURCES OF ENERGY

An alternative way of classifying sources of energy is to classify them as either **renewable** or **non-renewable.** A renewable source of energy is one which is replaced at the same rate as it is used. The renewable sources of energy are based on energy from the sun. Solar energy, for example, is direct electromagnetic radiation from the sun and can be used by plants (and converted to chemical energy) or by man (to heat water for domestic use, to warm swimming pools etc.). The use of solar energy in this way should not cause any depletion in the total amount of solar energy available provided that the output from the sun is considered to be constant and infinite. Thus there is a theoretical supply which will remain constant for the foreseeable future.

A non-renewable source of energy represents a store of energy (usually in the form of chemical energy) which was built up over a very long period and cannot be replaced to any realistic extent. The non-renewable energy sources are represented by the fossil fuels: coal, lignite, peat, petroleum products, natural gas, oil shale, and tar sands. These fossil fuels are the result of the conversion of solar radiation into chemical energy (in the form of plant and animal tissues) over a 50-million year period. It is likely that man will use up this store of energy over a period of 150 years in the case of petroleum products or over about 800 years in the case of coal (Committee on Resources and Man 1969). Because of this inbalance between the rate of formation and the rate of usage they are, effectively, non-renewable.

Current primary energy sources for the UK (1974) and the USA (1970) are shown in Table 1.1. It can be seen that there is at present a large reliance on non-renewable sources of energy. The reason for this is the relative ease of use of these energy sources.

Table 1.1
Primary energy sources (Percentage of total used)

	UK (1974)	USA (1970)
Nuclear	3.0	0.3
Hydro	0.5	4.3
Natural gas	14.3	38.8
Petroleum	51.7	38.1
Coal	30.3	21.5

Sources

USA data: Berthouex & Rudd (1977).
UK data: Department of Energy (1974).

1.7 THE CURRENT DOMESTIC ENERGY MARKET

In section 1.1 it was seen that 20–25% of primary energy supplies are used in the home. The source of energy is oil, solid fuel, natural gas, and electricity. Table 1.2 shows a breakdown of the UK domestic energy market (Department of Energy 1981) in terms of these fuels.

Table 1.2
Energy supplied to the domestic sector 1980
(millions of therms)

		%
Solid fuel	3313	(20.9)
Gas	8439	(53.5)
Oil	1093	(6.9)
Electricity	2939	(18.7)
Total	15784	

It must be remembered that this table represents a static picture. Over the past hundred years the balance between energy sources has been constantly changing. Most noticeable has been the decline in the use of solid fuels and the increasing proportion of gas and electricity.

This energy is used for a variety of purposes – heating, cooking, lighting, etc. Figures on the relative usage of energy for domestic activities are given in Table 1.3 for the US and for the UK. These are only estimates and represent average values. The usages of individual consumers obviously will vary a great deal depending on many factors including income, life style, and size of family.

Table 1.3
Use of energy in the home
(Percentages of domestic consumption)

	US	UK
Space heating	58	63
Water heating	15	21
Air conditioning	4	–
Cooking	6	9
Other uses	17	7

Sources
US data based on the year 1968 (Stamford Research Institute 1972).
UK data based on the year 1975 (Brundrett *et al.* 1977).

The consumption of non-renewable fossil fuels has increased at an alarming rate over the last century. This has been associated with higher standards of living, mostly in the Western world, and the increasing use of oil-based fuels for transport. This has meant that Governments have had to develop energy policies to handle the changes which will be required in the types and amount of fuel used (Francis & Peters 1980).

1.8 FUELS FOR THE FUTURE

It is impossible to be prescriptive about changes which will occur in the supplies of fuels for the future. There are many unknown factors which make predictions, based on current known reserves of fuel and extrapolations of historic fuel-use data, prone to error. The purpose of this section is to give some indication of likely changes in the supply of fossil fuels, which currently provide about 80% of the energy used in the World, and to discuss the likely contribution of some alternative energy sources to these resources.

1.8.1 Fossil fuels

Fossil fuels represent the most important group of non-renewable sources of energy (see section 1.6). For the year 1971 they represented about 80% of the World's consumption of energy, of which 35% was crude oil, 26% coal and lignite and 17% natural gas. (Cook 1976); These are global figures; different countries use a different fuel mix, dependent on indigenous energy resources and on the life style of its inhabitants. Per capita annual energy consumption in the USA is about 900 MJ, an average value for Western Europe is 440 MJ, and for underdeveloped countries 30 MJ.

Details of the extraction, preparation, distribution and use within the home of the fossil fuels are given in Chapter 2 (gaseous fuels), Chapter 3 (liquid fuels), and Chapter 4 (solid fuels); this section is mainly concerned with an estimate of the expected production cycles of these fuels. Fossil fuels are normally classified into one of three groups (solid, liquid, or gas), but in fact there exists a whole spectrum of products which have properties between the three groups. For example, liquefied petroleum gases are transported as a liquid but burned as a gas. Similarly tar is intermediate in properties between heavy (viscous) crude oils and bituminous coal.

A feature of non-renewable fuels is that their extraction follows a production cycle as shown in Fig. 1.2 (Committee on Resources and Man 1969). The characteristic curve shows an initial slow growth in the use of the fuel, leading to a period of exponential growth, as the technology for burning the fuel is improved and as more sources of the fuel are found. Once the rate of extraction is greater than the rate at which new sources of the fuel are discovered, production reaches a peak before going into decline. The area under the graph represents the total (cumulative) production of that fuel. In reality, because of a number of

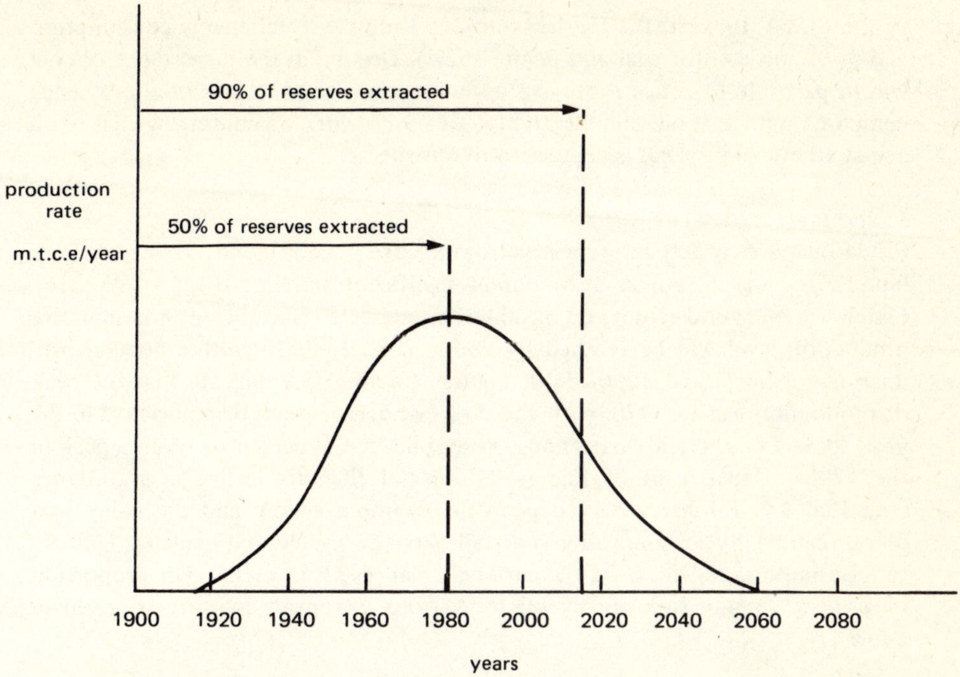

Fig. 1.2 – Production cycle for a non-renewable fuel (e.g. liquid fuel).

factors, it is unlikely that any fossil fuel will follow such an idealised curve. Once a fuel becomes scarce, economic factors come into play, which mean that sources which were once considered to be uneconomic because of mining or transport costs may become a viable proposition. Currently, oil wells and mines are closed at a stage when there are still considerable quantities of fuel left underground. At some time in the future, when the costs of fossil fuels have increased, it may be economic to reopen these mines or oil wells to extract these residual quantities of fuels.

Solid fuels

Current best estimates for world supplies of solid fuel indicate that the cumulative production will reach the 90% value by the year 2380 with US reserves reaching the same level by 2440 (Committee on Resources and Man 1969). The UK has an estimated 45 000 million tons of recoverable coal, which is sufficient to provide a significant contribution to UK energy needs for several centuries (Department of Energy 1976). Because of the relative abundance of solid fuels (reserves of solid fuels, excluding tar sands and shale, are ten times greater than oil reserves), it is likely that the decline in the relative importance of coal may be reversed. In 1969 the World's energy consumption was estimated as 6500 million tons of coal equivalent (m.t.c.e.) of which 2300 m.t.c.e was coal and lignite (35%). In 1950 the figures were 2607 m.t.c.e. total energy consumption and 1605 m.t.c.e. for coal and lignite (62%). Decline in the importance of coal was in part due to concern over air pollution and also to its lack of convenience compared with gas, oil, and electricity. Research work on converting coal to oil and synthetic natural gas is being actively pursued.

Liquid fuels

Crude oil varies widely in properties from tar-like viscous liquids to low-viscosity liquids, and many sources also contain significant amounts of dissolved gases. Estimates of world crude oil production indicate that the 90% cumulative production level will be reached by about 2020 to 2030, with a peak of production in the period 1990–2000. Figures for the USA indicate that the peak of production was in 1970, with the 90% cumulative point being reached in the year 2000. For the UK, indigenous oil supplies are expected to reach a peak in the 1990s (Department of Energy 1976) and then to decline in accordance with Fig. 1.2. Europe is very dependent on imported oil, and the rising cost of oil, caused by an increasing shortfall between supply and demand, is likely to have important effects on the national balances of payments. The proportion of primary energy supplied by oil for various European countries is given in Table 1.4.

Table 1.4
Proportion of primary energy supplied by oil in five European countries

Italy	79%
France	70%
Holland	58%
Germany	58%
UK	52%

Source: Hawthorne (1975).

Natural gas

Predictions of natural gas production for the USA (Committee on Resources and Man 1969) indicate a peak production in 1980 with the 90% level being reached by about 2020. UK estimates are for a peak production to be achieved between 1985 and 1990 (Madeley 1978). Consumption of natural gas depends largely on the availability of indigenous supplies. In Holland natural gas contributes 38% of the total energy consumption, 33% in the USA, 13% in the UK, and 1% in Japan. Proved reserves of natural gas throughout the World are given by Valais & Heigel (1977).

Other fossil fuels

Throughout the World there are large deposits of tar sands. This is sand impregnated with viscous crude oil which cannot be extracted easily. A technology is being developed which will facilitate the extraction of this crude oil. The tar sand deposits of Canada represent the equivalent of 10% of the total World crude oil supplies, and it is likely that these will be a significant source of fuel in the future. Oil shale, where the oil is so viscous as to form a solid, represents an even greater problem in terms of extraction. There are significant deposits in Africa, Asia, Europe, N. America, and S. America (Committee on Resources and Man 1969), although, based on current technology, many of these are marginally economic in terms of use. However, with the increasing cost of other fossil fuels it is likely that they will make a significant contribution by the year 2000.

1.8.2 Nuclear energy

Nuclear energy contributes at present only a small proportion of the World's energy supply, although it has been predicted that nuclear energy will supersede coal as the main energy supply to the USA electrical power generating industry (Reay 1977). There are several ways in which the energy from the nucleus of an atom can be released. All of these methods have in common the fact that this energy is released in the form of heat, which is then used to generate electricity. Details of the process of electricity generation from nuclear energy are given in Chapter 5.

Present nuclear power stations use **fission** (the breaking up of nuclei) to produce the release of energy, and the fuel used is *uranium*. This in its natural state is composed of three isotopes, uranium 234 (less than 0.1%), uranium 235 (0.7%), and uranium 238 (99.2%). For the generation of nuclear energy the important isotope is uranium 235, since this is **fissile**, which means that it can be made to break down readily to give a variety of fission products, neutrons and large amounts of heat. One gram of uranium releases as much energy as 2.7 tonnes of coal. The reaction is carried out in a **burner reactor.**

It is possible to convert uranium 238 (the most abundant isotope) into plutonium 239 which, because it is fissile, can be used as an energy source. Similarly, thorium 232 can be converted into uranium 233, which is also fissile. These are the types of reactions which occur in a **breeder reactor** (fast reactor).

The future for burner reactors is closely linked to the availability and cost of uranium ores. Known reserves of uranium ores are sufficient to support existing and planned reactors for the next 10 to 20 years (Department of Energy 1976, Committee on Resources and Man 1969). As supplies become scarce, the cost will increase compared with coal-fired power stations. Breeder reactors are still at a development stage and are not likely to be used in commercial situations until 1990 (Cook 1976). These reactors should use uranium 60 times more efficiently than burner reactors. The 250 MW prototype fast reactor at Dounreay, Scotland, had generated a significant amount of electricity by 1975 (Department of Energy 1976).

There is a possible alternative to nuclear fission as a means of releasing nuclear energy, this process is known as **nuclear fusion**. The controlled use of fusion reactions is still at an early developmental stage, and the research is highly speculative. It may prove to be impossible to produce the required conditions for the generation of electricity in power stations.

Nuclear fusion is the process responsible for the energy output of the sun. The reaction involves the fusion of the nuclei of light elements to produce heavier ones. In its normal form, the nucleus of hydrogen contains a proton only. Isotopes of hydrogen containing 1 proton and 1 neutron (deuterium) and 1 proton and 2 neutrons (tritium) are known. In all water supplies, including sea-water, there exists a ratio of 1 deuterium atom for every 6500 hydrogen atoms. The fusion reactions which are of most interest are those involving the fusion of two deuterium atoms. An alternative process involves the use of lithium as a means of generating tritium, which can then be fused with deuterium, giving helium.

Looking first of all at the availability of dueterium, 1 cubic kilometre of sea-water contains the equivalent deuterium to 269 billion† tonnes of coal (Committee on Resources and Man 1969). Thus the sea represents a vast reserve of energy, if the technology of the process can be solved. Some idea of the magnitude of the energy reserves in the sea is given by the fact that a reduction in level of deuterium in the sea by 1% would be equivalent to 500 000 times the energy content of the total fossil fuel supplies of the World (including supplies already consumed and known reserves). As far as the reactions involving lithium are concerned, scarcity of lithium makes this a less plentiful supply of energy. In the sun matter exists at temperatures of 100 million °C or more, in the form of a **plasma**. In the sun the plasma is held together by gravitational forces. Terrestrial research into alternative ways of holding this plasma together has been concentrated on the use of magnetic fields.

Research is continuing in the USSR, Europe, and the USA on developing this technology, but the process is still a long way from the prototype reactor stage. The research is very costly and also very speculative, but the rewards for success are very high (Department of Energy 1976). An additional advantage of the fusion process over fission is that it will not result in any long-life radioactive wastes to be stored or processed.

1.8.3 Solar energy

Very large amounts of energy arrive at the surface of the earth from the sun in the form of **solar energy**. Some of this energy is converted into chemical energy via the photosynthetic processes of plants: of this a proportion is used by

† USA billion (10^9).

man as fuel, both in the form of food and for combustion to give heat. The amount of solar energy used in this way is only a minute proportion of the total solar energy falling on the earth's surface. When considered as an energy source, solar energy can be utilised in a number of ways (Madeley 1978):

(a) conversion to electricity by solar cells
(b) space and/or water heating via solar collectors
(c) photosynthetic processes
(d) as an energy source for biomass growth (growth of plants to give chemical energy for use as fuel – Pankhurst 1980).
(e) conversion to hydrogen (see 1.8.11).

Many processes within this classification are the subject of research and may, in the future, provide significant amounts of energy at an economic cost. As far as domestic utilisation is concerned, the main application is the use of solar energy to heat water for domestic use and as a part of space-heating systems. Without substantial improvements in conversion efficiencies and costs it is unlikely that electricity from solar cells will be used for electricity generation. Schemes have been proposed for the use of orbiting satellite solar collectors, which could transmit energy down to the earth by microwave energy, as a means of overcoming fluctuations in supply caused by cloud.

The feasibility of using solar energy for water and space heating at a given geographical location depends on (a) the level of solar energy, (b) the match between seasonal variations in available solar energy, and (c) the demand by members of the household for hot water and space heating.

Fig. 1.3 gives a comparison of mean summer and winter levels of radiation in a number of countries. It must be remembered that within countries such as the USA there are large variations. As for the match between supply and demand,

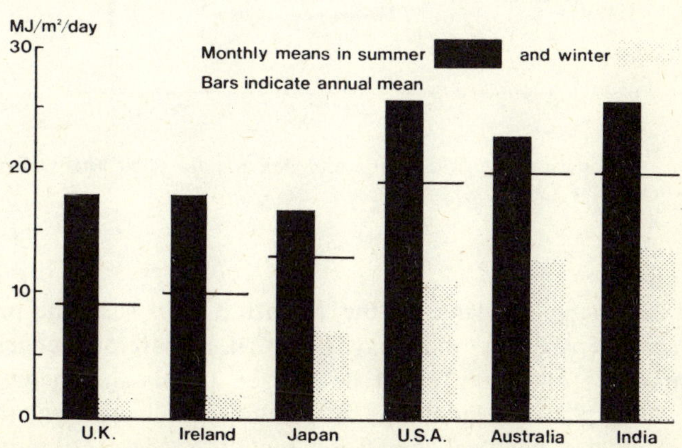

Fig. 1.3 – Comparison of available solar energy in several countries (Madeley 1978).

Fig. 1.4 indicates that, for space heating applications at least, there is a poor match. This can lead to design problems. For further discussion of solar energy see sections 9.7 and 10.3.3.

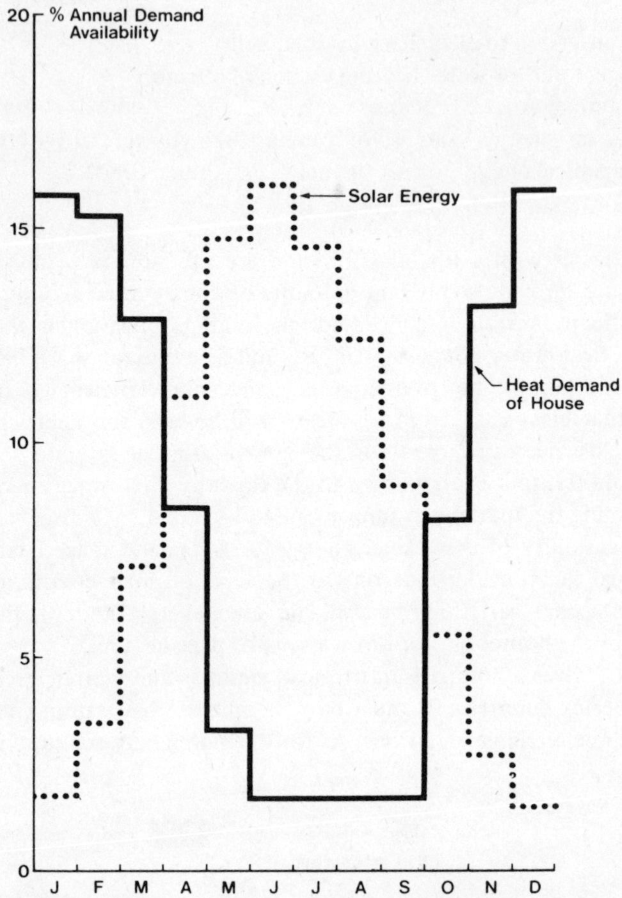

Fig. 1.4 – Comparison of availability and demand for solar energy for space heating – (Madeley 1978).

1.8.4 Wave energy

The theoretical energy available in the Atlantic Ocean is of the order of 70 kW/m of wave (Department of Energy 1976). It is therefore conceivable that significant amounts of energy could be extracted from the energy in waves. Research is taking place on the means by which the mechanical energy of waves can be converted into electrical energy (see section 5.2.5). Until more information on this is available it is difficult to establish the cost-effectiveness of wave energy. There is a reasonable match between the demand for electricity and the availability of wave energy (see Fig. 1.5). Problems are likely to arise, however, because of short-term fluctuations which would make control of the electrical supply difficult. It has been estimated that by the year 2025 wave power could be providing the UK with 50 m.t.c.e. of energy (Department of Energy 1976).

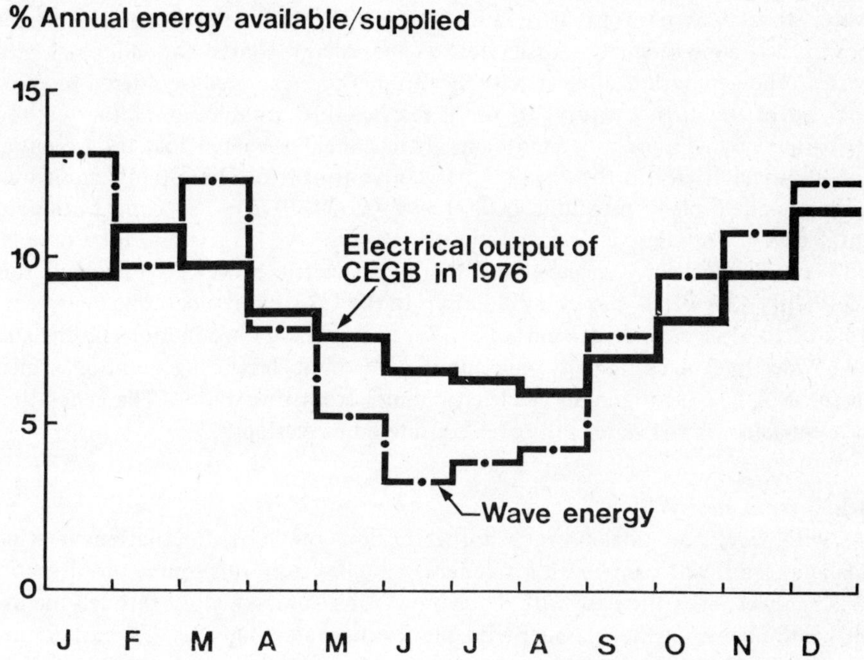

Fig. 1.5 – Comparison of wave energy availability with demand for electricity – (Madeley 1978).

1.8.5 Tidal Power

The difference in the height of water between high and low tides involves considerable amounts of potential energy. This energy could be used to provide electricity by the use of tidal barrages and water turbines (see section 5.2.5). A scheme based on the Severn estuary, England, has been proposed; but such a scheme would involve high capital costs and considerable environmental problems (Madeley 1978). Pilot schemes have been built in the Rance estuary (France) and in Russia. It has been estimated that the Severn estuary could provide 1.7 MW, but this is small when compared with a conventional power station. The

Bay of Fundy (Canada) provides a number of suitable sites which could provide up to 30 MW. Estimates of various sites in France have a potential of 11 MW, and estimates for sites in Russia are put at 16 MW (Committee on Resources and Man 1969). However, there are a limited number of suitable sites.

1.8.6 Hydroelectric Power

The uses of water power can be traced back to the Middle Ages, when the water-wheel was used to drive mills. During the Industrial Revolution water power was superseded by fossil fuels as the energy source for industrial processes. The generation of electricity by this method was first considered towards the end of the 19th Century. Its use is restricted to suitable sites, either natural or formed by strategically sited dams. It has been estimated that the potential for hydroelectricity in the North American continent is 313 000 MW compared with installed plant providing 60 000 MW (Cook 1976). The contribution to total energy supplies varies from country to country. In Norway, for example, 99% of all electricity is generated by hydroelectric power stations, and this represents 42% of all energy production. In the USA hydroelectricity represents 16% of total electricity generation or 1% of total energy production. In England and Wales hydroelectricity is only about 0.1% of all electrical generation, whilst in the North of Scotland all electricity comes from this source. The generation of electricity via the water turbine is considered in section 5.2.3.

1.8.7 Wind energy

As with wave and tidal power, short-term and long-term fluctuations in wind make a significant contribution to energy supplies from this source problematic. Wind power, as is the case with water power, has a long history through the use of windmills. A further parallel with the use of water power is that current and future applications of wind power involve the generation of electricity. Two approaches to this have been considered: small localised generators and large generators connected into regional and national distribution systems. The Energy Technology Support Unit (Department of Energy 1976) have suggested a possible supply of 8 m.t.c.e. for the UK, based on 10 000 1 MW generators. Because of the siting of these, they are likely to cause severe environmental problems. One suggestion for overcoming this is to site the generators off-shore. The design of small localised wind-powered generators has improved considerably (Hickok *et al.* 1975), although the contribution of these to national supplies is likely to be small.

1.8.8 Geothermal energy

Geothermal energy consists of heat trapped in pockets within the layers of the earth's surface, often in the form of hot water or steam, but more generally as hot rocks at a greater depth. At present it is only the localised hot water and steam supplies which have been exploited as an energy source. Steam from geothermal source has been used for electricity generation in a number of countries – Italy, Japan, New Zealand, and the USA. In the case of deeper rocks a method of generating steam by drilling two parallel wells has been developed. Water is pumped down one well and hot water or steam is recovered from the other. Set against the output of a power station the quantities of electricity are likely to be small.

1.8.9 Heat Pumps

In section 1.4 the problem of wasted energy in the form of low-quality thermal energy was discussed. There is, however, one method available to upgrade this low-quality heat, to make it suitable for space and water heating. The machine for carrying out this conversion is known as a heat pump (Sumner 1976). Heat pumps and their applications to space heating are covered in section 9.6.

1.8.10 Energy from waste

Large volumes of waste material are produced from homes, from agriculture, and from industry, and much of this material is conbustible. Currently, only a small proportion of this energy is recovered (Anderson & Tillman 1977). If this energy is to be recovered it is more likely to be on an industrial scale for process heating or for district heating (see section 9.5) rather than for domestic use.

1.8.11 Hydrogen

Energy is difficult to store and transport, but one practical method of doing this is to convert it to hydrogen which can then be liquefied for storage and distribution. Heat can be converted into hydrogen by the use of thermochemical reactions (the heat could come from sources such as solar energy). Electricity can also be converted into hydrogen via the electrolysis of water (World Energy Conference 1978).

1.9 ENERGY CONSERVATION

As the availability of abundant supplies of fossil fuels declines, accompanied by an increase in the cost of all types of fuels, the efficient and economic use of energy will become increasingly important. Table 1.5 summarises some of the areas where energy conservation can be achieved. It was shown in section 1.4 that in reality energy is not lost from a system but that it is converted from one form to another. Because of this, energy conservation is concerned therefore with maintaining the quality of heat and preventing it from being dissipated as heat at a temperature close to that of its surroundings. It is also important to maximise the efficiency of energy conversion processes, since any energy lost will be converted into low-quality thermal energy.

Table 1.5

Energy losses connected with the domestic use of fuels

Location of loss	Cause of loss
1. Energy losses during manufacture and distribution	1.1 Process inefficiency 1.2 Distribution losses
2. Energy losses in the home	
2.1 Losses from equipment	2.1.1 Equipment inefficiency 2.1.2 Incorrect equipment selection 2.1.3 Incorrect equipment usage
2.2 Energy losses from fabric of home	2.2.1 Standard of construction 2.2.2 Standard of maintenance 2.2.3 Poor heating system design 2.2.4 Incorrect heating system usage

In considering energy conservation measures it must be remembered that over one half of energy used in the home is for space heating (Table 1.3). For energy conservation measures to be cost effective initially they should be concentrated on those aspects of the home which use large amounts of energy, i.e. space and water heating. An important component of energy conservation is the education of consumers, through the development of 'energy literacy' (Mathews 1978). This must be the role of government, fuel (utility) companies, and independent consumer organisations. Conservation must be seen as an overall policy, with consideration being given to fuel efficiency from source to final consumer. The picture seen by the final consumer is often distorted when compared with the overall situation, but it is the consumer who makes important decisions on the selection and utilisation of fuels and equipment. As an example a gas water boiler has a lower efficiency than an electric one and therefore may seem less desirable from an energy conservation standpoint. However, when the efficiency of electricity generation is taken into account the position changes: electricity has a lower overall efficiency than gas for water heating. This information can be given to consumers through the structure of energy tariffs. Traditionally an energy tariff was based solely on recovering the operating costs for providing that energy, including the cost of raw materials, production, distribution, marketing, research, development, and investment in new plant and machinery. Increasingly, however, governments are interested in controlling the relative costs of alternative energy supplies to reflect the national interest.

REFERENCES

Anderson, L. L. & Tillman, D. A. (1977), *Fuels from waste,* Academic Press.

Berthouex & Rudd (1977), *Strategy of pollution control,* Wiley.

Brundrett, G. W., Leach, S. J., Parkinson, M. J., Pickup, G. A. & Rees, N. T. (1977). Research into energy conservation in dwellings, *Gas Engineering and Management,* p. 430-441.

Committee on Resources and Man (1969), Freeman.

Cook, E. (1976), *Man, energy and society,* Freeman.

Department of Energy (1974), Energy Flow Chart, HMSO.

Department of Energy, (1976), Energy research and development in the UK, Energy Paper No. 11, HMSO.

Francis, W. & Peters, M. C. (1980), *Fuels and fuel technology,* Pergamon.

Hawthorne, W. R. (1975), Energy a renewed challenge to engineers, *Proc. Inst. Mech. Eng.,*/Dec., p. 189.

Hickok, F. (1975), *Handbook of solar and wind energy,* Cahners.

Madeley, G. D. (1978), Future energy resources, *Gas Engineering and Management,* **18**, p. 262-274.

Mathews, W. (1978), *Journal of Consumer Studies and Home Economics,* **2**, p. 99-118.

Department of Energy (1981), *Digest of Energy Statistics,* HMSO.

Pankhurst, E. S. (1980), Biogas, *Gas Engineering and Management,* **January**, p. 3-13.

Reay, D. A. (1977), *Industrial Energy Conservation,* Pergamon.

Stamford Research Institute (1972), *Patterns of energy consumption in the US.*

Sumner, J. A. (1976), *Domestic heat pumps,* Prism.

Valais, M. & Heigel, M. (1977), *The Gas Industry World,* Graham & Trotman.

Watt Committee on Energy (1977), Energy research and development in the UK, Report No. 1.

World Energy Conference (1978), *World energy – looking forward to 2000,* IPC Science and Technology Press.

BIBLIOGRAPHY

American Institute of Physics (1975), *AIP Conference Proceedings No. 25,* Chapter 3, Energy Indoors in Efficient Use of Energy.

Brandwein, P. (1968), *Energy, its forms and changes,* Harcourt, Brace and World Inc.

Davis, A. J. & Schubert, R. D. (1974), *Alternative energy sources in building design,* Van Nostrand.

Folley, G. (1976), *The Energy Question,* Penguin.

Ling, G. (1975), *Solar energy, Department of Energy, Paper No. 6,* HMSO.

Reay, D. A. & MacMichael, D. B. A. (1979), *Heat Pumps; Design and Application,* Pergamon.

Sabady, P. (1978), *The Solar House,* Newnes-Butterworth.

Schoen, R. S., Hurshberg, A. S. & Weingart, J. M. (1975), *New Energy Technologies for Buildings,* Ballinger Publishing Co.

Summonds, C. I. (1974), *The Ecology of Natural Resources,* Edward Arnold.

Szokolay, S. V. (1975), *Solar Energy and Buildings,* Architrctural Press.

Watt Committee (1981), *Assessment of Energy Resources, Report No. 9.*

CHAPTER 2

Gaseous fuels

2.1 INTRODUCTION

There are three common types of gaseous fuel currently in use in the home. These are natural gas (NG), liquefied petroleum gas (LPG) and manufactured gas. Of these, NG is now the most important source in most countries. However, since the availability of these supplies beyond the turn of the century is in doubt, researchers are developing processes for the manufacture of substitute natural gas (SNG) which will have the same characteristics as NG but which will be manufactured from oil and/or coal.

2.2 MANUFACTURED GAS

When coal is heated in closed retorts (excluding air and in this way preventing combustion) to high temperatures, a large number of useful products are obtained. These products include coke, tar, ammonia liquor and gas. Each of these materials finds its uses: the coke is used for iron-smelting; the liquid by-products are used for the manufacture of fertilisers, pharmaceuticals, road tars, pitch etc. and the gas is used as a source of thermal energy. Until the widespread discovery of supplies of NG, most gas supplied by the gas industry was manufactured in this way. Nowadays, the only manufactured gas produced is that which arises as a by-product of coke manufacture and this is used by industrial consumers near to the coke works.

2.3 NATURAL GAS

2.3.1 The source of natural gas

Petroleum (Chapter 3) and natural gas are thought to have been formed from the fossils of minute marine animals such as plankton. As these animals died they fell to the bed of the sea and were mixed with sand to become trapped in sedimentary rocks. Under the influence of high pressures and temperatures and, in the absence of air, the organic matter of these animals underwent anaerobic decay processes. This resulted in a range of hydrocarbons being formed, with chain lengths ranging from molecules, such as methane, which are gases at atmopsheric pressure, to larger molecules which are tar-like solids. In between these extremes is a large proportion of liquid hydrocarbons which make up the bulk of crude oil. In many cases these hydrocarbons, as they were formed, would rise to the surface of the earth, where the gaseous fractions would evaporate leaving deposits such as the tar lakes of Trinidad. However in certain circumstances the upward progression of the oil and gas was prevented by the presence of an impermeable barrier, known as a **cap rock**. This is associated with several types of geological structure as shown in Fig. 2.1.

These structures always have three components: an impervious **cap rock** prevents the upward rise of oil and gas. The upward pressure comes from **water** or **brine**, above which is the gas and oil. The third component is the **reservoir rock** which holds the oil and gas. This rock is porous containing many small holes caused by the sedimentary nature of the rock, and by the cavities formed from fossilised animals or by geological fractures.

Often gas and oil are found together but there are examples of natural gas fields with no crude oil and vice versa. When exploring for new oil and gas deposits, geologists look for rock structures associated with those shown in Fig. 2.1 (Cook 1976) but their identification does not always lead to the discovery of oil and gas. This can only be determined by test drilling, which is used to confirm the presence of oil or gas and also to determine the economic viability of extraction from any finds.

The southern part of the North Sea contains natural gas deposits only whilst the northern part contains oil deposits. The gas deposits are associated with deep coal fields. When a source of natural gas has been found by an exploratory drilling well it is replaced by a fixed well. This varies depending on whether the field is to produce both oil and gas and by its physical location (on land or offshore). In the case of offshore drilling platforms, a single platform can drill a number of wells covering an area of several hundred square metres (Macartney 1969).

2.3.2 Transportation, preparation and storage of natural gas

Natural gas can be transported by pipeline or in tankers. Pipelines are used to transport gas from offshore wells, along the sea-bed to coastal terminals. Similarly pipelines can be used for land wells. To facilitate the economic transport of gas by tanker (sea, road or rail) it must first be converted into a liquid by cooling it to -160 °C.

Prior to distribution to the consumer through the gas distribution pipeline system, the gas has to be processed. This involves:

(1) removal of hydrogen sulphide (not always necessary)

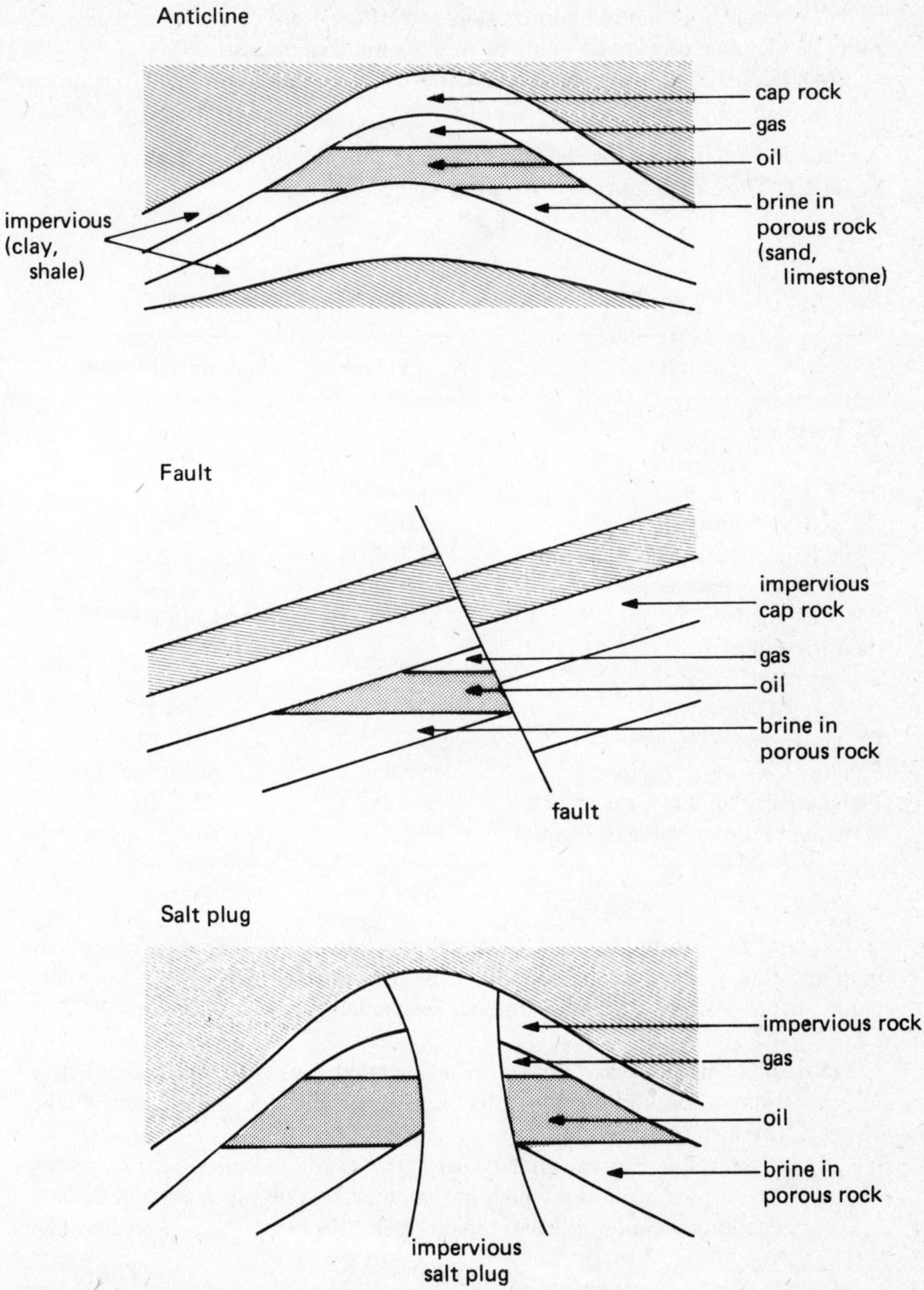

Fig. 2.1 – Gas and oil traps.

(2) recovery of liquid hydrocarbons by refrigeration
(3) blending to give the required combustion characteristics
(4) odourisation as natural gas deposits often have no smell (Densham *et al.* 1971).

A typical composition for the processed gas and its properties are recorded in Table 2.1. These data however vary from well to well.

Table 2.1
Typical properties of Natural Gas and Commercial Propane

Property		Natural Gas	Commercial Propane
Composition			
CH_4	methane	94.4 %	0.1 %
C_2H_6	ethane	3.1 %	5 %
C_3H_8	propane	0.6 %	82%
C_4H_{10}	butane	0.2 %	10%
C_5H_{12}	pentane	–	2%
Specific gravity		0.59	1.40–1.55
Calorific value,			
gross (MJ/m^3)		38.8	93.2
net (MJ/m^3)		35.0	86.2
Wobbe number (MJ/m^3)		50.6	78.8–74.9
Sulphur		negligible	negligible
Flammability limits (% gas in air)		5–15	2.2–10
Maximum burning velocity (mm/s)		325	390 (for propane)

Natural gas is stored to allow for fluctuations in demand. Storage may either be long term or short term to cover daily and seasonal variations in customer requirements. Various storage techniques are available:

(1) Liquid natural gas: storage in refrigerated tanks or a hole dug below ground, the soil forming the side of the hole is maintained in the frozen state by the liquid gas.
(2) Underground storage in aquifiers: the gas is pumped into an underground rock structure which has an impermeable cap rock under which is a porous, water filled reservoir rock. The water is displaced by the gas.
(3) Storage in transmission pipelines: a very large volume of gas is held in the transmission pipelines which are operated at a high pressure.

2.4 LIQUEFIED PETROLEUM GAS (LPG)

Most LPG comes from the refining of crude oil (section 3.3). During the refining process a number of chemicals are obtained which are gases at normal temperatures and pressures but which can be liquefied by the application of pressure alone, without the need for refrigeration. The major hydrocarbons which make up LPG are propane (C_3H_8) and butane (C_4H_{10}). The commercial gas on sale to the consumer (propane) is not pure and has a typical composition as shown in Table 2.1

LPG is transported by bulk road or rail tanker and is stored in a tank on the premises of the customer. For domestic consumers with only one appliance the gas is delivered in portable cylinders which when empty, are returned for refilling.

2.5 DOMESTIC INSTALLATIONS FOR GASEOUS FUELS

2.5.1 Natural gas

Gas from the transmission system is taken to a house via a service pipe. In the transmission system gas is at a high pressure (70 bar) but this varies depending upon the demand for gas. The pressure is reduced in several stages using governors (regulators). These are used to control the supply of gas at a low pressure when fed from a supply at a higher and fluctuating pressure as shown in Figs. 2.2 and 2.3.

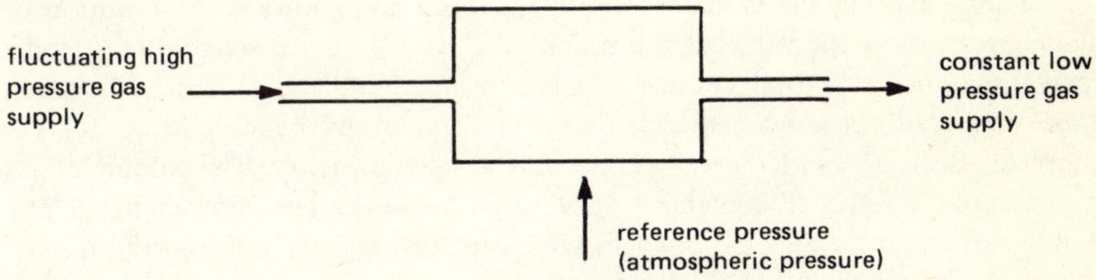

Fig. 2.2 – The function of a governor.

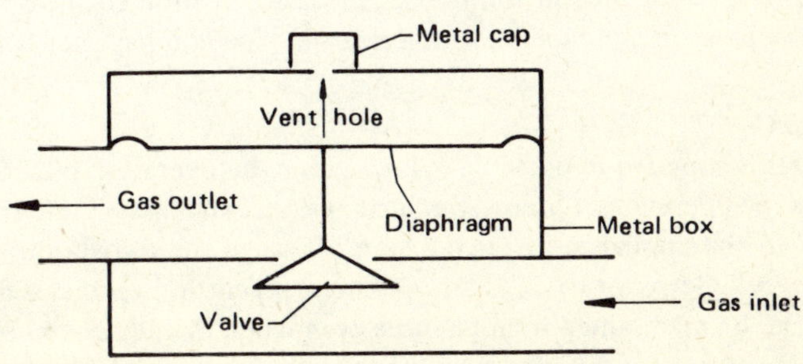

Fig. 2.3 – Cross-section of a gas governor.

Where the service pipe enters a house it is fitted with an on/off valve and a meter (which has its own governor). The on/off valve enables the supply of gas to the house to be shut off at a point near to where the gas enters the premises. A gas meter measures the volume of gas consumed in the house and it may be one of two types: a pre-payment meter or a credit meter. The volume of gas used (as measured by the meter) is converted into a quantity of heat for payment purposes (see Fig. 2.4).

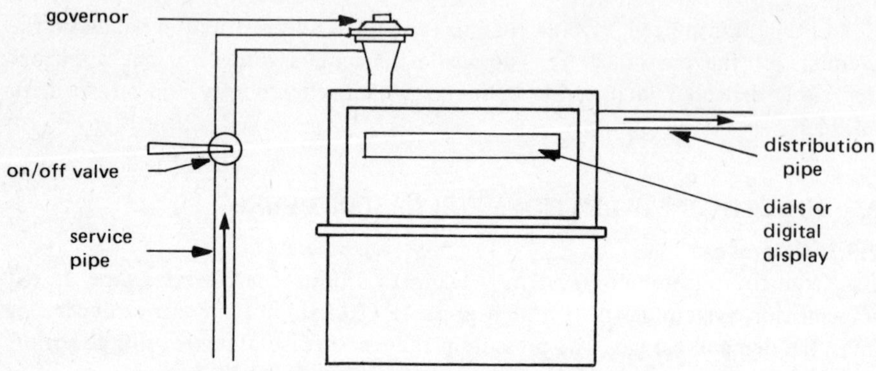

Fig. 2.4 – Diagram of a gas meter and associated fittings.

A pre-payment meter has a coin mechanism and allows a consumption of gas equivalent to the value of the money received by the mechanism. A credit meter measures the total volume of gas consumed and the difference between successive readings is used to calculate the volume and hence cost of the gas supplied. Both types of meter operate on the same principle. The volume of gas is measured by the filling and emptying of bellows. The movement of the bellows operates mechanical linkages which in turn register the consumption of gas on clock-type dials or a digital indicator.

Distribution pipes of copper or iron are used to transport the gas from the meter to each appliance in the house. Final connection to the appliance can be by rigid copper or iron pipe or with a flexible pipe and bayonet connection.

2.5.2 LPG

Where gas is supplied in portable cylinders for the operation of a single appliance there is no permanent domestic installation. The gas cylinder is connected directly to the appliance through an on/off valve and a governor. One or several portable cylinders can be used to operate a piped domestic system, so that the supply can be maintained to appliances even when cylinders are being changed.

Where the consumer has a bulk tank, this is sited outside the premises. The bulk tank is fitted with an isolating on/off valve. The pressure of the gas leaving the tank is reduced by a governor, situated outside or inside the house. As with natural gas, the distribution of gas is through rigid pipes but the final connection can be made with flexible piping.

2.6 THE COMBUSTION PROPERTIES OF GAS

2.6.1 Combustion

When methane, the major component of natural gas, burns in the presence of sufficient air to ensure complete combustion the reaction can be expressed as

$$CH_4 + 2O_2 = CO_2 + 2H_2O.$$

The combustion of propane and other hydrocarbons, hydrogen etc. can be represented in a similar way. In practice the oxygen which is used in the combustion process is accompanied by the other consituents of air. These take no part in the reaction, but they are heated and form a part of the combustion products. Since each mole of oxygen is accompanied by 3.76 moles of nitrogen, the process of combustion in air can be represented as:

$$CH_4 + \underbrace{2O_2 + 7.52N_2}_{\text{theoretical air}} = \underbrace{CO_2 + 2H_2O + 7.52N_2}_{\text{combustion products}}.$$

This equation represents the case where the air present is just sufficient to burn completely the methane to carbon dioxide and water. When this is the case, the volume of air required is referred to as the **theoretical air**.

If less than the theoretical air is present, combustion of the methane will be incomplete leading to the production of carbon monoxide and carbon particles in the combustion products. This is undesirable because of (i) the toxicity of carbon monoxide (ii) the carbon particles give a smoking flame (iii) incomplete combustion represents a loss of efficiency. In practice, to prevent this situation arising, more air is used than is theoretically necessary for complete combustion of the gas. The surplus air is known as the **excess air.** The amount of excess air must be limited since it represents a loss of efficiency. Also, a resultant increase in the volume of combustion gases (Milson and Kirk 1980) leads to an increase in the loss of heat in the combustion products. The heat is lost to the atmosphere and is not used usefully in the appliance.

2.6.2 Calorific Value

The calorific value of a fuel is a measure of the total thermal energy obtained when it is burned. For a gas this is expressed as thermal energy obtained from a unit volume of the gas, i.e. in SI units MJ/m^3. (In Imperial units it is given as Btu/ft^3). Two calorific values are used, net and gross. The gross value includes the heat in the condensed water vapour from the combustion products. The net calorific value excludes this heat. In all domestic appliances the water vapour in the combustion products is **not** allowed to condense and therefore the net calorific value is used for comparing fuels. Table 2.2 shows typical net calorific values for the common domestic fuel gases.

Table 2.2
Net Calorific Values for Gaseous Fuels

Natural Gas	35.0 MJ/m^3
Manufactured Gas	18.9 MJ/m^3
Propane	86.2 MJ/m^3

The gross calorific value of a gas is used by the Gas Board/Utility to calculate the amount of thermal energy consumed by a customer from the volumes measured by the meter. If the pressure of a gas varies, so will its calorific value and this is why a gas meter needs a governor to standardise the pressure of the gas at the inlet of the meter.

The Wobbe number is also used as a specification for the gas and guarantees the heat output of an appliance irrespective of fluctuations in the density of the gas supply. It is calculated by dividing the gross calorific value by the square root of the specific gravity of the gas. Control of this parameter by the gas industry as well as gross calorific value, is important in terms of consistency of performance of consumers' appliances.

2.6.3 Flammability Limits

These define the proportions of gas and air within which a gas/air mixture will burn. Fig. 2.5 shows the flammability limits for natural gas.

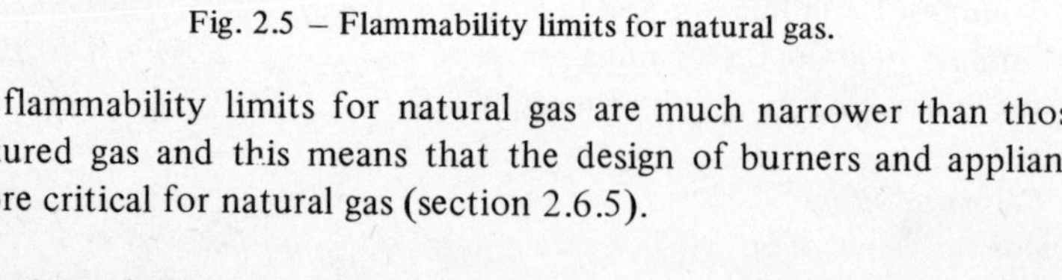

Fig. 2.5 – Flammability limits for natural gas.

The flammability limits for natural gas are much narrower than those for manufactured gas and this means that the design of burners and appliances is much more critical for natural gas (section 2.6.5).

2.6.4 Burning velocity

The flammability limits define those mixtures of gas and air which will support a flame. Within these limits the stability of the flame on a burner will vary and an important factor in this context is the burning velocity. The burning velocity is measured as the rate at which a flame travels down a tube filled with a stationary air/gas mixture. As can be seen from Fig. 2.6 the burning velocity is at its lowest when the air/gas ratio is close to the flammability limits. The burning velocity is an important factor in flame stability since, if the velocity of gas leaving a burner jet is greater than the burning velocity, the flame will **lift-off** the burner. Conversely, if the velocity of the gas is less than the burning velocity, then the flame will **burn-back** down the burner pipe. This is discussed in section 2.6.5.

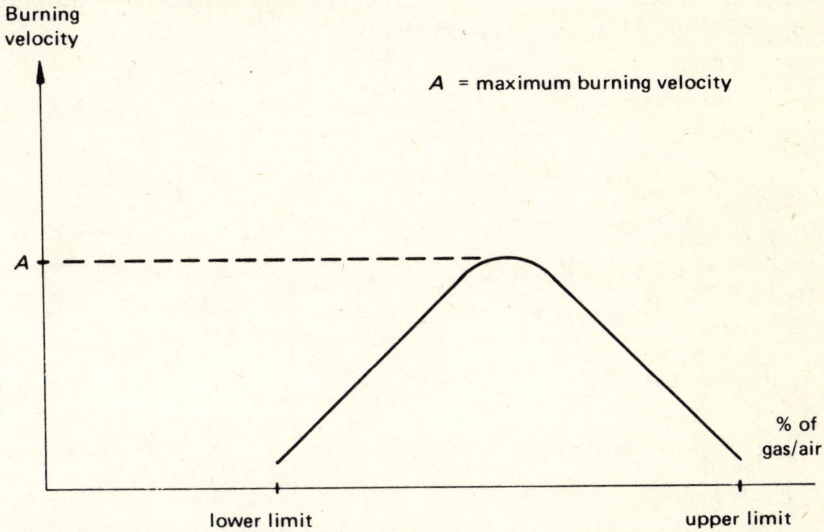

Fig. 2.6 – Burning velocity for a gas: air mixture.

The inter-relationship between a number of factors affecting the combustion characteristics of a gas can also be illustrated by means of a Fuidge diagram (Fig. 2.7). This diagram shows the range within which a satisfactory stable flame is formed.

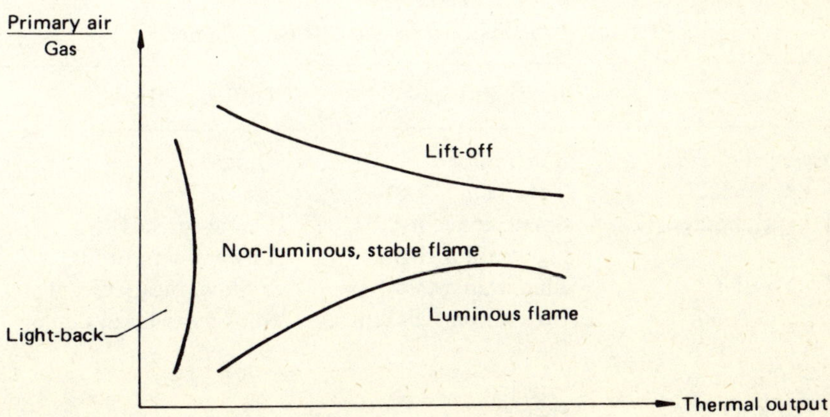

Fig. 2.7 – Fuidge diagram.

2.6.5 The design of gas burners

It has been shown that combustion is a reaction between oxygen (from air) and a fuel. A prerequisite for the reaction is that the fuel and air should be intimately mixed. This is relatively easy for a gaseous fuel. There are two basic types of flame. One is where a proportion of the air for combustion is mixed with the gas prior to combustion: an *aerated* flame. The second type is where there is no pre-mixing of air and gas: a *diffusion* flame. The two types of flame are represented by a Bunsen burner (Fig. 2.8), (Pearson 1980).

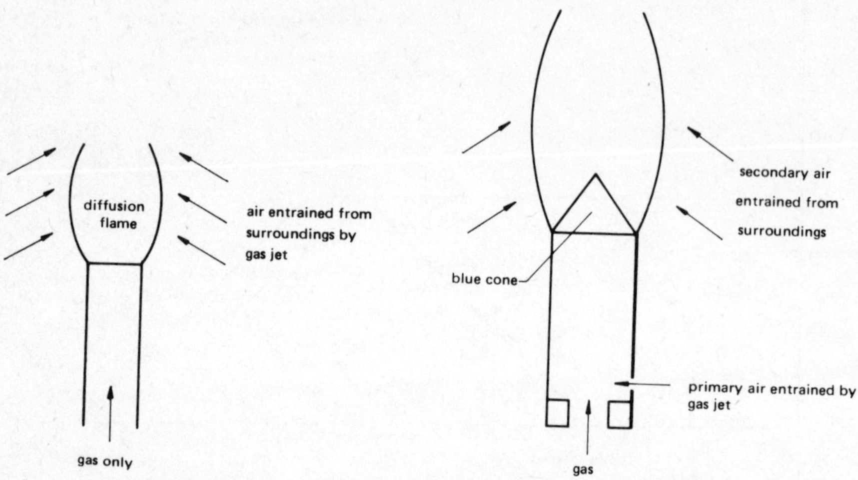

Fig. 2.8 – Aerated and diffusion flames using a Bunsen burner.

The characteristics of the two types of flame are given in Table 2.3. Specific appliance burners are described in the appropriate chapters later in this book.

Table 2.3
Characteristice of aerated and diffusion flames

	Aerated flame	Diffusion flame
Noise level	Some noise	Quiet
Lighting back	Yes	No
Heating characteristics	Small, concentrated high output flame	Flame spreads Lower heat output
Colour of flame	Blue, transparent	Yellow, incandescent
Soot	No soot particles in flame	Soot particles in flame
Susceptibility to blocked jets	High	Low

As an alternative to the use of a flame, flameless catalytic combustion may be used. The reaction takes place on a heated fibrous pad containing a catalyst. Initially the pad must be heated to 200 °C, but once the reaction is activated the catalyst is maintained at 400 °C (Pearson 1980).

2.7 THE REMOVAL OF COMBUSTION PRODUCTS

The combustion products leaving a flame contain heat and consist of nitrogen, carbon dioxide, oxygen and water vapour. These must be removed from a room, since, if they were allowed to build up, they would lead to discomfort for the occupants of the room (see Chapter 11). In addition to the maintenance of conditions acceptable to people, it is also important that, because of dilution by combustion products, the oxygen supply to the gas burner is not affected. If, through the build up of combustion products in a room with a gas burning appliance, the oxygen supply to the flame falls below the level required for complete combustion, toxic carbon monoxide can be produced or the flame may be quenched. A depletion in the oxygen content of air supplied to a flame is known as **vitiation** (Smith & White 1976).

If an appliance produces small quantities of combustion products only, or is used intermittently, the normal room ventilation eliminates the combustion products (providing that the normal ventilation is not impaired by draught sealing). Where large volumes of combustion products are produced, these must be removed by a flue. The division between these appliances requiring a flue and those which can operate satisfactorily without one is normally related to the heat output of the appliance and to the size of the room in which the appliance is situated. Flues are not normally needed for cookers, small gas fires and instantaneous water heaters but are needed for storage water heaters and for heaters which are part of a central heating system (Billington 1972).

A flue is designed to extract the products of combustion from an appliance and to remove them from the household environment, normally to the air surrounding a building (Gibson 1969). It must be able to do this safely under all weather conditions. If the design of a flue allows wind to blow the combustion products back down the flue into the appliance, the flame may be quenched or the combustion products may spill into the room (Fig. 2.9).

The danger of flame quenching may be overcome by fitting a draught divertor on the flue (Fig. 2.10). This consists of a cone-shaped fitting on the flue with a gap allowing exit or entry of gases from the flue into the room. There is also a baffle which causes partial blocking of the flue at this point. Under normal conditions combustion products flow up the flue and around the baffle, drawing in air which dilutes the combustion products. Under adverse wind conditions combustion products are blown through the diverter into the room instead of being blown into the appliance. They still escape into the room however.

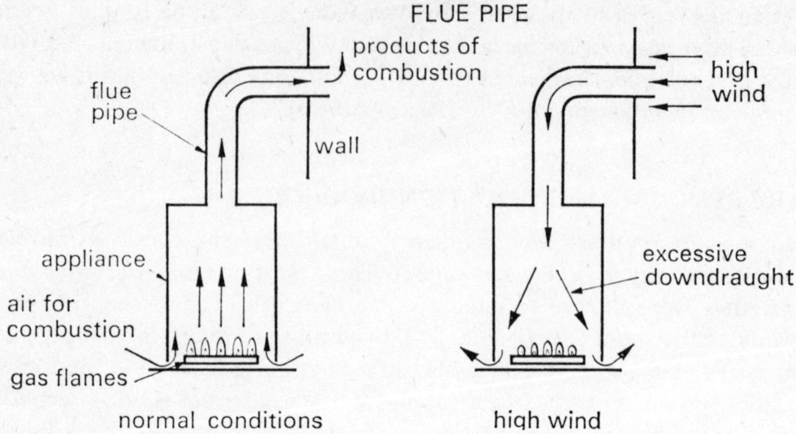

Fig. 2.9 – Effect of high wind on a poorly designed or situated flue.

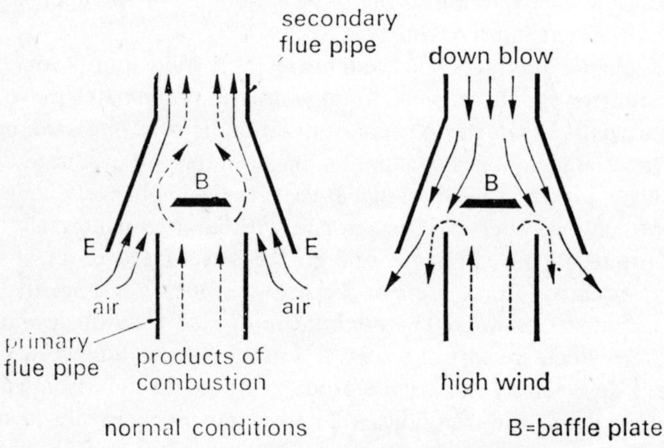

Fig. 2.10 – Draught diverter under normal and adverse wind conditions.

Because of the problems of providing stable conditions for combustion and to minimise the risk of combustion products entering a room, a **balanced flue** is often used. This is sometimes referred to as a room-sealed appliance. In this type of flue, both air for combustion and combustion products do not come into contact with air in the room in which the appliance is situated. Air for combustion is drawn in through a wall terminal into the combustion region of the appliance and, after combustion and having passed over a heat exchanger, leaves the appliance through the same terminal. Because air for combustion and combustion products enter and leave by the same terminal, the effects of wind are cancelled out, producing stable combustion conditions (Fig. 2.11). The design of a terminal is very important since there must be no risk of combustion products being drawn in with the air for combustion (Brady and Kerac 1969, Bennett 1979).

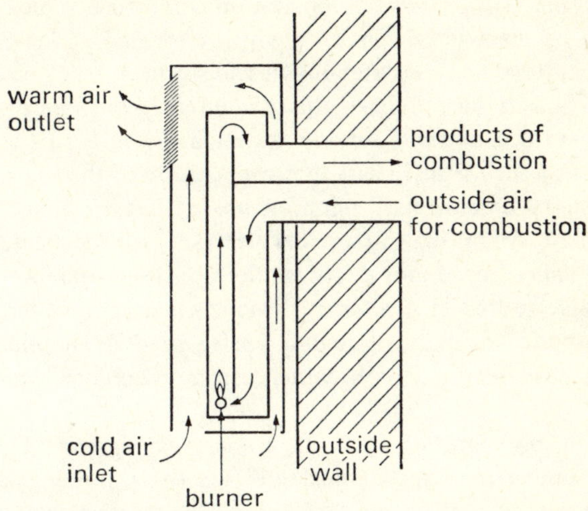

Fig. 2.11 – Balanced flue room sealed appliance.

2.8 FLAME IGNITION

Automatic gas ignition is becoming more generally used with gas burning appliances because of the increased convenience this gives to the user. In some cases automatic ignition is used in conjunction with a pilot flame or it can be used to ignite the main burner, which eliminates the waste of energy associated with a permanent pilot flame.

2.8.1 Pilot flame

A permanent pilot is a small flame, adjacent to the main gas burner, which burns continuously, so that, when the gas supply to the main burner is turned on, the pilot ignites the gas flow immediately. Pilot flames are designed to have a small stable flame, which must not be extinguished in the often turbulent air flow close to the main burner. To prevent the jet from becoming blocked by dust or other particles, a filter in the gas supply pipe is necessary. On cookers the pilot must be sited so that spillage of food cannot cause a blockage of the jet. In addition to these design problems, pilot flames use gas continuously and this is regarded as a wasteful use of energy. Pilot flames may be ignited manually or automatically.

2.8.2 Automatic ignition

All of the methods available for the automatic ignition of gas utilise electricity – either using a heated filament or a spark. They may be used to light a pilot flame or the main burner directly.

A low voltage catalytic igniter uses a platinum wire which is heated to about 400 °C by passing an electric current through it, (0.4 amp at 1½–3 volts) supplied either from a battery or, via a step-down transformer, from the mains (service) supply. When gas is passed over this heated filament, localised combustion occurs at the surface of the filament causing a rise in temperature of the wire to over 1000 °C, which is sufficient to cause ignition of the gas.

Other systems of automatic ignition use an *electric spark,* which may be generated by battery, by mains (service) electricity, or mechanically. In the case of battery or mains ignition an electrical circuit is used to convert the electricity into high voltage pulses of electricity which, when applied across the gap of a spark plug, produce a chain of high-intensity sparks. The electircal circuit is activated by a push-button switch, which can be combined with a flame-failure device.

Mechanical spark-ignition systems use a *piezo-crystal.* This is a material which, when subjected to mechanical pressure (by squeezing) generates a high voltage, which can be applied to a spark plug. The mechanical pressure is applied by a plunger on the appliance (Fig. 2.12) or it can be incorporated in the on/off control valve to produce a spark when the applicance is turned on.

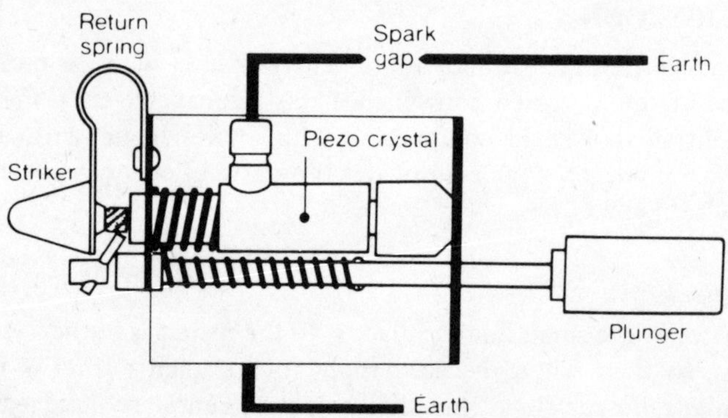

Fig. 2.12 – Piezo-electric igniter.

With all of the methods of automatic ignition using a spark, the siting of the spark plug is very critical. It must be close to the gas jet (pilot or main) if it is to ignite the air/gas mixture before it becomes diluted with air outside its flammability limits. However, it must not be so close to the flame that it, or its electrical connecting wire, becomes overheated by the flame (Fig. 2.13). An alternative approach is to conceal the electrode within the burner (Pearson 1980).

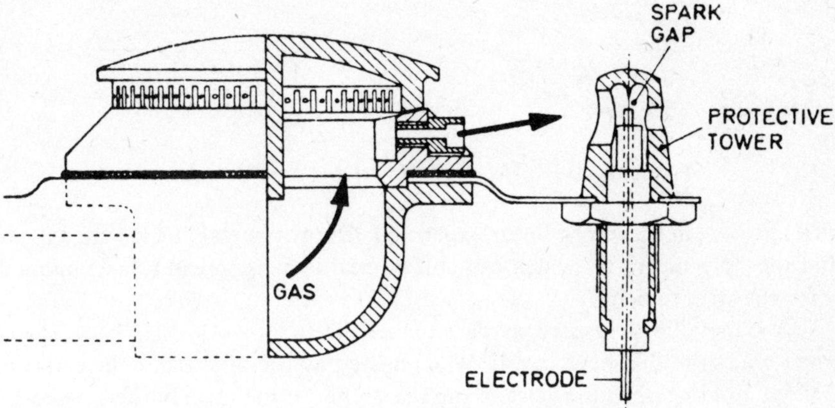

Fig. 2.13 – Spark igniter attached to the burner.

2.9 FLAME-FAILURE DEVICES

Increasingly flame failure devices (FFD) are used on gas burning appliances, particularly in cases where the appliance is under the control of a timer. If the pilot light of a gas central heating appliance is extinguished, when the timer causes the main burner to be turned on, a dangerous situation will exist. A FFD can be used to isolate the main burner from the gas supply if the pilot is extinguished. FFD's can be used in conjunction with automatic ignition systems as a way of ensuring that the gas supply to a burner is terminated if ignition has been unsuccessful.

Flame failure devices operate by sensing the heat given off by a flame and converting this into a mechanical force sufficient to hold a gas valve open. Three different methods are used for converting the heat of a flame into the physical movement of a valve. These are bimetallic strip, thermo-electric and vapour pressure devices.

2.9.1 Bimetallic strip FFD

A bimetallic strip is made of two metals which have different coefficients of linear expansion. The strip responds to a change in temperature by bending as shown in Fig. 2.14.

When a metal is heated it expands and the degree of expansion can be expressed as the coefficient of linear expansion. This is the increase in length as a proportion of the origianl length caused by a rise in temperature of 1°C. When the bimetallic strip is heated the brass expands more than the iron because

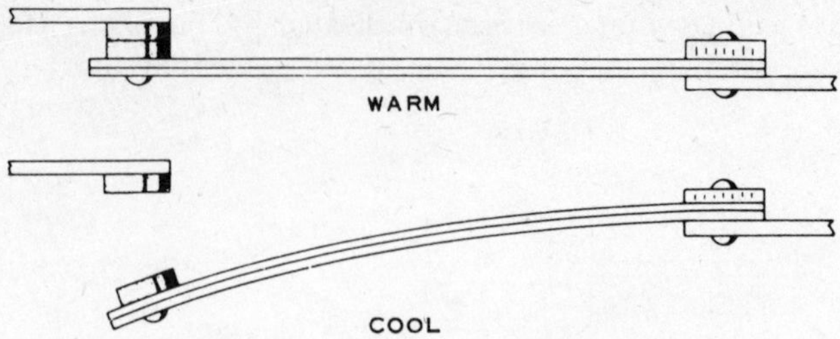

Fig. 2.14 – The effect of heat on a bimetallic strip.

of its higher coefficient of linear expansion. The two metals are fastened together and the only means by which this differential expansion can be accommodated is for the strip to bend.

A bimetallic strip used as part of a FFD is shown in Fig. 2.15. The strip when cold is in the shape of a U. When heated by the pilot flame the ends of the U close, holding open the valve in the gas supply to the main burner.

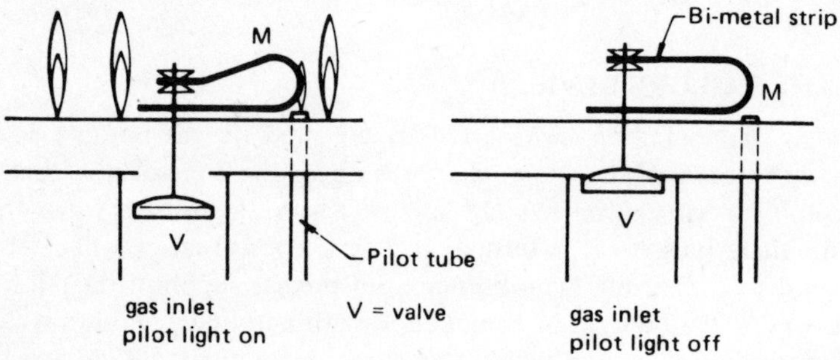

Fig. 2.15 – A Bimetallic strip FFD.

2.9.2 Thermoelectric FFD

If the junction of two different metals is placed in a flame a small electrical voltage is generated across the ends of the two metals, as shown in Fig. 2.16. This is known as the **thermoelectric** effect. The electricity produced is sufficient to generate a magnetic field if it is applied to the coil of a small electro-magnet.

A thermo-electric flame failure device is shown in Fig. 2.17. Initially the gas valve (V) is closed. It can be opened manually by the push button (B). This allows gas to flow to the burner which can then be ignited. The flame from the burner heats the thermo-couple junction (T) and this generates sufficient electricity to activate the electro-magnet (M). This holds open the gas valve after the push button is released. If the flame is extinguished the magnetic field will decay allowing the spring (S) to close the valve. These devices can be used with main

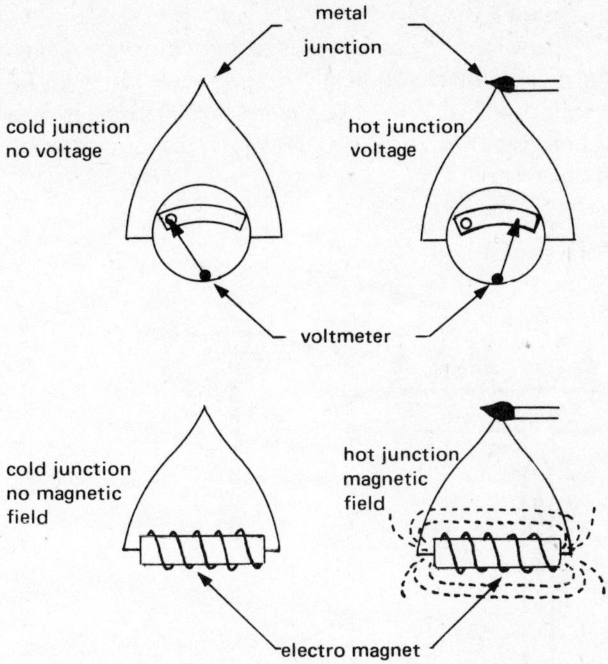

Fig. 2.16 – The thermo-electric effect.

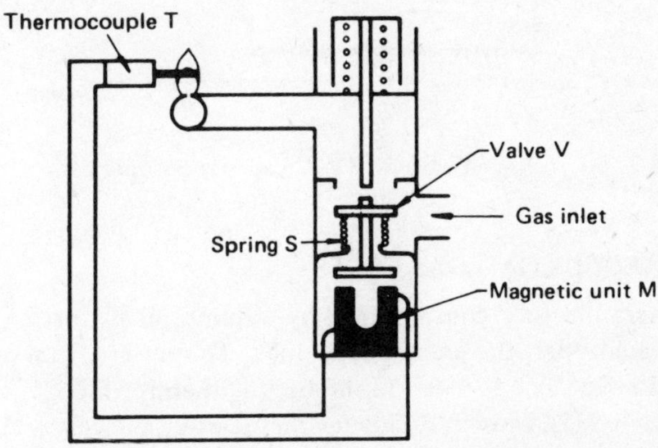

Fig. 2.17 – Thermo-electric FFD.

burners or with pilot lights. In the latter case, mechanical operation of the push button allows gas flow to the pilot but not to the main burner. When the push button is released, and providing that the pilot has ignited, gas flow to the main burner is allowed.

2.9.3 Vapour pressure FFD

The heat from a flame can be used to change a liquid to a vapour and, if this liquid is in a sealed tube, there will be an increase in pressure. An FFD utilising this principle is shown in Fig. 2.18. The vapour pressure tube is connected by a capillary (a fine bore tube) to a bellows or capsule, the movement of which can be made to hold open a gas valve.

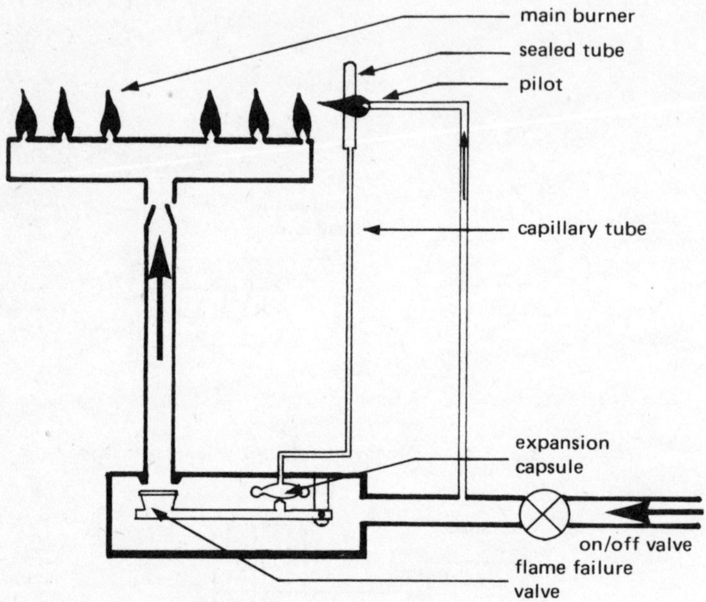

Fig. 2.18 – A Vapour pressure FFD.

2.10 CALCULATING GAS COSTS

Gas meters measure consumption by volume, in m^3 or ft^3, but the charge for fuel is based upon the energy consumed. To convert from one to the other, the gross calorific value is used. In the UK the **therm**, which is 100,000 Btu (equivalent to 94.9 MJ), is used. Following metrication, the calculation will be based on the **metric therm** (equivalent to 100 MJ). Thus,

$$\text{therms used} = \frac{\text{volume (ft}^3) \times \text{gross calorific value (Btu/ft}^3)}{100\,000}$$

$$\text{or metric therms} = \frac{\text{volume (m}^3) \times \text{gross calorific value (MJ/m}^3)}{100} .$$

The charge for gas is based on a tariff structure. This is made up of a number of different charges such as a *standing* charge and a charge per therm. The latter can be a *variable charge,* composed of a high rate for the first block of therms used in an accounting period, and a lower rate for subsequent units.

REFERENCES

Bennett, W. J. (1979), Flues, past, present and future, *Gas Engineering and Management,* **October**, p. 515-533.

Billington, N. S. (1972), Ventilation of home kitchens, *H & V E,* **April,** p. 512.

Brady, K. J. D. and Kerac, D. (1969), Design optimisation of balanced flue terminals, *IGE Journal,* **9,** p. 239-256.

Cook, E. (1976), *Man, Energy and Society,* Freeman.

Densham, A. B., Gibbons, P. A., Glidewell, M. D. & Palmer, R. (1971), The odorisation of natural gas, *IGE Journal,* **11**.

Gibson, J. (1969), Domestic chimney design, *J. Inst. Fuel,* **42,** p. 155-160.

Macartney, I. (1969), Production of North Sea Gas, *IGE Journal,* **9,** September, p. 535-551.

Milson, A. & Kirk, D. (1980), Chapters 2 and 11, Principles of design and operation of catering equipment, Ellis Horwood.

Pearson, E. J. (1980), Burning gas in the domestic market, *Gas Engineering and Management,* **June,** p. 247-256.

Smith, E. J. & White, I. F. (1976), The effects of vitiation of combustion air on the flame stability of domestic multipoint aerated burners, *J. Inst. Fuel,* **44,** p. 30-34.

BIBLIOGRAPHY

Peet, L. J., Pickett, M. S., Arnold, M. G. & Wolf, I. H. (1970), *Household Equipment,* Wiley.

Plant, E. M. (1971), *Gas and its Domestic Uses,* Women's Gas Federation.

Taylor, B. (1973), *Le Fevre's Domestic Utilisation of Gas,* Ernest Benn.

Valais, M. & Hiegel, M. (1977), *The Gas Industry in the World,* Graham & Trotman.

CHAPTER 3

Liquid fuels

3.1 INTRODUCTION

Liquid fuels originate from crude petroleum and are separated from it during the refining process. There are two main liquid fuels of importance to the domestic market: kerosine (or paraffin) and domestic fuel (or gas oil). These materials are not used exclusively for domestic heating.

3.2 THE ORIGIN OF PETROLEUM

It was seen in section 2.3.1, that Natural Gas is thought to have originated from small marine animals trapped in layers of sedimentary rock. Crude petroleum is associated with similar, geological structures as those described for natural gas (Fig. 2.1). Unlike natural gas, crude petroleum requires complex processing before it is in a suitable form for the consumer. This process is known as **refining.** Because most of the world's crude petroleum sources (other than those of USA, Russia and the North Sea) are located thousands of miles from the consumer, it has to be transported over long distances. Often the oil is not refined at the oil well but is transported as crude oil to the country of consumption. This reduces the import cost of the oil (compared with the refined product). It is transported overland by pipeline and over sea by tanker.

3.3 REFINING CRUDE PETROLEUM

Crude oil consists of a mixture of hydrocarbons of varying structure and chain length. There are also small amounts of organic chemicals containing sulphur, nitrogen and oxygen. Mixed with these organic chemicals there are usually small amounts of salt, sediment and water. The actual composition of crude petroleum varies, depending on its geological source. Typically it contains 85–90% carbon, 10–14% hydrogen and between 0–4% sulphur. It can be green, brown or black in colour with a specific gravity between 0.73 and 1.00. Crude oils are blended from various sources dependent upon the market demand for the different products.

At the refinery the crude petroleum is separated and converted into a range of materials using both physical and chemical methods. These products of refineries vary depending on the characteristics of the crude petroleum and the demand for petrochemicals (see Fig. 3.1).

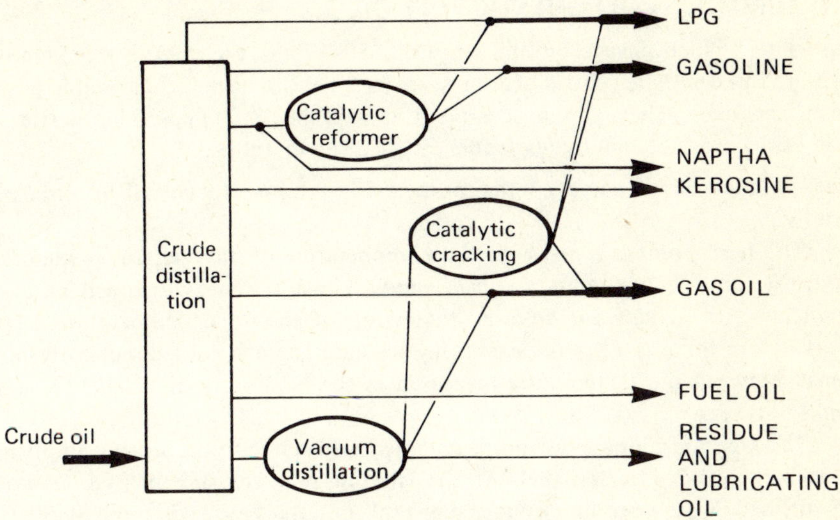

Fig. 3.1 – A flow diagram of a typical refinery.

The major physical process which takes place in a refinary is **distillation,** the separation of crude oil into a range of products, based upon a difference in their boiling points. Steam and crude oil are pumped into the base of a tall column and fractions taken off at various heights up the column. Fractions from high up the column have a low boiling point (less than 30°C) and fractions nearer the base of the column have a high boiling point (over 350°C). Normally five or six fractions are produced by the primary distillation process, most of these products then being subject to further refining processes, such as the removal of sulphur compounds.

The low boiling fractions are compounds of one to four carbon atoms (chiefly propane C_3H_8 and butane C_4H_{10}) and these are used for the manufacture of LPG (see section 2.4). The **kerosine** fraction is composed chiefly of C_{10} to C_{13} compounds. The next lowest fraction on the column is **gas oil** containing compounds of C_{14} to C_{22} carbon atoms.

The proportions of the major fractions recovered from crude oil varies with its source. For example crude oil from Kuwait contains up to 14% LPG whereas that from Venezuela contains only 2%. In general, oils from the Middle East have a low viscosity with a high proportion of the lighter (higher boiling point) fractions, while oils from South America have a high viscosity with a high residue content. North Sea crudes tend to be composed of low boiling, light oils with a low sulphur content. The important products from a domestic viewpoint are LPG, kerosine and light fuel oils.

3.4 KEROSINE AND DOMESTIC FUEL OILS

Kerosines, which have a boiling range of 150–250 °C, are used in free standing heaters and domestic central heating systems. Fuel oils have higher boiling points than kerosines. They can be classified as in Table 3.1. Typical properties of domestic kerosines and domestic fuel oil are given in Table 3.2.

The significance of some of the properties of liquid fuels will be discussed briefly.

The **flash point** is a measure of the temperature of the fuel, under standardised test conditions, required to give a vapour above the fuel which will ignite on application of a flame. It gives an indication of the upper temperature safety limit before there is a fire hazard during storage. There is not a direct correspondence between the temperature measured in the laboratory and that in storage tanks.

The **freezing point, pour point** and **viscosity** relate to the storage and handling characteristics of the fuel. At the freezing point crystals of hydrocarbons are formed, which can block pipes and tank outlets. Below the pour point the oil does not flow when held vertically for five seconds. Viscosity is a measure of the ease with which fuel oil can be pumped from a delivery vehicle into a storage tank and from the storage tank into the appliance. Freezing point, pour point and viscosity do not usually pose a problem for domestic fuel oil except under very severe weather conditions. Viscosity is important in the case of appliances which use a wick, since viscosity counteracts the capillary action which is responsible for drawing oil up the wick. Viscosity also determines the ability of an oil to atomise (break up into small droplets). Domestic fuel oils are easily handled and burned without the use of expensive equipment.

The **calorific value** was defined in section 2.6.2. The presence of *sulphur* in the fuel is undesirable because

(a) it produces an unpleasant odour in the fuel
(b) it causes corrosive deposits in chimneys and flues
(c) it makes the fuel unsuitable for direct combustion in ovens because of contamination of the product
(d) sulphur dioxide in the combustion products make it unsuitable for flueless appliances
(e) the sulphur compounds are corrosive and may affect tanks and pipes.

Kerosines are likely to produce a smoky flame if combustion is incomplete. The **smoke point** is the maximum flame height beyond which a smoky flame

Table 3.1

Classification of fuel oils

		Increasing viscosity ↓	Increasing sulphur content ↓	Increasing boiling point ↓
D	Domestic fuel oil (Gas oil)			
E	Light fuel oil			
F	Medium fuel oil			
G	Heavy fuel oil			

Fuel oil D is the only one of interest to the domestic consumer.

Table 3.2

Typical properties of Kerosine and Domestic Fuel Oil

Property	Kerosine: Premium Grade	Kerosine: Regular Grade	Domestic fuel oil
Specific gravity (15.5/15.5°C)	0.78–0.79	0.78–0.79	0.84
Flash point °C	45–50	45–50	77 (min 55)
Freezing point °C	−42	−47	
Pour point °C	−48	−51	−12
Kinematic viscosity at 37.8°C (10^{-6} m^2/s)	1.42	1.45	1.6 to 6.0
20.9°C (10^{-6} m^2/s)	1.85	1.87	
Gross calorific value MJ/kg	46.4	46.5	45.3
Net calorific value MJ/kg	43.6	43.7	42.3
Sulphur content %	0.02	0.06	1.0 max
Smoke point mm	40	33	
24 h char value mg/kg	2	9	
Water content %			0.05 max
Sediment %			0.01 max
Ash %			0.01 max
Conradson carbon %			0.2 max
Hydrocarbon type analysis:			
paraffins %	59.5	58.0	
naphthenes %	36.0	33.0	
aromatics %	4.5	9.0	

results when the fuel is burned on a standard burner, Similarly char builds up on wicks causing rapid deterioration of the wick, leading to higher maintenance costs. The **char value** is a measure of char build up on a wick and is associated with high boiling point compounds which do not vaporise and hence char, leaving a deposit on the wick.

The **water content, ash, sediment** and **carbon residue** are used as a part of the specification for fuel oils. Water, if present, will sink to the bottom of a fuel tank, and over a period of time, gradually builds up. It therefore needs to be drained off periodically, and, the higher the water content, the more frequently will this need to be done. Sediment can cause blockages in pipes if it is allowed to accumulate. It collects with water in the base of the storage tank. Ash is the residue after combustion of any mineral matter in the oil. This needs to be minimised or it will build up in the combustion space. Carbon residue represents the amount of unburned carbon which is formed in the flame. Solid carbon particles are formed in the flame which gives the flame its radiating properties. If particles strike a surface before combustion is completed, then a soot deposit results: this needs to be cleaned off periodically.

The aim of the oil industry is to avoid possible problems by manufacturing their products to specifications which minimise problems for the consumer. Fuel oils supplied to these specifications and used on correctly maintained equipment will give a satisfactory performance.

Liquid fuels are a mixture of hydrocarbons together with small amounts of other materials. The major hydrocarbons present can be one of four types:

(a) straight chain hydrocarbons

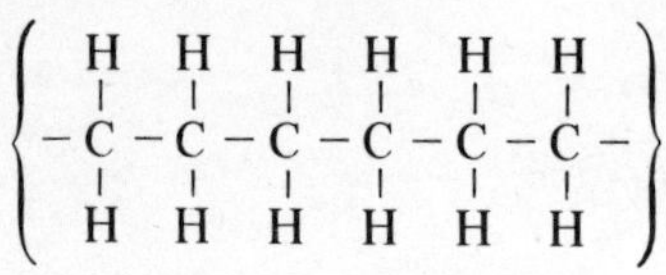

(b) branched hydrocarbons

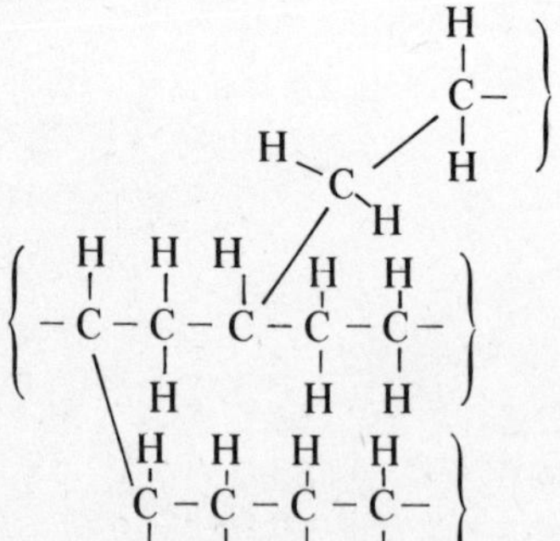

(c) naphthene derivatives

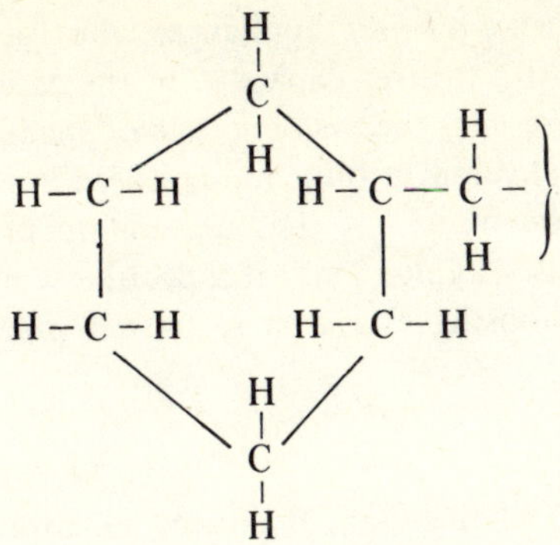

(d) benzene derivatives (aromatics)

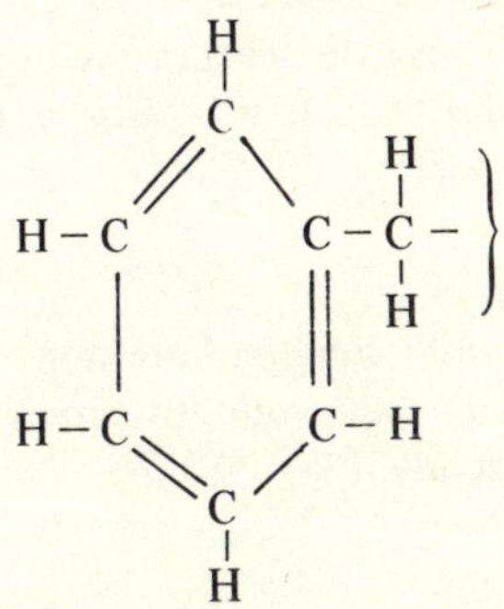

Naphthenes and aromatic hydrocarbons have lower freezing points than straight and branched molecules. However, aromatics have a tendency to produce a smoky flame, more char and combustion products with a strong smell. Straight chain hydrocarbons have a higher smoke point and produce less char, but this is dependent on the ratio of carbon to hydrogen. The smoke point of branched hydrocarbons and naphthene derivatives are intermediate between those of straight chain molecules and benzene derivatives. Normally, a blend of the four types of hydrocarbon is required to give a product with easy handling properties and good combustion characteristics.

Premium grade kerosine is used in appliances which do not have a flue. It has a lower benzene derivative content and a lower sulphur content. This means that the combustion products can enter the domestic environment without causing objectionable odours (from sulphur dioxide) and have a lower tendency to produce smoke and charring of the wick. Reference to the properties of Premium and Regular grade fuel oils in Table 3.2 clearly shows these differences in composition and properties. Regular grade kerosine and domestic fuel oil are used in flued appliances such as central heating boilers.

3.5 STORAGE AND HANDLING OF DOMESTIC FUEL

For households with single kerosine-burning appliances, the fuel may be purchased in cans and drums of up to 20 litres capacity. In the case of households where the fuel oil or kerosine represents the major source of space heating, water heating or cooking, the fuel is delivered in bulk and is stored in a tank. The size of this tank is related to the rate of use of the fuel and to price advantages obtainable from seasonal and bulk buying. Typical capacities would be between 1 m^3 and 3 m^3. High density housing estates may have a piped distribution system from a central tank.

3.5.1 Siting of the storage tank

The tank should be within 30 m of the road. If greater distances than this are involved, a permanent feed line may be needed. It should also be close enough to the appliance to enable the oil pump on the appliance to provide a satisfactory flow of oil. It should not however be closer then 2.5 m. In general, the tank should not be higher than 3 m above the level of the burner and should be separated from the appliance by a solid wall. If it is outside, the tank should be given protection from wind, rain and frost.

3.5.2 The storage tank and its fittings

Storage tanks are normally made of mild steel and are normally raised off the ground, both to protect the tank from damp and to provide sufficient liquid head to aid flow of fuel along the pipeline. Fig. 3.2 shows a storage tank and its fittings.

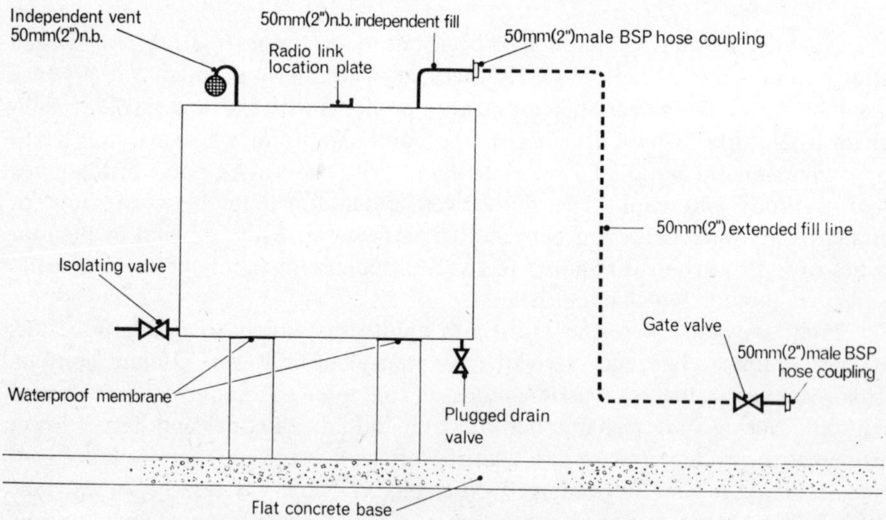

Fig. 3.2 – An oil storage tank and its fittings.

The fill pipe is for connecting the hose of the delivery vehicle to the tank and must have an appropriate coupling. In some cases, for example where the tank is raised above ground level or where it is not within easy access of the delivery vehicle, the fill pipe must be extended.

A remote control point enables a device to be used which automatically terminates pumping of fuel, by the road tanker's pump, into the storage tank once it is full (referred to as "radio link" on Fig. 3.2).

The vent pipe allows air to leave the tank as it is filled and re-entry of air as it empties. It also ensures the venting of the more volatile components which evaporate during storage.

The isolating valve is the connection point for the domestic installation being situated above the base of the tank to minimise the risk of water or sediment being drawn into the burner. The isolating valve is manually operated and is a safety device to cut off the tank from the domestic installation (see section 3.7).

The drain valve is used to run off any sediment or water from the bottom of the tank.

The contents gauge in its simplest form is a plastic sight tube situated on the outside of the tank. With this the consumer observes the depth of fuel in the tank. It has an isolating valve at its base which is only opened when it is desired to estimate the contents of the tank. An alternative type utilises a float on the surface of the oil, connected mechanically to the calibrated scale of a dial. The position of the dial on the scale indicates the quantity of fuel in storage.

3.5.3 The domestic installation

Copper tube of 8–11 mm outside diameter is normally used to transfer oil from the storage tank to an appliance. The final section of the pipe may be a flexible tube of reinforced plastic. Two types of pipework are used: a one-pipe system (flow only) and a two-pipe system (flow and return). Two pipe systems are normally used with pressure jet burners (Section 3.6.4).

3.6 COMBUSTION OF LIQUID FUELS

3.6.1 Wick-type burner

This is used on small paraffin (kerosine) heaters (Fig. 3.3). Paraffin is drawn up through the wick by capillary forces from a reservoir. It evaporates from the top of the wick, mixes with air and burns with a blue flame. The wick is raised or lowered to control heat output. Maximum output is limited because if the wick is turned up too far, there is insufficient air for combustion and smoke is produced. A chimney is required. The wick needs occasional trimming for efficient combustion.

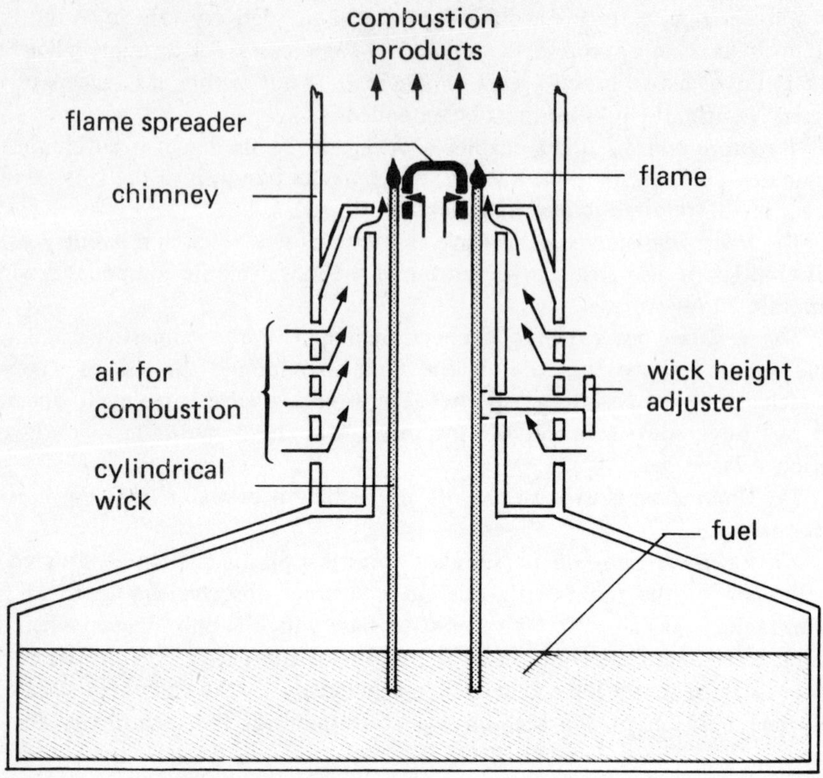

Fig. 3.3 – A wick type paraffin heater.

3.6.2 Drip-feed type burner

As with the wick-type this is also for small paraffin-burning heaters (Fig. 3.4). Paraffin is fed to the burner from an inverted fuel container standing in a bowl. The supply of fuel is controlled by a valve in the fuel line. The flame impinges on a metal mask making it hot. A polished reflector is placed behind the metal mask and therefore this type of heater resembles a radiant fire.

3.6.3 Vaporizing burner

This type of burner can be used with kerosine or domestic fuel oil. In it the fuel is vaporized prior to combustion, that is, it burns as a vapour (gas). The heat from the flame provides the energy necessary for the vaporization of new fuel. For initial ignition, enough energy must be provided to vapourize a small quantity of fuel and so establish the flame. There are three different types of this burner:

(1) pot-type burner
(2) sleeve-type burner
(3) rotating 'wall-flame' burner.

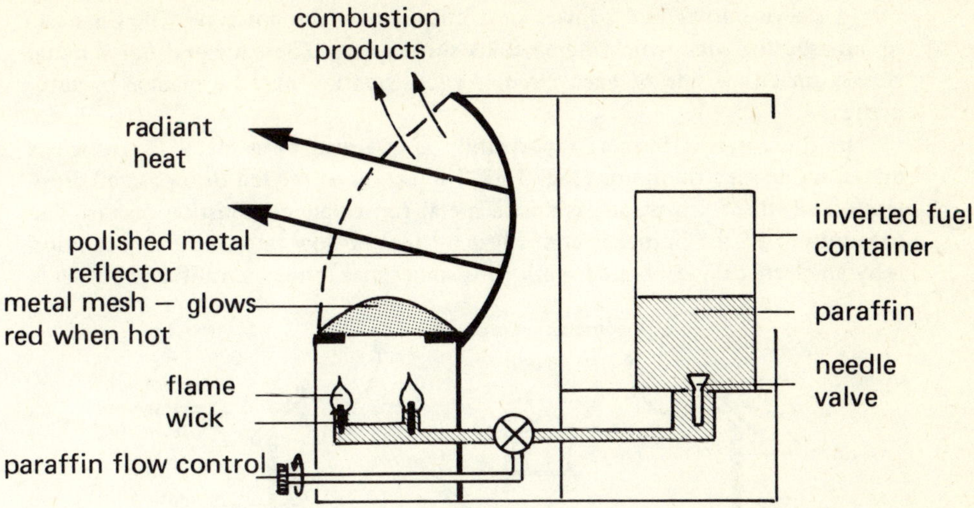

Fig. 3.4 – A drip-feed type paraffin heater.

A pot-type burner is illustrated in Fig. 3.5. The oil forms a thin layer over the base of the burner. The sides of the burner have perforated holes for the air supply. The flame is sited at **fire rings** on the walls of the burner. Flame ignition may be manual or automatic. An electric fan may be used to supply the air to the burner.

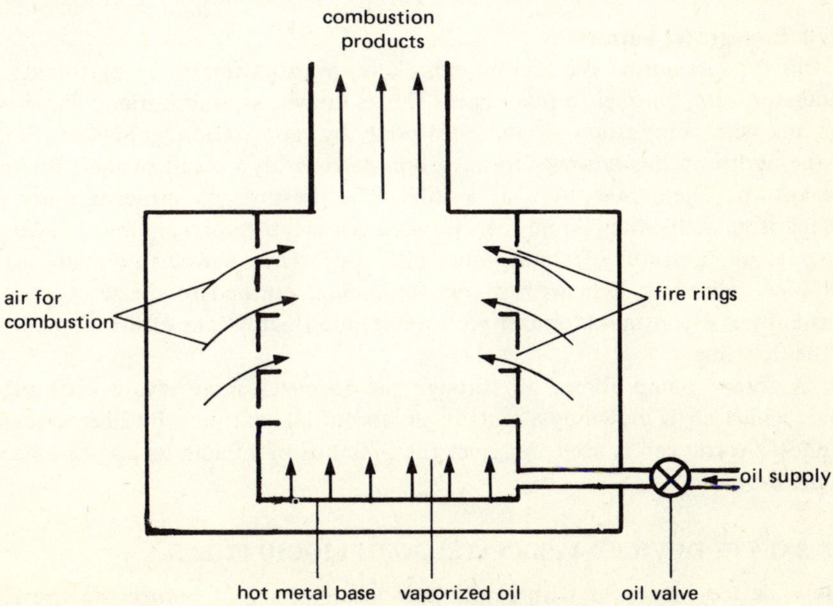

Fig. 3.5 – Pot-type vaporizing burner.

A sleeve burner has a lower heat output than the pot type. The oil soaks up an asbestos wick which increases its surface area. There are perforated metal sleeves on either side of each circular wick. Ignition may be manual or automatic.

In the rotating burner, vaporization is achieved by a metal fan which is driven by an electric motor (Fig. 3.6). The action of the fan throws small droplets of oil, dispersed in air, against a metal rim where combustion occurs. The heat output of the burner is controlled by the oil flow rate to the fan. Ignition is by an electrically-generated spark. It is sometimes termed a **wallflame** burner.

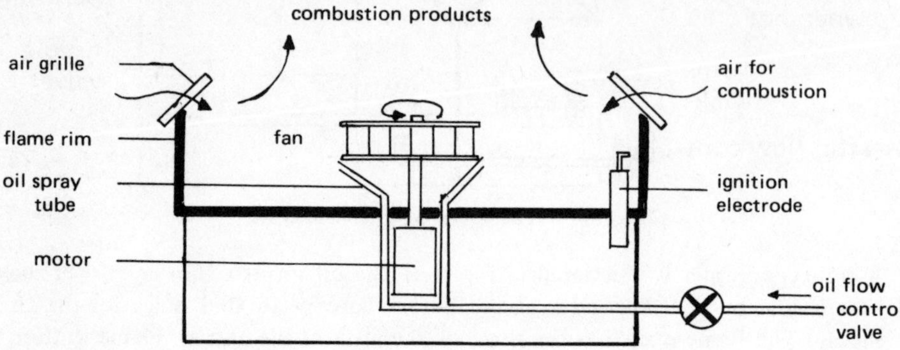

Fig. 3.6 – Rotating type vaporizing burner.

3.6.4 Pressure-jet burners

In this type of burner the fuel oil is broken up into a fine spray by forcing it, under pressure, through a small hole. This is known as **atomization.** The droplets mix with combustion air and burn partly by vaporization (gas phase). Some of the hydrocarbons however form carbon particles as a result of their thermal breakdown. These then burn as a solid. The pressure jet burner is noisy by comparison with other liquid-fuel burners. An electric rotary pump is used to generate oil pressures of about 7 bar (100 psi). These burners normally use a two-pipe delivery system with excess fuel being returned to the storage tank. This ensures a constant atomizer pressure despite fluctuations of the oil pressure in the flow line.

A rotary pump forces oil through the atomizer spray into a blast tube, down which air is blown by a centrifugal fan. Initial ignition is by electric spark. A photoelectric cell is used to detect the presence of a flame, as part of a flame failure device.

3.7 SAFETY DEVICES ASSOCIATED WITH LIQUID FUELS

Safety devices are used with liquid fuel installations, to reduce the fire risk associated with the use of liquid fuels. One group of devices is used to isolate the storage tank from the pipework within the house should the temperature of the environment surrounding the boiler rise above 65–70°C. A sensor monitors the environmental temperature close to the oil-burning appliance. This can either be a vapour-pressure phial (Fig. 2.18) or a fusible link (Fig. 3.7). The fusible link is made of two strips of metal welded together by a low melting point metal. It is held under tension. If the temperature rises, the weld melts, allowing the tension in the wire to be released, automatically closing a valve in the oil supply line. The aim of the device is to cut off the flow of oil to the appliance when there is a fire in the vicinity of the appliance. In this way, the fuel does not add to the fire problem.

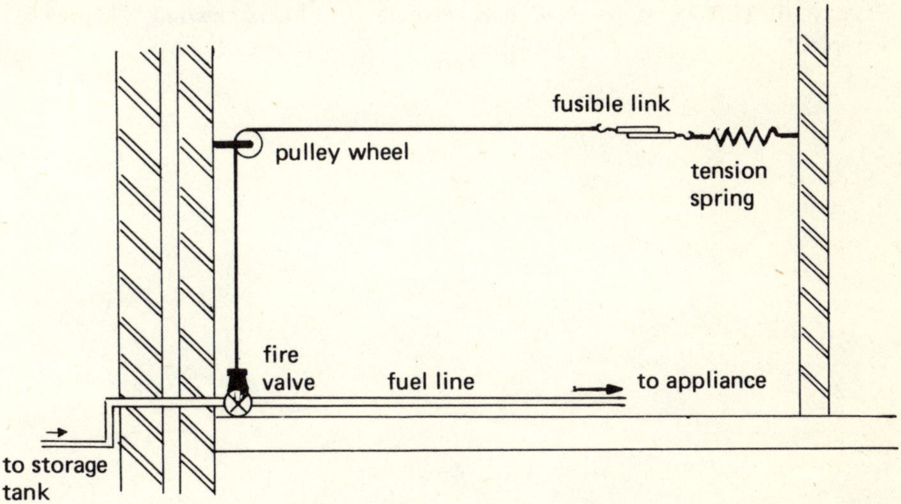

Fig. 3.7 – A fusible link isolating valve.

In some cases both a fusible link and a vapour-pressure phial are used. The fusible link is close to the valve and the phial is at a distance from it, but connected to it by a capillary tube.

Flame failure devices are used with oil burners for the same reason as those for gas burners (section 2.9). In the case of pressure jet burners photoelectric flame failure devices are used. These detect the light output from the flame and convert it into an electric current. This current operates a relay which controls the electricity supply to the fuel pump. Alternatively a bimetallic strip device may be used. This is mounted in the stream of the hot flue gases. If the flame fails, the temperature of the gases around the bimetallic strip falls, and the safety device is operated.

The use of portable oil burning appliances can be a fire hazard and because of this the design of these appliances should incorporate a number of safety features. They must for example be self-extinguishing if knocked over (see section 20.2.2).

BIBLIOGRAPHY

Brewster, H. R. (1970), Oil in the domestic heating market, *Domestic heating,* **3**, (3), p. 3–9.

Francis, W. & Peters, M. C. (1980), *Fuels and fuel technology,* 2nd Ed. Pergamon.

BP Marketing Ltd. (1973), *Storage and handling of domestic fuels.*

Locklin, D. W. (1969), *Small Oil Burners ASME Conference Proceedings, Los Angeles.*

McMullan, J. T., Morgan, R. & Murray, R. B. (1977), *Energy Resources,* Arnold.

Slome, J. (1968), *Domestic oil fired central heating,* Arco Publications.

Stanley, T. (1969), *More small bore heating,* Crosby Lockwood, Chapter 5, p. 125.

CHAPTER 4

Solid fuels

4.1 INTRODUCTION

Solid fuels originate from vegetable matter. Some vegetable materials, such as wood and straw, can be used directly as a fuel, in which case they can be considered to be a renewable fuel. The rate at which they can be used is related to their rate of growth together with the stock of living trees. In general, however, the amount of wood and vegetable materials used directly as a fuel source is small when compared with the total usage of solid fuels. Most of these were originally vegetable material which has been transformed under specific conditions of pressure and temperature and over long (geological) periods of time. This process does not result in a single end product, but in a range of products varying from **peat** to **anthracite.** Peat represents a vegetable material in its early stages of change, whilst anthracite represents the most advanced state of change from the original material. The extent to which this change has taken place determines the *rank* of the coal.

Table 4.1

Classification of solid fuels by rank

Low rank	Peat
↓	Lignite (brown coal)
↓	High volatile bituminous coal
↓	Low volatile bituminous coal
High rank	Anthracite

4.2 THE CONVERSION OF VEGETABLE MATTER TO COAL

Coal is a sedimentary rock, but it is unique because it consists mainly of organic material. Its major components are carbon, hydrogen and oxygen, with small amounts of nitrogen and sulphur. In addition, inorganic material is always found in coal, not as part of its chemical structure, but as a mixture with the organic material.

The formation of coal is thought to have started with the accumulation of vast quantities of dead and decaying woody materials; this was submerged in water because of an abrupt subsidence in the area. This could be in the form of a swamp, fresh water lake or sea. In all of these cases the decaying vegetable matter was covered by a layer of sand and silt, the typical components of sedimentary rocks. Because of the layers of sand, silt and water above the vegetable material, air was excluded, resulting in anaerobic decay of the material, i.e. by micro-organisms in the absence of air.

Two types of accumulation of woody material have lead to the formation of two types of coal deposit.

4.2.1 *In Situ* Accumulation

Extensive forests grew on swampy ground and the trees and other vegetable material died and lay where they had once grown. Over very long periods of time this material slowly sank into the swamp, causing the build up of a thick layer of vegetable material. This layer was covered with silt and sand deposits from the overlying water. This process of growth, death and subsidence took place several times, resulting in parallel uniform coal seams which were low in inorganic content (Fig. 4.1).

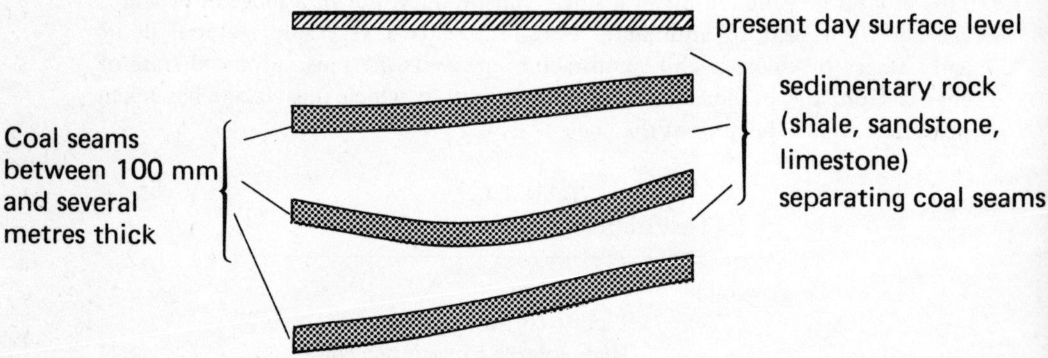

Fig. 4.1 – Cross-section of *in situ* coal field.

4.2.2 Drift Accumulation

In this situation dead vegetation drifted onto moving water which was part of a river system. The material accumulated at a point where there was little movement of water. Here it was eventually submerged and buried, resulting in thick localised deposits with a basin-like structure (Fig. 4.2).

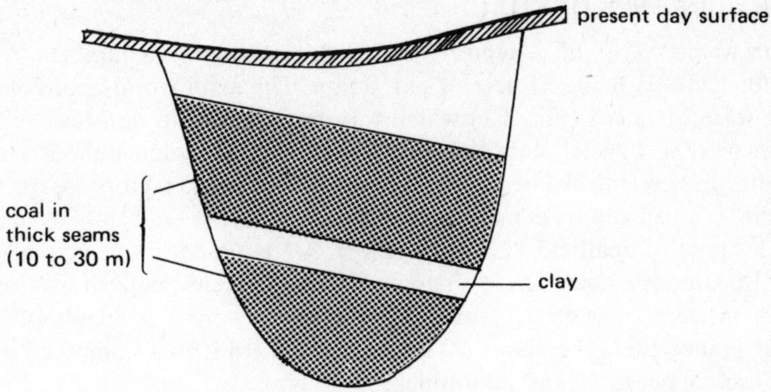

Fig. 4.2 – Cross-section of 'drift mechanism' coal field.

4.2.3 Coalification

The conversion of an accumulation of layers of submerged vegetable matter into coal is known as **coalification.** Initially the buried partially decayed matter would have a consistency like that of peat. Further changes to the material were the result of elevated temperatures and high pressures (caused by the weight of overlying sedimentary rocks), coupled with the passage of millions of years. This resulted in a loss of water from the tissue and chemical changes leading to a reduction in the proportion of oxygen and hydrogen and an increase in the proportion of carbon. Peat represents the first stage in the conversion; it is found near the surface. Lignite or brown coal is more compact and drier than peat, but the remains of woody material can still be seen in it. Bituminous coals are much drier than lignite and, other than the occasional fossil, the vegetable material cannot be seen with the naked eye. Anthracite has a high carbon content and is hard, black and bright in appearance. Table 4.2 shows the changes in the proportions of carbon, hydrogen and oxygen with the progress of coalification.

Table 4.2

Principle components of solid fuels (% dry matter)

Solid Fuel	Carbon	Hydrogen	Oxygen
Peat	60	6	34
Lignite	67	5	28
Bituminous coal	88	6	6
Anthracite	94	3	3

4.3 OCCURRENCE OF COAL

The economic value of a deposit of coal depends upon a number of factors, the principal one being its ease of extraction. The main world coal fields occur in the form of seams (Fig. 4.1) which vary in thickness from a few millimetres to four metres. Typical seam thickness for British coalfields is about 1 to $1\frac{1}{2}$ m. In addition to seam thickness, other important economic factors are the number of seams in a bed and its depth below the surface.

The type of coalfield illustrated in Fig. 4.2 is found in the southern hemisphere. A famous example is in Victoria, Australia where a deposit of brown coal 300 m in depth has been found, interupted only by thin bands of clay. In central France there are about 50 small coalfields which are composed of 3 or 4 seams each between 10 and 20 m thick.

4.4 EXTRACTION AND PREPARATION OF COAL

There are two major methods used for the extraction of coal, underground mines and strip, or surface, mines. Underground mining is expensive and dangerous. It is used when the coal is sufficiently far below ground level to make strip mining impossible. Strip mining is used where the seems are relatively close to the surface and where the overlying surface soil can be removed before the coal is extracted.

Following the mining process, coal is washed and graded by size. Washing removes dust, dirt and shale from the coal, making it more acceptable to the consumer. The washing process uses the fact that shale and dirt have a higher density than coal and therefore sink in the washing liquid whereas coal floats. Screens are used to separate the coal into pieces of different size ranges, for example 25 to 50 mm, each size relating to the end use of the coal.

In the UK domestic users have their solid fuel delivered by road, either in bulk or, following automatic or manual bagging, in sacks. An alternative is the use of specially designed road vehicles which transport the fuel in bulk, the fuel being bagged and weighed when the vehicle reaches the customers' door. Coal and other solid fuels are also available in small pre-packaged quantities for customers requiring small volumes of fuel or for those who do not have adequate storage facilities.

4.5 DOMESTIC STORAGE OF SOLID FUELS

The storage volume for solid fuels depends upon a number of factors, such as the type of fuel used, the rate of usage, frequency of delivery, the cost of a unit quantity of fuel and discounts for bulk delivery.

Coal and solid fuels are sold by weight and this can be related to storage volume using the bulk density of the fuel. The bulk density of a solid fuel is its mass in lump form per unit volume (including the air volumes between lumps). Anthracite, coal and smokeless coals have a bulk density of approximately 700 kg/m^3 whereas cokes and coke-like smokeless fuels have a bulk density of approximately 500 kg/m^3 (Bradbury, 1973). In general, it is advantageous to have as large a fuel store as possible since it is then possible to buy a significant proportion of the household's annual consumption during off-season periods when there is a price advantage. A large store will also lead to less frequent deliveries and a lower risk of running out of fuel during long or unexpected periods of inclement weather.

The store volume will depend on the number of appliances which utilise the fuel and whether the fuel is being used for space heating, water heating or both. Typical domestic fuel bunkers range in size from $\frac{1}{4}$ m^3 to $1\frac{1}{2}$ m^3.

The siting of a solid fuel store is very important. In addition to aesthetic considerations, it should provide easy access both for delivery of the fuel and for its removal before transport into the house. The store should be under cover otherwise the fuel will absorb moisture during rain and snow. The increased moisture content will adversely affect its combustion properties. Also it should be possible for the householder to remove fuel from the bunker and transfer it into the house in the dry. It must be remembered that the greatest demand for solid fuel will be during periods of unpleasant weather conditions. The store can be made from prefabricated concrete panels or built from bricks and building blocks. In some cases, where fuel requirements of appliances differ, two or more compartments may be needed to segregate the different fuels.

4.6 COMPOSITION AND CHARACTERISTICS OF COALS

Coal does not represent a single material. Coals from different locations show marked variation in composition and properties which affect their quality and suitability for different applications. Coal as it is mined consists of three parts:

(a) *organic matter* derived from the original vegetable material;
(b) *moisture* derived both from the vegetable material and also from the water which covered it;
(c) *mineral matter* from the sedimentary silt which covered the vegetable material and mixed with it during coalification.

4.6.1 Organic Matter

The major components of the organic matter are carbon, hydrogen, oxygen and sulphur. An important factor affecting the characteristics of coal is its volatile matter. This is the proportion of the coal which is given off as gas and smoke when it is heated in the absence of air and under specified heating conditions. The volatile matter is important for two reasons. Firstly the gaseous proportion of the volatile matter determines (a) the ease with which a solid fuel can be ignited and (b) the rate at which it burns. This property is known as the *reactivity* of a coal. The second factor related to the volatile matter is the amount of smoke produced on combustion, which can lead to atmospheric pollution (see section 4.8). Coals with a low volatile matter are those which have been subjected to the highest degree of coalification, that is, of highest rank such as anthracite, which has less than 10% volatile matter. This compares with 30–40% for some bituminous coals. The sulphur content is also important because of its potential for causing atmospheric pollution.

A further property of coal is related to its ability to fuse together forming a solid piece of coke on heating in the absence of air. This in known as the *caking* power of the coal and is an important factor in determining the application for which any given coal is suitable.

For domestic use in open fires, a coal of high volatile matter (greater than 32%) and weakly caking properties is required. In closed stoves, where combustion conditions are more closely controlled, a low volatile (9–14%) coal is satisfactory. These coals (anthracite and steam coal) are also the only naturally occuring smokeless fuels.

Table 4.3
Calorific values of solid fuels

Solid fuel	Net Calorific Value MJ/kg (dry and free of mineral matter)
Anthracite	34
Welsh dry steam coal	33
Bituminous coal	33–36 (increases with increasing rank)
Smokeless fuels	29–33
Coke	29

The calorific value (see section 2.6.2) of a solid fuel is due to its organic matter. Typical values are recorded in Table 4.3. These values are for dry coal, free of mineral matter, that is for the organic matter.

4.6.2 Moisture Content

In addition to the water associated with coal as it exists underground, water may be found in a sample of coal which is the result of the washing process or of rainfall. Since coal is sold by weight this additional water is considered to be an important factor. It is often referred to as *free water* since it is easily removed by evaporation under normal temperature and humidity conditions. Some moisture remians after the free water has been removed because coal is a porous material.

4.6.3 Mineral Matter

The amount of mineral matter varies typically from 3 to 10% although in some cases it may be much higher or lower. It is usually measured as the *ash*, which is the proportion of a sample of coal remaining after all of the organic material has been removed by combustion at a given temperature. The higher the ash of a coal, the more frequently will a domestic appliance have to be emptied. Also if ash is heated to a high temperature it may fuse together forming large lumps of *clinker* which can be a problem in appliances which have automatic and semi-automatic ash removal and in long burning appliances. Here the clinker will build up over a period of time and may eventually affect the supply of air to the fuel and can be difficult to remove. A domestic solid fuel should have an ash of no more than 5–6%.

4.6.4 Size

The size of fuel burned on a grate should be fairly uniform. Otherwise small particles fill up the voids (air spaces) between the large particles. This reduces the rate of air flow through the bed of burning fuel on the grate and in this way reduces the rate of combustion.

4.7 COMBUSTION OF COAL

Before combustion of coal can occur two conditions must be satisfied. Firstly the fuel must be heated to a specific temperature before ignition will occur. This is known as the **ignition temperature.** Typical values are 300°C for wood, peat and bituminous coal, 400°C for anthracite and Welsh dry steam coal and 700°C for coke and carbon. Secondly there must be an adequate supply of air which is in close contact with the coal. Initially, when the coal is heated volatile matter is evolved leaving coke, the major component of which is carbon. The volatile matter contains inflammable gases which burn above the coal on the grate. This aids the ignition of the coal. The coke burns as carbon, that is

$$2C + O_2 + 3.76\,N_2 = 2CO + 3.76\,N_2$$

$$2CO + O_2 + 3.76\,N_2 = 2CO_2 + 3.76\,N_2$$

Overall: $$2C + 2O_2 + 7.52\,N_2 = 2CO_2 + 7.52\,N_2\,.$$

The conversion of carbon to carbon monoxide occurs in the fuel bed and the combustion of carbon monoxide above the fuel bed. Sticks, fire lighters or a gas ignition system may be used to achieve the high temperature required for initial ignition.

Reactive coals are much easier to ignite and burn more freely then low-reactivity fuels. For this reason combustion conditions are much more critical for smokeless fuels than for bituminous coals (see section 4.10). In all cases a good flow of air through the fuel bed and up the flue is essential. Initially for ignition to occur a good flow of air is required but once combustion has been achieved the flow of air can be reduced to control the rate of combustion. Therefore some means of controlling the supply of air to the fuel bed is required (see Fig. 4.3). Air for combustion enters through an adjustable slot. For ignition this is opened fully until a good rate of combustion is established. The size of the gap can then be reduced to control the rate of combustion. In the case of appliances designed for use with low reactive fuels, a fan is sometimes used to provide forced draught ensuring a good supply of air to the burning fuel.

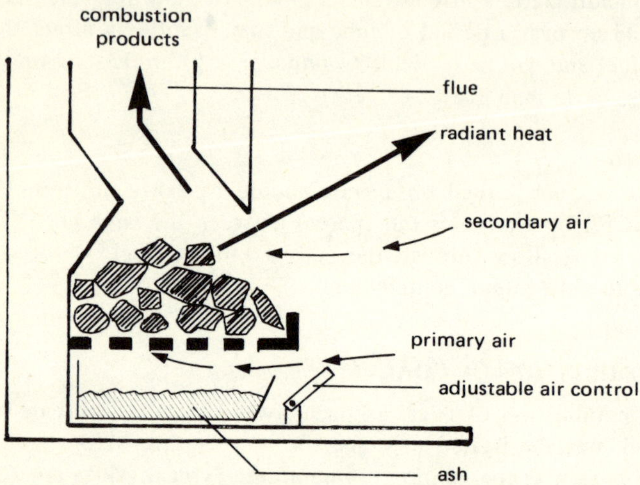

Fig. 4.3 – Combustion of a solid fuel in an open fire.

Specifications for solid fuels often incorporate levels of performance under controlled combustion conditions. In the case of open fires (see section 9.2.1.1) the radiant heat output is very important, particularly the time to reach a given heat output after ignition and also the rate of recovery of heat output after the fire has been refuelled. A typical test to measure these properties is shown in Fig. 4.4. In this test the fuel is ignited under specified conditions and the time taken to achieve a defined level of heat output (A) measured. Following this the peak level of heat output (I) is recorded and, when the output falls to a specified level (B) an additional quantity of fuel is added to the fire. The second peak (II) is then recorded and the fire refuelled for a second time when its heat output falls again to (B). The third radiation peak (III) is then measured.

Another important property of solid fuels is their ability to burn at low levels of combustion (for example overnight) and then to be capable of rapid recovery to the normal level of heat output. A solid fuel for use in a domestic hot water boiler should be capable of operating for periods of 10–12 hours at a rate of combustion of 15% of the maximum value.

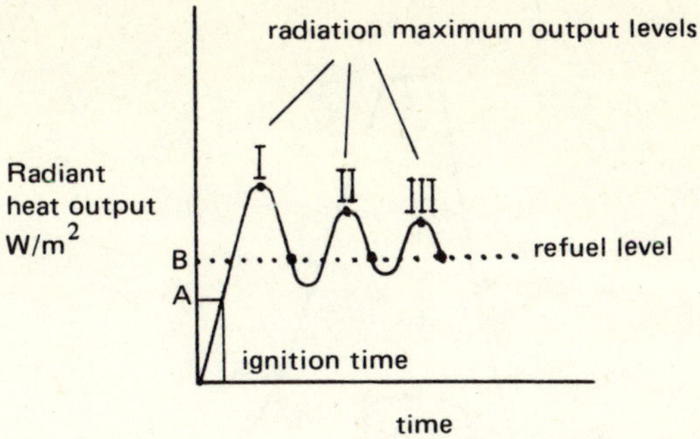

Fig. 4.4 – Three peak radiation test for a domestic solid fuel.

4.8 ATMOSPHERIC POLLUTION CAUSED BY THE COMBUSTION OF COAL

During the period between the 5th and 9th of December 1952, London was enclosed in still cold fog, and smoke was also trapped at ground level resulting in very dense smog. (Smog is the term used to describe the mixture of smoke and fog.) It is estimated that this smog was responsible for 3 500 to 4 000 deaths. In particular, deaths from bronchitis increased 9.4 fold, pneumonia by 4.5, tuberculosis by 4.5 and heart disease by 2.5. Most of these deaths were in elderly people who had a history of respiratory illnesses (see Fig. 4.5). This lead, in the UK, to serious concern about the effects of smoke on the population's health.

Pollutants found in the atmosphere are of two types: *primary pollutants* are those which result from direct combustion of the fuel and *secondary pollutants* which are formed in the atmosphere by chemical reaction. As an example, combustion of some sulphur containing fuels can lead to the emission into the atmosphere of sulphur dioxide and hydrogen sulphide. These are primary pollutants. The sulphur dioxide and hydrogen sulphide can react in the atmosphere to produce sulphur trioxide, sulphuric acid and sulphates (Seinfeld, 1975).

The major pollutants associated with the combustion of coal are smoke and sulphur dioxide. The smoke contains dust, grit and organic compounds, some of which have carcinogenic properties, notably benzpyrene. *Threshold limits* have been set for the major atmospheric pollutants such as those shown in Table 4.4.

Table 4.4

Limits for major pollutants (USA) (Sienfeld 1975)

Pollutant	Clean Air	Polluted Air
Sulphur dioxide	<0.01 ppm	0.02–2 ppm
Carbon dioxide	310–330 ppm	350–700 ppm
Carbon monoxide	<1 ppm	5–200 ppm

ppm = parts per million

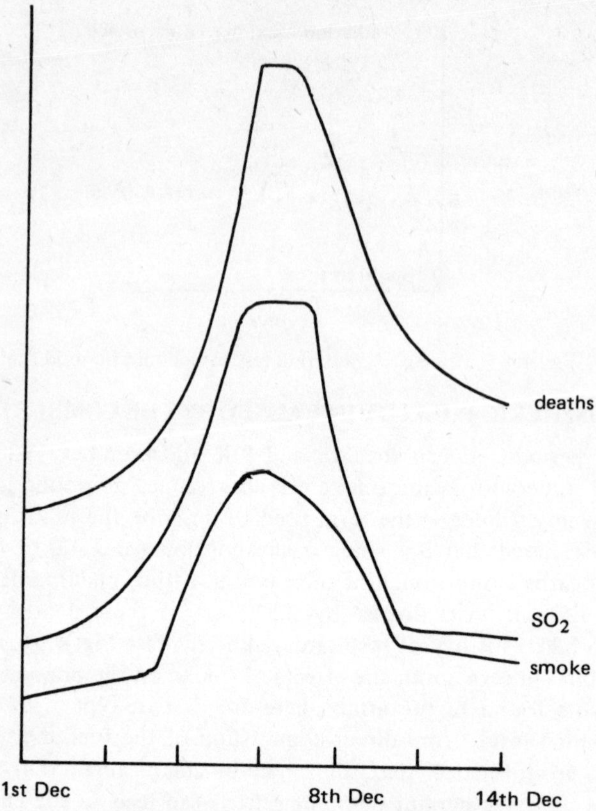

Fig. 4.5 – Deaths and pollution levels in London during December 1952.

These air pollutants have a number of undesirable effects. For example, they reduce visibility and lead to the formation of fog which reduces the amount of solar energy reaching ground level. Combustion products from the burning of fuels can affect temperature distributions and wind patterns. The pollutants can also affect the growth of vegetation. Many, such as sulphur dioxide, are corrosive. Finally they can have harmful effects on health. Many of these effects, particularly at low levels of pollution, are long term and it is impossible to be certain about what represents a safe level for these materials. In the UK, which has a history of burning coal domestically, almost half of the smoke pollution in towns and cities was from domestic fires before legislation was introduced.

4.9 CLEAN AIR LEGISLATION

Legislation has been passed in many countries to regulate the amount of pollutants which may be emitted into the atmosphere. In the UK the Clean Air Acts of 1956 and 1968 give local authorities control over the emission of dark smoke, grit, dust and fumes and the heights of chimneys.

Control over smoke is possible in two different ways. Firstly, it is an offence to emit dark smoke (except under certain specified temporary conditions). Secondly the local authority can establish *smoke control areas.* In these areas only specified fuels (which do not produce smoke) and appliances may be used.

In the USA the Clean Air Act of 1967 designated air quality control regions, giving greatest priority to the most heavily polluted areas. It was amended in 1970 to include all regions as air quality control regions. The Environmental Protection Agency establishes maximum levels of pollutants particularly sulphur oxides, particulates and nitrogen oxides. The latter is a pollutant caused by the internal combustion engine.

The effects of clean air legislation can be seen in Fig. 4.6 which shows the increase in sunshine levels measured at the London weather centre brought about by more effective pollution controls.

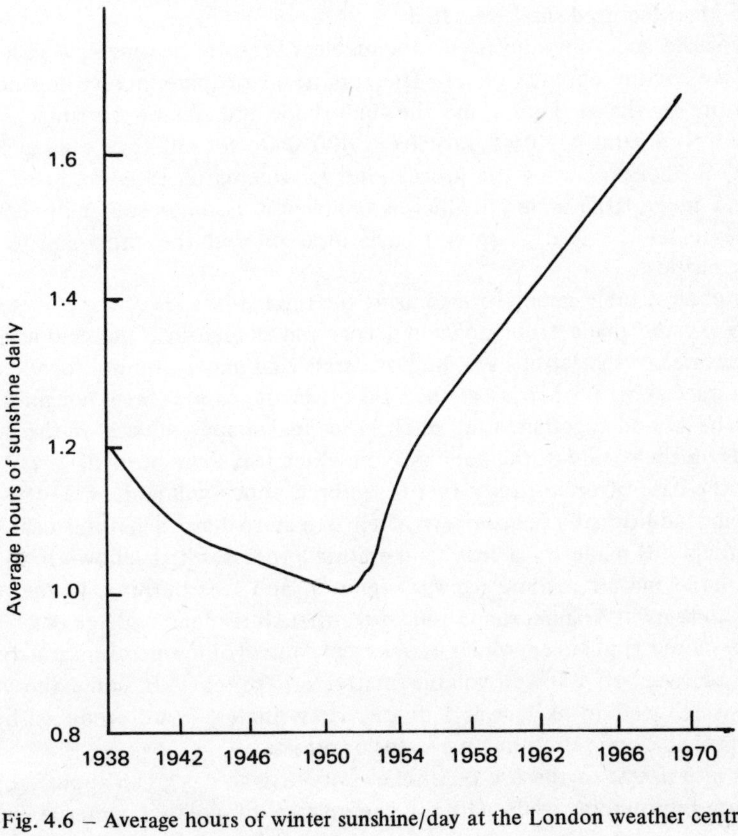

Fig. 4.6 – Average hours of winter sunshine/day at the London weather centre.

4.10 SMOKELESS FUELS

One of the most significant effects of the Clean Air Act in the UK has been a switch from the use of bituminous coals to smokeless fuels resulting from the designation of *smoke control areas* by local authorities. One disadvantage of smokeless fuels is that they have a low reactivity and therefore they are more difficult to ignite and burn. For this reason special appliances are often required to burn smokeless fuels. In particular, it is important that the correct fuel is selected to suit a given appliance. Some coals, as mined, have a low volatile matter and these are known as natural smokeless fuels. These represent only 3% of the UK coal output. Others are manufactured from bituminous coals so that the demand for smokeless fuels can be met.

4.10.1 Natural smokeless fuels

In the UK there are two main classes of natural smokeless fuel: anthracite and Welsh low volatile coal. Both of these coals do not ignite easily and burn slowly; they are better suited to closed appliances than to open fires.

4.10.2 Manufactured smokeless fuels

High volatile coals are converted to smokeless fuels by heating coal in a closed retort, i.e. in the absence of air. The conditions of manufacture depend upon the nature of the coal used and the final shape and size of the smokeless fuel required. Temperatures used vary from 400°C up to 950°C or even as high as 1050°C for cokes. During the process, the volatile matter is reduced to less than 10% and the volatile material which is removed is used as a source of chemicals. In many cases the coal gas produced is used to heat the retort and for other heating purposes.

Smokeless fuels can be formed using the original shapes of the pieces of coal or they can be made from powdered coal and shaped into pieces during their manufacture. In the latter case the powdered coal may be bound together using the natural caking properties of the coal or, in the case of weakly caking coals, it may be bound together using pitch. The final shape and size of the product depends on the nature of the appliance on which it is to be burned.

In the case of open fires a fuel is required which will ignite easily and burn freely and additionally, consumers often like it to have a natural coal shape. These fuels are made by a low temperature process which allows it to retain 10% volatile matter making for easy ignition and free burning. In the process the coal retains its original shape and size, although its density decreases.

For closed appliances, which have better control of temperature and draught, smokeless fuels of a lower volatile matter are suitable. In some the volatile matter is reduced to as low as 1 or 2%. These materials are produced by high temperature carbonisation in coke ovens or retorts.

An alternative to the use of smokeless fuels is to design an appliance which can burn bituminous coals without the release of volatile material into the atmosphere. This can be achieved by collecting and passing the combustion products through the fire bed. At these high combustion temperatures many of the components of the smoke are oxidised to carbon dioxide.

4.11 WOOD AND CHARCOAL

With the rising cost of fuels there has been a resurgence of interest in the use of wood burning appliances. Because of their low bulk density an appliance burning these fuels needs more frequent filling than an equivalent solid fuel burning appliance. They also require a supply of chopped dry wood of a suitable type. Flue conditions can often be critical in these appliances.

Charcoal, which is manufactured by heating wood in a closed retort, consists almost entirely of carbon and is used in barbeque grills.

REFERENCES

Bradbury, J. (1973), *Solid fuel in the home*, 2nd Edition, Women's Solid Fuel Council.

Seinfeld, J. H. (1975), *Air pollution – physical and chemical fundamentals*, McGraw-Hill.

BIBLIOGRAPHY

Clarke, J. W., Weissman, W. & Hammer, M. J. (1977), *Water supply and Pollution control*, 3rd Edition, IEP.

Francis, W. & Peters, M. C. (1980), *Fuels and fuel technology*, 2nd Edition. Pergamon.

Potter, N. M. & Martindale, J. R. (1966), Modern Developments in Smokeless Fuels, *The Mining Engineer*, **Dec.** pp. 195–205.

Royal Commission on Environmental Pollution (1971), *First Report*, HMSO.

CHAPTER 5

Electricity

5.1 INTRODUCTION

The energy associated with electricity is related to the distribution of electrical charges. If one area has an excess of negatively charged electrons and a second area has a deficiency of electrons, a flow of electrons from the first area to the second will result in a release of energy. One way in which this flow of electrons can occur is along a **conductor** which connects the two areas.

An example of a good electrical conductor is copper. The copper atom contains only one electron in its outer shell and this electron is held by relatively weak forces, and can move easily from one atom to another. Within a piece of the metal, copper electrons will be in constant random movement between the atoms, as shown in Fig. 5.1.

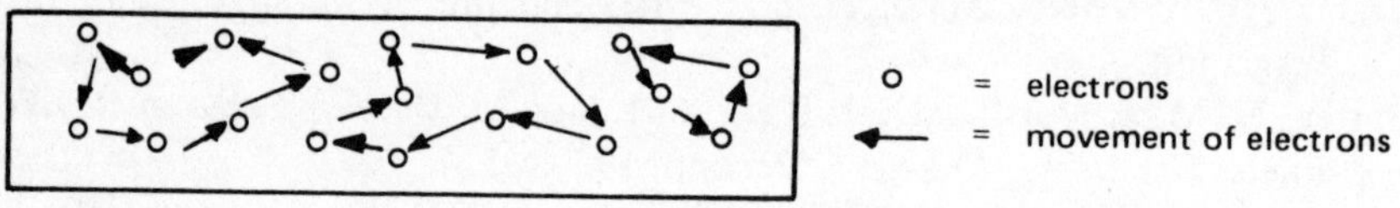

Fig. 5.1 – Random movement of electrons within a copper bar.

If one end of a copper bar has a defieciency of electrons and the other end a surplus of electrons, there will be a general movement of electrons from the end with an excess of electrons to the end with a deficit, producing a flow of electrons. This flow of electrons is known as a **current.**

A battery has two terminals, one of which has a surplus of electrons and the other a deficit. This uneven distribution of electrons is produced by a chemical reaction taking place inside the battery. If a piece of copper wire is connected between the two terminals (Fig. 5.2), electrons will flow along the piece of copper wire until the uneven distribution of electrons between the terminals no longer exists. At this point, when the two terminals have equal numbers of electrons, the battery is said to be 'flat'.

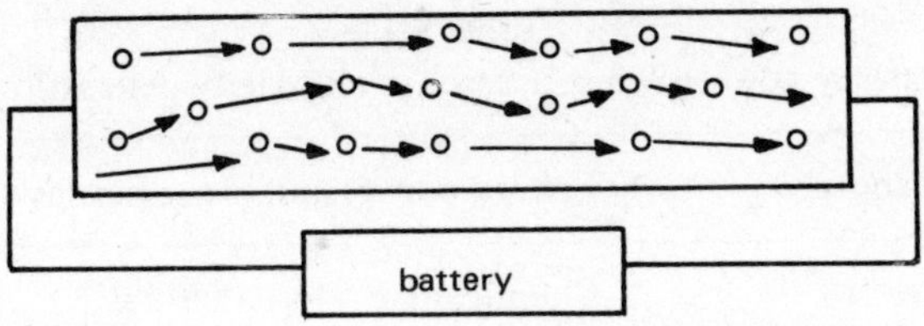

Fig. 5.2 – Flow of electric current in a copper bar.

The atoms of good electrical conductors have only one or two electrons in their outer electron shells. Examples of good conductors are copper, silver, gold, aluminium, brass, platinum, acids and water. In the case of acids and water the electrical charges are carried by positively charged hydrogen ions (H^+).

Materials which do not conduct electricity easily are known as **insulators.** A feature of these materials is that their atoms have full, or almost full, outer shells. In this case the electrons are tightly bound into the atom and as a consequence are much less free to move. Examples of good insulators are plastics, glass, rubber, dry paper, dry air and mica. Insulators such as plastic and rubber are commonly used to insulate electrical cables and appliances. Under certain circumstances insulators can **breakdown** and will then conduct electricity. It is therefore important that insulators used in electrical appliances are completely safe under the conditions in which they will be used. By the term 'breakdown' we mean that the material ceases to act as an insulator and becomes a conductor. The electrical current (I) flowing in a conductor is measured in **amperes** or **amps.** The difference in electrical charge between any two points, which is the driving force for the flow of electrons, is known as the **voltage** or **electro-motive force** (E) and is measured in **volts.**

The relationship between the voltage at two ends of a wire, and the current flowing along the wire is given by **Ohm's Law.** This states that for a conductor at a constant temperature, the voltage across the conductor is proportional to the current flowing through it. This may be written as:

$$\frac{\text{Volts } (E)}{\text{Amperes } (I)} = \text{a constant} = \text{resistance } (R)$$

The resistance of a conductor is measured in ohms (Ω).

5.2 GENERATION OF ELECTRICITY

5.2.1 Energy conversions in electricity generation

Electricity can be regarded as a secondary fuel since it is always generated by an energy conversion process, as shown in Table 5.1.

Table 5.1
Energy conversion processes in electricity generation

Primary energy source	Energy conversion processes involved
Fossil fuel	Chemical to thermal Thermal to mechanical Mechanical to electrical
Nuclear	Nuclear to thermal Thermal to mechanical Mechanical to electrical
Hydro-electric and Tidal Power	Mechanical (potential) to mechanical (kinetic) Mechanical (kinetic) to electrical
Wind power and Wave power	Mechanical (kinetic) to electrical
Solar energy	Electromagnetic to electrical
Geothermal energy	Thermal to mechanical Mechanical to electrical

The principal methods used for the generation of electricity are from fossil fuels (chiefly coal), nuclear energy and hydro-electricity. The balance of these depends mainly on a country's indigenous energy supplies. For example in the USSR, Norway, Sweden and Switzerland a large proportion of electrical energy is generated by hydro-electric turbines. In many industrial countries fossil fuels form the major energy source. In England and Wales for example, over 75% of all electricity is generated from coal with only 10% from oil and 0.5% from gas. In the UK as a whole approximately 15% of electrical energy is from nuclear power stations. In the USA 50% of electricity generated is from coal, 20% from natural gas, 20% from oil, with approximately 5% each from hydroelectric and nuclear power stations.

5.2.2 The steam powered turbine

All electricity power stations which use fossil fuels (except gas – see section 5.2.5) and many nuclear power stations, convert heat, from the chemical or nuclear reaction, into steam. This steam is then used to drive a steam turbine. It is not possible to use the combustion products directly to drive a gas turbine because of the formation of ash and because of the corrosive nature of the combustion products from oil and coal. The steam turbine converts the pressure of the steam (a form of potential energy) into a turning movement (a form of kinetic energy). This rotational movement of a shaft produced by the turbine is then used to drive an electrical generator. The process is illustrated schematically in Fig. 5.3. The efficiency of this multi-stage energy conversion process is low, varying from 25 to 36%. The waste energy, usually in the form of warm water, is discharged into rivers and the sea. By making use of this waste heat it is possible to improve the overall efficiency of the process to 75 or 85% (Reay, 1977). A number of ways have been suggested for the recovery of this waste heat such as its use for heating greenhouses, in fish farming and for the supply of domestic hot water and space heating (combined heat and power system). The latter is discussed in more detail in section 9.5. Alternatively the waste heat could be used as the energy source for a heat pump (Sumner 1976).

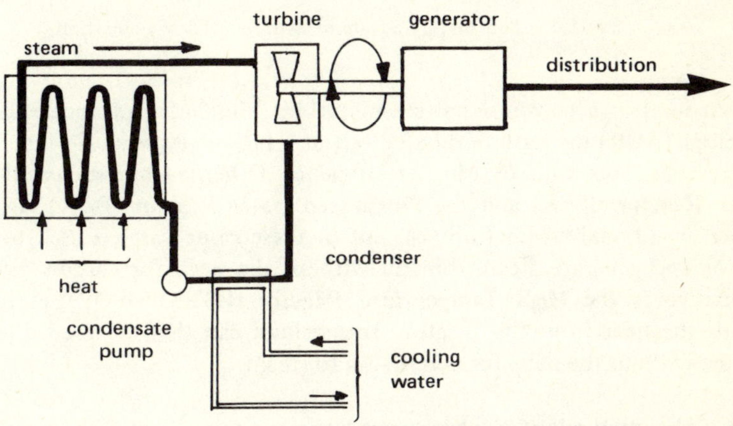

Fig. 5.3 – Schematic diagram of steam powered electrical generation.

5.2.3 Water turbines

Water turbines are used in hydro and tidal electricity generation. The flow of water through the turbine is used to rotate the shaft of a generator. The water for hydro-generation is supplied from a reservoir situated above the turbines (Fig. 5.4).

Surplus electricity can be used to drive the attached generator of a water turbine in reverse to elevate water to the reservoir level. This water can then be used to drive the turbine to produce more electricity when required. The use of these pumped storage systems is common in mountainous areas (Overman 1974).

5.2.4 Nuclear electricity generation

The first generation of nuclear reactors in the UK (MAGNOX) use carbon dioxide to transfer heat from the reactor to a boiler which generates steam. This

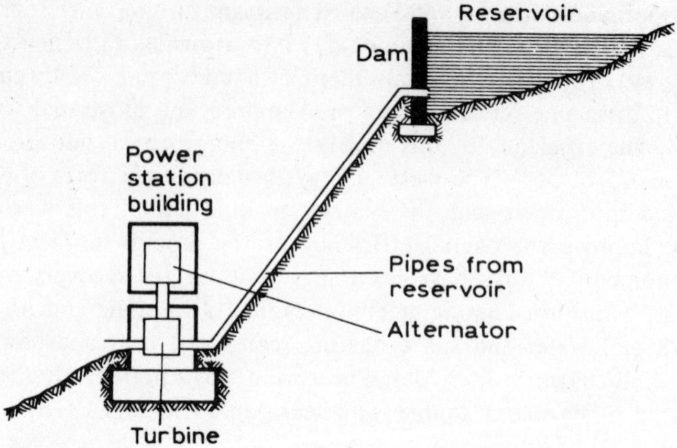

Fig. 5.4 – Schematic diagram of hydro-electrical generation.

is then used in a conventional steam turbine. Similarly **Advanced Gas Cooled Reactors** (AGR) use carbon dioxide but at a higher temperature than the Magnox reactor. This leads to a higher efficiency. Other reactors such as the **Boiling Water Reactor** (BWR) and the **Pressurised Water Reactor** (PWR) also operate with conventional steam turbines, but in these cases water is used to cool the reactor and generate steam directly without the need for carbon dioxide. An alternative is the **High Temperature Reactor** (HTR), which uses helium to absorb the heat from the reactor. This helium can then be used directly in a turbine without the need for conversion to steam.

5.2.5 Other methods of electricity generation

Gas turbines are fired directly with fossil fuels such as natural gas and LPG because these fuels do not produce ash and/or corrosive combustion products. They start up faster than steam turbines and cost less to build than coal, oil and nuclear power stations. However, they have higher running costs because of their use of premium fuels. For these reasons they are used to supply electricity at times of peak demand.

All methods of generating electricity so far described utilise a turbine for the generation of a rotational force which is used to drive the electrical generator. The same principle applies in the case of wind power, where a *windmill* can be considered as an air turbine (Hickok 1975). Wave power research is still in progress to design a reliable system for the conversion of the kinetic energy in waves into electrical energy. Most of these systems convert the wave energy into pressure energy (commonly pressurised air) which is then used to drive a turbine. An alternative is to use the difference in water levels between the peaks and troughs of the waves to drive a turbine directly.

5.3 TRANSMISSION OF ELECTRICITY

5.3.1 Transmission losses

Electricity is transported from the power station to the consumer by a high voltage cable. The reason for using a high voltage is that it reduces the energy loss resulting from heat generated in the cable. The cable, although a good conductor, still has some resistance to the flow of electrons and because of this some heat is generated. When an electric current I flows along a wire there is a conversion of electrical energy into thermal energy caused by the resistance R of the wire to the flow of electrons. The amount of power P lost as the electricity passes down the wire depends on two factors: the magnitude of the current and the resistance of the wire. The power loss is given by:

$$P = I^2R \ .$$

This power loss is undesirable because of its effect on the overall efficiency of the power supply system. The heat in the transmission line also reduces its life. This heating effect can be reduced by minimising both the current flowing and the resistance of the wire.

The resistance can be minimised by using wire of large diameter made from a material of low resistance and by keeping transmission distances as short as possible. Current can be minimised by increasing the voltage. The amount of power in watts W transmitted along a cable is equal to the product of voltage V and current I that is,

$$P = VI \ .$$

By using a transmission voltage of 400 kV the current can be reduced to less than a thousandth of that at 250 V. A number of different voltages are used for the electricity distribution system. In the UK, older parts of the system operate at 132 kV but newer installations use 275 kV or 400 kV. In the USA, transmissions are typically at 138 kV and 345 kV. Examples of transmission voltages as high as 500 kV are common and 750 kV is used in some cases.

5.3.2 Alternating current (a.c.) electricity

In section 5.1 an electric current was explained in terms of a flow of electrons. The electrons were seen to be flowing in one direction along a conductor. This kind of current is called a **direct current** (d.c.). Its characteristic is that current and voltage remain constant with time.

If the voltage across the conductor were to change as shown in Fig. 5.5 the current would flow first in one direction along the wire and then in the reverse direction. This is known as an **alternating current** (a.c.).

Reference to Fig. 5.5 shows that the current and voltage are zero at A, rise to a maximum positive value at X and then fall back to zero. The current then flows in the opposite direction until it reaches a maximum negative value at Z

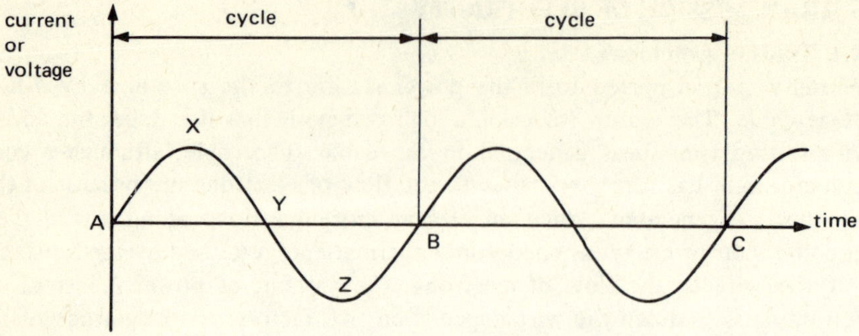

Fig. 5.5 – The cycle of an alternating current.

and then decreases once more to zero at B. This process is then repeated continuously and each change from A to B is known as a cycle. An a.c. supply is characterised by the number of these cycles occurring every second. In the UK electricity is supplied at 50 Hz (1 Hz = 1 cycle/second) and in the USA the supply is commonly at 60 Hz with some areas having supplies at 50 Hz.

In addition to the frequency, electricity supplies are also characterised by a voltage. Since over one cycle the voltage of the positive part of the cycle is equal in magnitude but opposite in sign to the negative half of the cycle, the average voltage is zero. To obtain an avergage value the *root mean square (RMS) value* is used. This is the average value of the voltage over one cycle, ignoring the sign of the voltage.

$$\text{Root mean square voltage} = 0.707 \times \text{maximum voltage} \ .$$

An a.c. electrical supply consists, in its simplest form of two wires: a live (line) wire and a neutral wire. This is referred to as a single phase two-wire system and is used in the UK. The voltage across the two lines is 240 V. An alternative single phase system, in common use in the USA, is a single phase three wire system. In this system there are two lines each at 115–120 V compared with the neutral wire (Fig. 5.6).

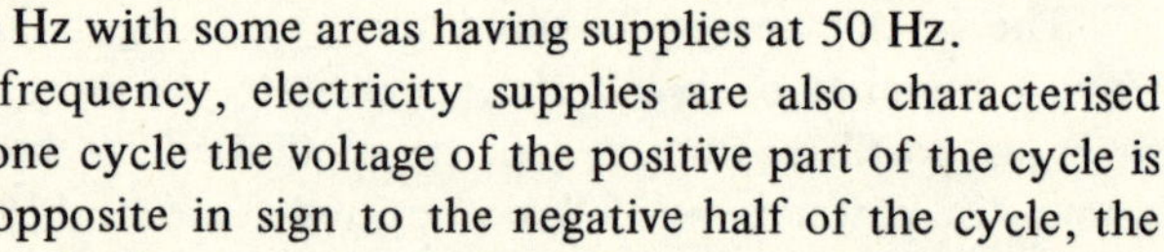

Fig. 5.6 – Three-wire single phase system.

In almost all cases domestic electrical supplies are of the single phase type. Larger consumers such as shops, offices and factories are supplied with what is known as a three phase supply (Hall 1977).

5.3.3 The electrical distribution system

After it leaves the power station, electricity is transported as a high voltage alternating current. Its voltage is reduced in a number of stages to bring it down to the 240 V required by the domestic consumer.

The electricity is brought into the home through a service cable which can be either underground or overhead. In the UK the majority of houses have underground service cables. In the USA overhead supplies are normally used, but new buildings often have underground supplies.

5.4 SUPPLY OF ELECTRICITY WITHIN THE HOME

Following its entry into the house the service cable feeds into an electrical distribution circuit. Between the point of entry of the service cable and the distribution circuit there is normally an isolating switch, a fuse or circuit breaker (section 5.4.2), a meter for measuring consumption (section 5.4.1) and a provision for *earthing* or *grounding* of appliances (section 5.4.3). The exact arrangement of these, and of the distribution system, varies from one country to another. Even within a country there are variations because of the age of the installation and also regional variations. Codes of practice for electrical installations are provided by the National Electrical Code (NEC) for the USA and by the Institute of Electrical Engineers (IEE) for the UK. Distribution systems are discussed in section 5.4.4.

5.4.1 Measurement of consumption

At a point inside or outside a house where responsibility for the wiring changes from the supply company to that of the consumer, the consumption of electricity is metered. This is measured by the number of **kilo-watt hours** (kWh) or **units** used in an accounting period.

One kWh is the amount of electricity consumed by an appliance rated at 1000 W when it is used for one hour. Meters for measuring the consumption of electricity have dials or digits (Fig. 5.7). Both types of meter produce a cumulative total of electricity consumption. The consumption for any period is given by the difference between the reading at the end of the period and that at the beginning.

In the case of White meters, used in the UK in conjunction with differentiated day and night rate tariffs, the meter has two sets of digits resdings. A time clock is used to switch the meter and house wiring between the two meters at pre-set times of day (see Fig. 5.16).

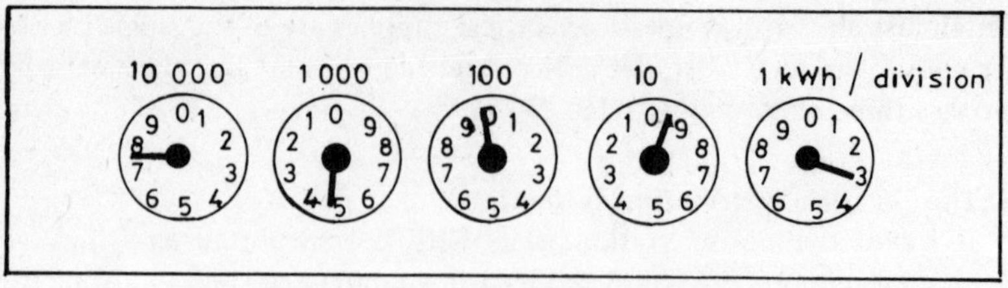

(a) dial type meter: reading 74993 kWh
(These meters may have 1/10 kWh dial)

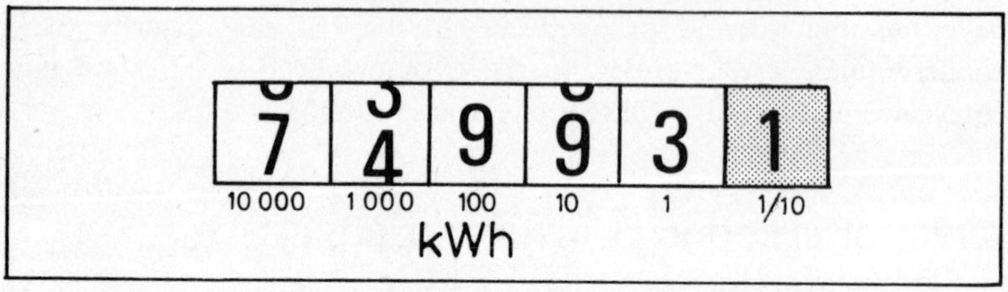

(b) digit type meter: reading 74993 kWh

Fig. 5.7 – The displays of electric kWh meters.

5.4.2 Electrical overload protection

If the flow of electricity through a wire becomes excessive, the wire will overheat. This can cause damage to the insulation of the wire leading to a risk of a short circuit or even a fire.

To prevent this happening, devices are included in an electrical circuit which will isolate a part or the whole of a circuit which is drawing excessive current. This device is connected into the live wire in such a way that when excessive current flows through it, there is a break in the circuit. These overload protection devices are either **fuses** or **circuit breakers.**

A fuse is a piece of wire designed to carry a stated current load but if the current increases above this level the fuse element becomes hot and melts. Since the fuse forms an electrical link between the supply and the appliance, once it melts the appliance is disconnected from the supply. Once a fuse has *blown* it must be replaced, after identifying the cause of this current overload. Ideally a fuse would melt instantaneously when the current rises, but in practice there is a time delay during which the fuse wire heats up. In practice *whether or not a fuse blows* depends upon both the magnitude of the current overload and its duration. Thus, for example, a 15 A fuse will blow immediately if the current rises to 30A, but it could survive a 20 A overload for several hours.

There are many different designs of fuse such as the rewirable fuse (UK), the cartridge fuse (USA and UK) and the plug fuse (USA) (Fig. 5.8). It is important that a replacement fuse is of the same current rating and with the same characteristics as the original. It is also important that a fuse is not replaced with a non-fusible link such as a piece of wire (renewable and cartridge fuse) or a small coin (plug fuse).

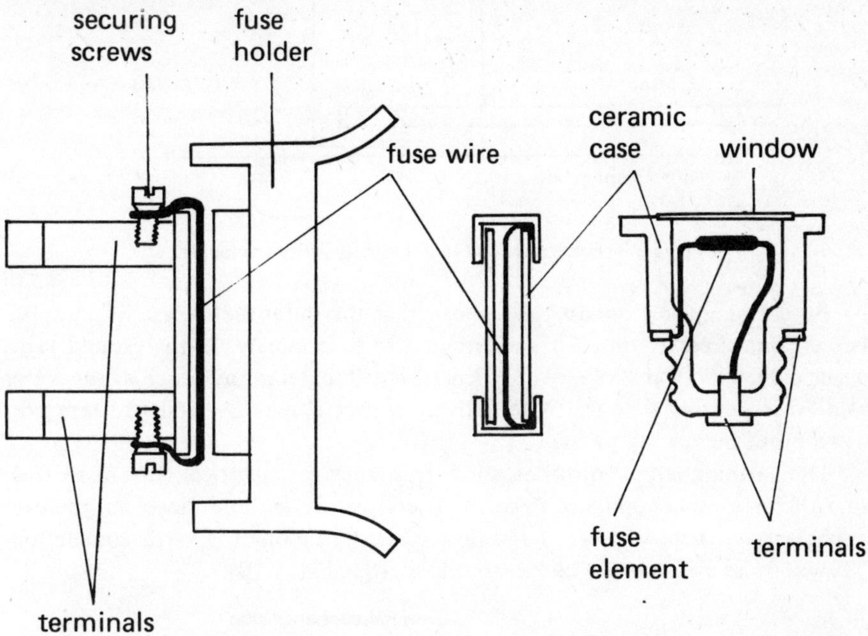

Fig. 5.8 – Appearance of common types of fuse.

A circuit breaker is an electro-mechanical device which operates in the same way as a fuse but which can be reset once the source of the current overload has been identified. Most circuit breakers use either bi-metallic strips or electromagnets to detect and react to overloads. Once a circuit breaker has disconnected the circuit it must be reset manually.

5.4.3 Equipment earthing (grounding) and double insulation

It is possible for a situation to develop in which the metal casing of an appliance becomes live due to a breakdown in insulation or if a wire becomes detached (Fig. 5.9). If a person then touches this metal casing it is possible for current to flow through that person. This can occur if the person is completing an electrical circuit to earth (ground) through gas or water pipes or if the person is standing on a solid floor.

This situation can be prevented in one of two ways: by **earthing (grounding)** the equipment or ensuring that no part of the equipment which can be touched by a user can become live (**all insulated** and **double insulated** appliances).

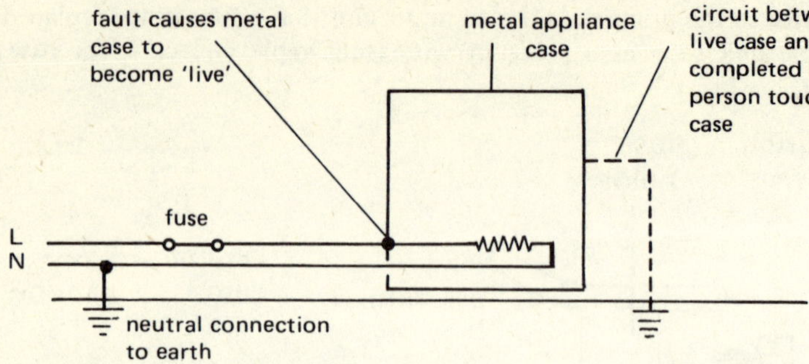

Fig. 5.9 – Electrical flow in unearthed faulty appliance.

By earthing (or grounding) we mean that the outer metal case of an appliance is connected to an electrical circuit which terminates in the ground. This can be a metallic water pipe which has travelled underground, such as the water main which enters a house. Alternatively a special earthing rod (20 mm corrosion proof metal) can be used.

The earthing circuit provides a low resistance for electrical current to flow from the case of an appliance to earth. Therefore if a fault develops and the case becomes 'live' there will be a large flow of current along the earth connection. This will cause a fuse in the circuit to 'blow'. (See Fig. 5.10).

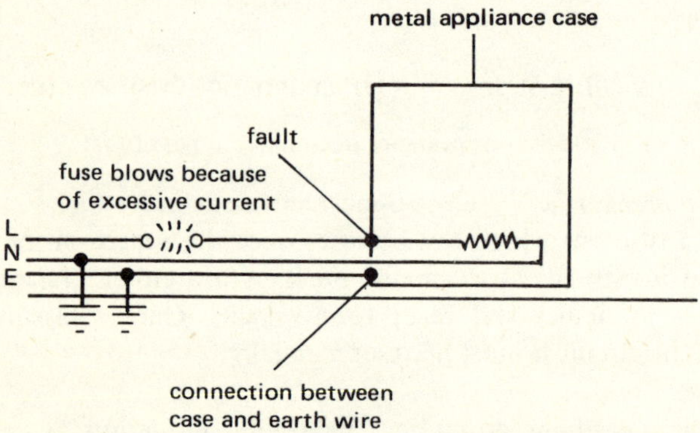

Fig. 5.10 – Electrical flow in earthed faulty appliance.

Appliances which do not need an earth connection are those which do not have any exposed or accessible metal parts **(all insulated appliances)** and those where any exposed metal parts are separated from any live metal parts by two layers of insulating material **(double insulated)**.

An **earth leakage circuit breaker** detects and responds to the leakage of small amounts of electricity to earth in the same way as a fuse or circuit breaker.

They respond to much smaller currents and react much more rapidly. The user is protected against defective or poor earthing. These devices can be used to protect the whole of the installation, a single circuit or a single appliance. The response to an earth leakage is to operate a double pole isolating switch which must be manually reset. They operate by detecting any difference in the current flowing in the live and neutral wires. Under normal operating conditions the flow of current in these two wires should be identical.

5.4.4 Household wiring circuits

Wiring circuits are of two types: radial circuits or parallel circuits. Schematic diagrams of these types of circuits are shown in Fig. 5.11. In a radial circuit each appliance or socket outlet has its own cable (containing live, neutral and earth wires) coming from a distribution board. A parallel circuit has several appliances or sockets connected on a single circuit which comes from a distribution board.

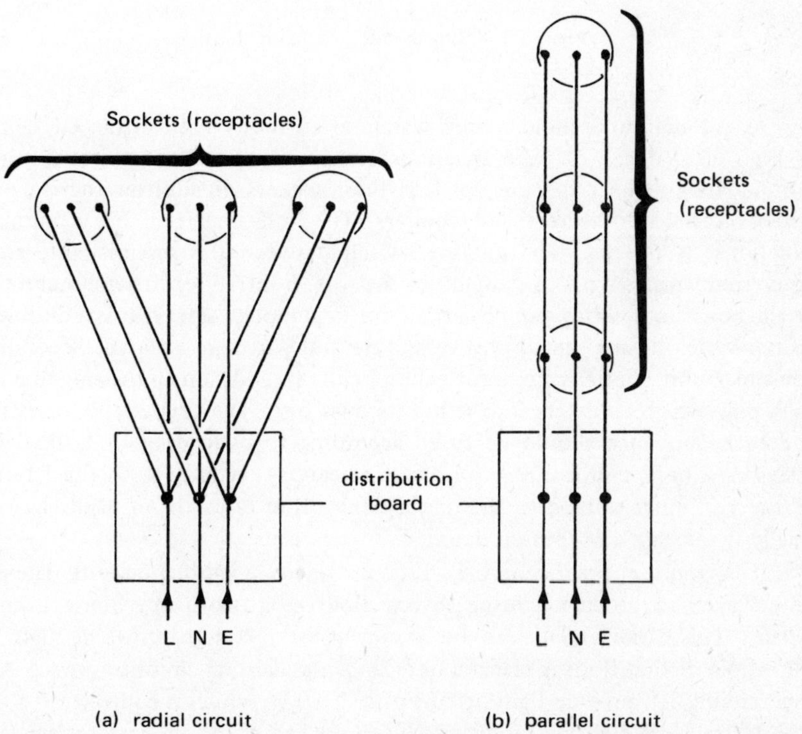

Fig. 5.11 – Radial and parallel circuits: schematic diagram.

A specialised form of the parallel circuit is the ring-main which has been in common use in the UK since 1950. A diagram of such a circuit is shown in Fig. 5.12.

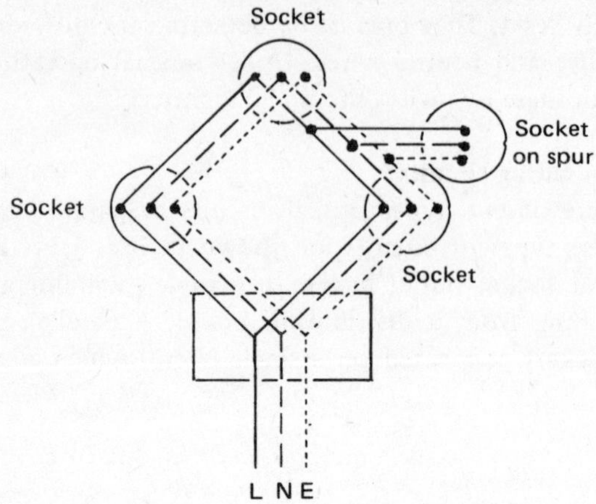

Fig. 5.12 – Ring circuit: schematic diagram.

A complete household wiring system as currently used in the UK is shown in Fig. 5.13. It can be seen that it is composed of two ring circuits, one for ground floor sockets and one for first floor sockets. In addition there are radial circuits for water heaters and cookers. The consumer unit has a number of functions. It provides an isolating switch between the service cable and the household wiring. With this switch in the 'off' position repairs and maintenance of the household wiring can be carried out in complete safety. It is a double pole switch which means that both live and neutral wires are isolated. Secondly the consumer unit contains fuses for each circuit. In addition to the ring-main fuse, each plug which connects into it has its own fuse. The ring circuit fuse is 30 A and each appliance should be fused according to its loading: 2 A for lighting, televisions, hi-fi, radios; 13 A for electric heaters, kettles, etc. A third function of the consumer unit is to provide a connection to earth for each circuit and finally it acts as a distribution board.

It is common practice in the UK to provide two lighting circuits. These may be arranged to provide lighting to one floor of a two-storey house from each circuit. This arrangement can be inconvenient if a fuse blows at night since the whole of that floor is blacked out. Since the lighting circuits have a 5 A fuse, each circuit can provide lighting of up to 1250 W, which is equivalent to twelve 100 W bulbs. A lighting circuit is shown in Fig. 5.14. An earth connection is provided, in addition to the live and neutral connections, for modern wiring circuits.

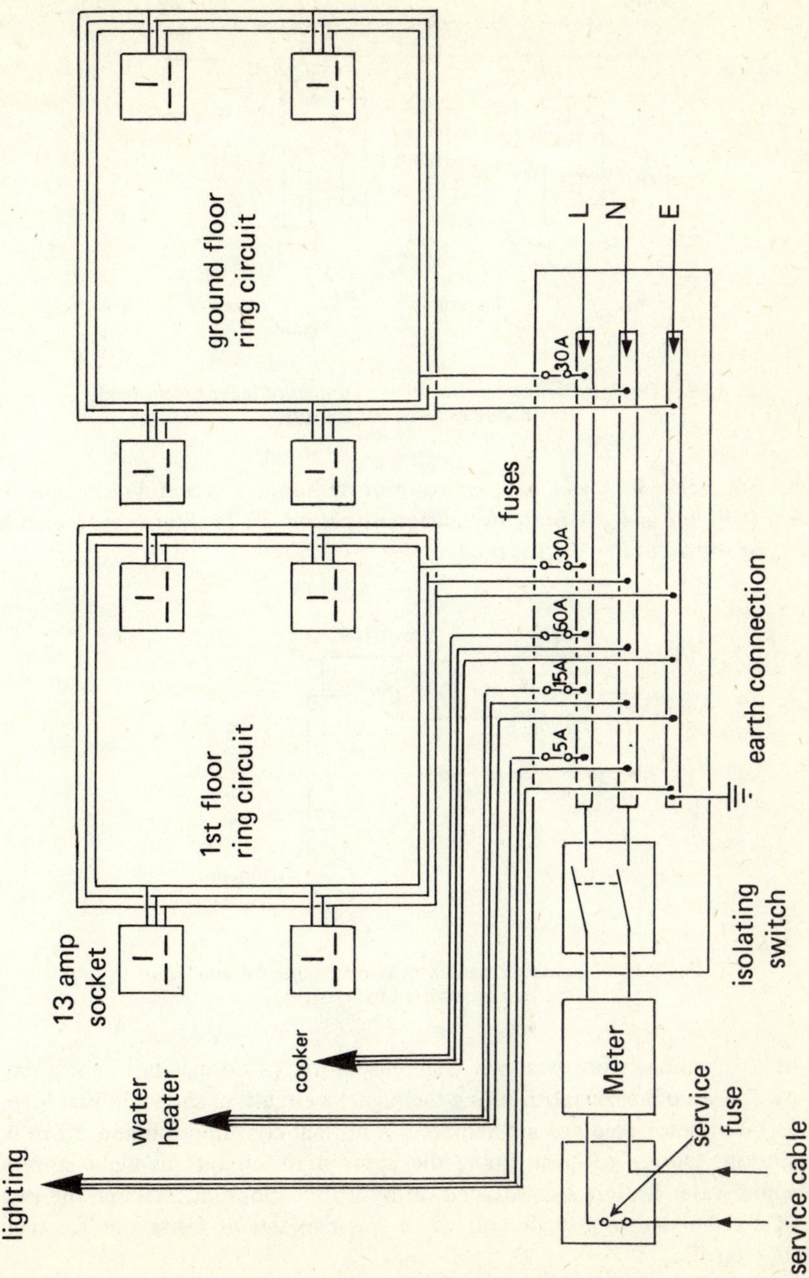

Fig. 5.13 – UK household wiring system.

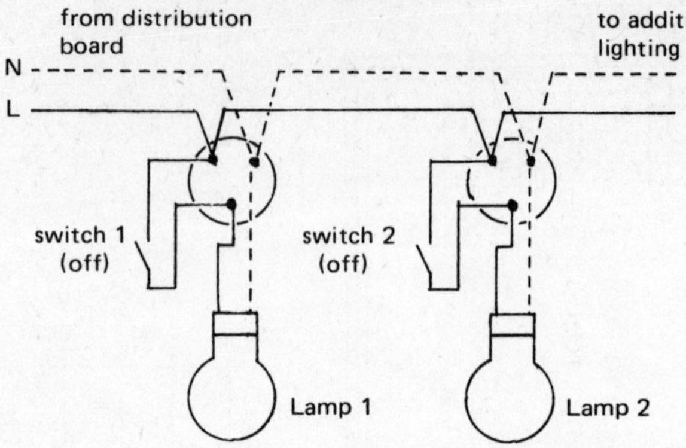

Fig. 5.14 – One way switches controlling a number of independent lights. (earth wire omitted for clarity).

In some cases, such as a hall, staircase or bedroom it is useful to be able to switch a light on and off from two different places. To facilitate this a circuit of the type shown in Fig. 5.15 is used.

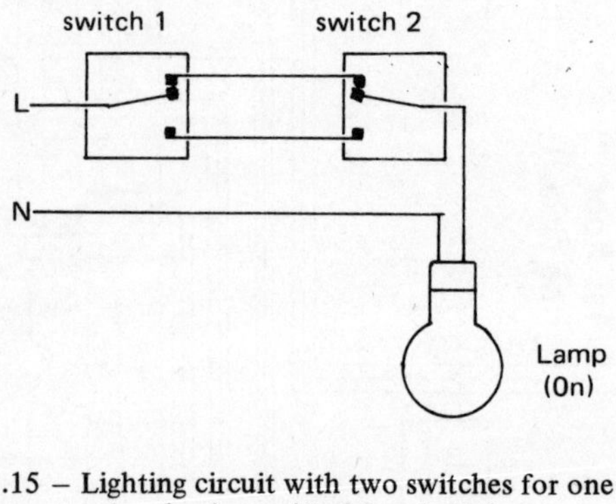

Fig. 5.15 – Lighting circuit with two switches for one bulb. (earth wire omitted for clarity).

For a consumer who uses two different tariffs (a normal tariff for a day time use and an off-peak tariff during the night) a circuit, as shown in Fig. 5.16, is used. Two meter readings are required, a normal consumption and off-peak consumption. During off-peak hours the appropriate circuits to night storage heaters and water heaters are switched on by a time clock. In practice the two meters are combined in a single unit which has two sets of digits, one for each of the two tariffs.

In the USA houses are supplied by a three-wire supply. In this system there is 115–120 V between either of the live (hot) wires and neutral and 230–240 V

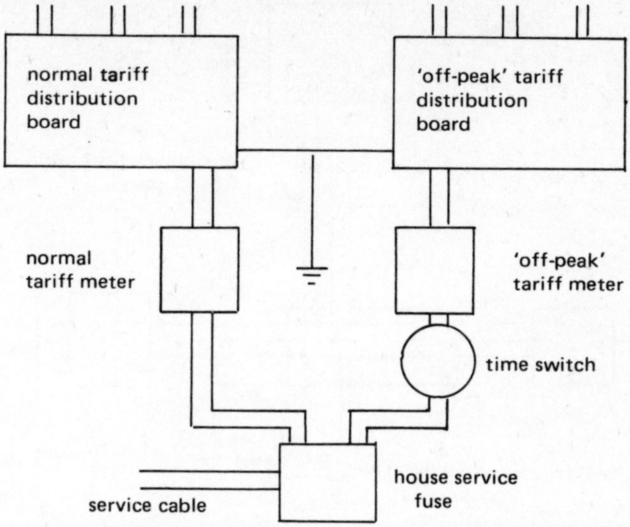

Fig. 5.16 – Arrangement of meters and time switch for domestic installations using off-peak heating.

between the two live wires (Fig. 5.6). Electricity is brought from the utility company's overhead or underground supply through service conductors. At the point of connection between the utility company's wiring and the household wiring is an isolator, commonly a switched fuse box. Prior to this is a meter for measuring consumption (kWh) which may be inside or outside the building.

In most cases the isolator is associated with a main distribution panel which, in addition to the main fuses or circuit breakers (100 A for each live wire), also contains fuses or breakers for the distribution circuits. The household wiring consists of a mixture of parallel and radial circuits. The main distribution panel supplies electricity to a number of branch distribution panels through a feeder cable (Fig. 5.17). As with the UK system there is an earth (ground) connection at the distribution board but, unlike the UK system, there is a direct connection between the earth and neutral lines. (In the UK the neutral line is earthed at the Electricity Board's sub-station.)

There are a number of different types of circuits in common use:

(1) *The general purpose branch* is a 115–120 V two wire parallel circuit, fused at 15 or 20 A and used for lighting and small appliance sockets (receptacles).

(2) *The appliances branch* is a 115–120 V two wire parallel circuit, fused at 20–30 A for appliance sockets. In new wiring this is being replaced with a three-wire circuit containing, in addition to the live and neutral

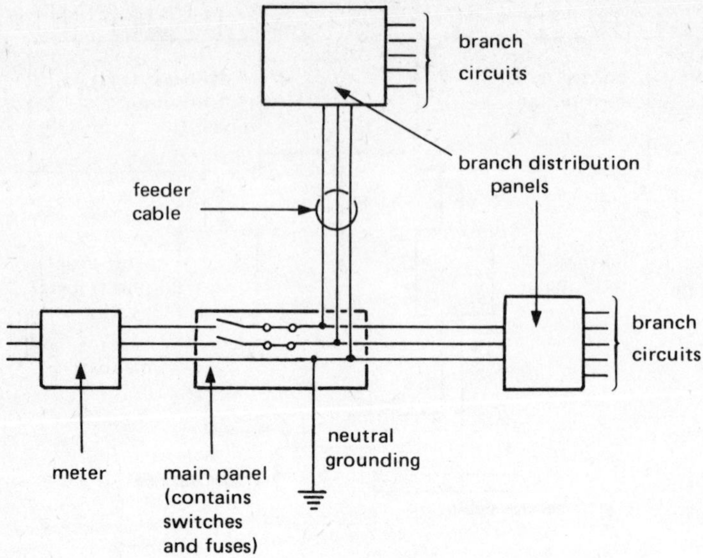

Fig. 5.17 – USA household distribution circuits.

wires, a ground wire. A newer method is to use a three-wire system connected to pairs of two-pin sockets (Duplex receptacles).

(3) *The individual branch* is radial wiring to an individual piece of equipment. It is used for appliances rated at 15 A or above, such as cookers (ranges), water heaters and room heaters. Each appliance is usually served by a three wire (120/240 V) supply. For example a range will have its own wiring circuit with a fuse rated at 35 A on each live wire. The third wire acts as a ground. Additionally some types of motorised equipment are provided with an individual branch because of the high current surge produced when they are switched on.

5.5 CALCULATING ELECTRICITY COSTS

As in the case of gas, electricity is charged on the consumption over an accounting period, which in the UK is 13 weeks (one quarter). The number of units (kWh) of electricity used, is measured by the difference between the meter readings at the beginning and at the end of the accounting period. Typical meters are shown in Fig. 5.7. The cost of electricity is usually composed of a Standing (fixed) charge and a Variable charge, which depends upon the quantity of electricity consumed in an accounting period.

In some cases the variable charge is made up of a single rate, in others there are two more different rates. For example the Yorkshire Electricity Board (UK) in 1979 charged 6.978p for each of the first (primary) 72 units and 2.768p for each additional (secondary) unit.

The cost of running an electrical appliance can either be estimated from its stated rating, or measured using a kWh meter. All appliances have a plate attached giving, amongst other information, their power rating in Watts or kilo-Watts. In the case of a simple piece of equipment, with no time switches, thermostats or other energy controls such as an electric kettle, the number of kWh consumed is equal to the power rating (in kW) multiplied by the time (in hours) for which the appliance is on.

Suppose that a kettle rated at 2.5 kW takes 2 minutes to boil two pints of water. Then:

$$\begin{aligned} \text{energy consumed (kWh)} &= \text{power rating (kW)} \times \text{time (hours)} \\ &= 2.5 \times \frac{2}{60}\,\text{kWh} \\ &= 0.083\,\text{kWh}\ . \end{aligned}$$

To assign a cost to this energy consumption it is first of all necessary to make some assumptions about the cost/unit if a consumer is on a variable tariff. Using the cost figures given above and a consumption of 500 units in a given accounting period (13 weeks), the average cost for 1 unit may be calculated as shown below.

$$\begin{aligned} \text{Cost of } 72 \text{ (primary) units} &= £72 \times .06978 \\ &= £5.02 \end{aligned}$$

$$\begin{aligned} \text{Cost of remaining 428 (secondary) units} &= £428 \times .02768 \\ &= £11.85 \end{aligned}$$

$$\begin{aligned} \text{Therefore average cost per kWh} &= \frac{5.02 + 11.85}{500} \\ &= £0.03374 \text{ or } 3.374\text{p}. \end{aligned}$$

A similar calculation for a consumption of 2000 units per quarter gives an average cost of £0.02920 or 2.92p. (It can be seen that this type of tariff structure is not in the interests of energy economy since units of electricity become cheaper the more that are consumed.)

Returning to the example of the kettle, and assuming it is used on average 3 times a day, it will be used 273 times in one quarter. Over this period the kettle will consume:

$$3 \text{ times/day} \times 91 \text{ days/quarter} \times 0.083 \text{ units/time} = 22.66 \text{ units}.$$

This will cost

(a) within a consumption of 500 units/quarter

£22.66 × 0.03374
£0.76

(b) within a consumption of 2000 units/quarter

£22.66 × 0.02920
£0.66 .

For an appliance with a thermostat or energy regulator, the estimation of energy consumption is not so straight forward. For example, although an oven is on for 2 hours, the heating elements will be switched on and off under the control of a thermostat (section 9.4.1). If the thermostat has an indicator light it is possible to measure the time when the light is on and to use this time, together with the rating of the oven heating element, to estimate energy usage and costs. However in cases such as these it is much more satisfactory to use a kilowatt hour meter to measure actual consumption as shown in Fig. 5.18.

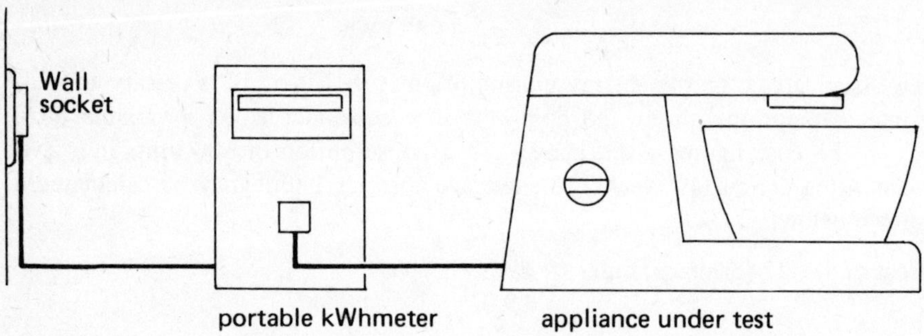

Fig. 5.18 – Measuring energy consumption for mixing machine.

5.6 THE CONNECTION OF AN APPLIANCE INTO A WALL SOCKET (RECEPTACLE)

Most portable appliances are connected to the electricity supply through a wall socket (receptacle) via a plug and flex (i.e. flexible cable). The type of plug chosen must suit the requirements both of the household supply and of the appliance. Similarly the flex used must be capable of carrying the current required by the appliance without overheating. It must also be capable of surviving the environmental conditions in which it will be used particularly the presence of water in the case of appliances used in the kitchen. Flex on portable appliances needs to be capable of withstanding repeated bending and twisting.

Many different types of wall sockets are used throughout the world. The International Electrotechnical Commission (IEC) is considering the adoption of international standards but progress is slow because of the wide variation in practices between countries (Consumers Association 1976, 1979). The greatest progress so far has been made in Europe with the adoption of the Europlug. This is a **moulded on** plug for use on appliances which do not need an earth. This plug will fit most European wall sockets (Fig. 5.19a).

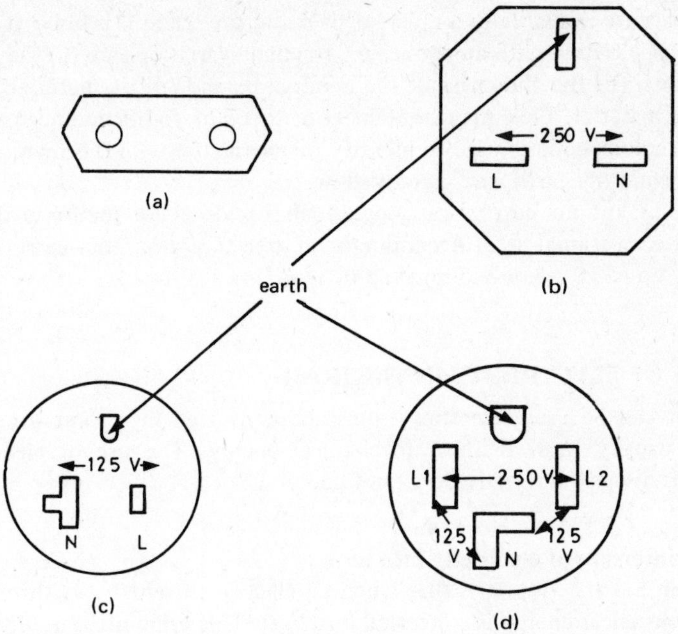

(a) Europlug (b) 3 pin 13 A (UK) (c) 3 pin 20 A (USA) (NEMA 5–20R) (d) 4 pin 30A (USA) (NEMA 14–30R)

Fig. 5.19 – Outlines of common types of appliance plugs.

In the UK the ring-main system uses a three pin fused plug for all portable appliances (Fig. 5.19b). For appliances which do not require an earth the same plug can be used with no connection made to the earth pin or an adaptor may be used which accepts a two-pin plug similar to the Europlug. The fuse in the three-pin plug is of the cartridge type (section 5.4.2) and the fuse should be chosen to suit the current requirements of the appliance:

2 A for lighting, television and appliances up to 500 W
5 A for appliances between 500 W and 1 kW
13 A for appliances between 1 kW and 3 kW.

In the USA a number of different types of socket are in use (McGuiness and Stein, 1977), the most common being a three-pin (two-pole plus ground) for appliances operating from 115 to 125 V and from 15 to 20 A (Fig. 5.19c). This plug is designated the 5-20R by the National Electrical Manufacturers Association (NEMA). Larger appliances need a 120/240 V supply at 30 A or 60 A. These may be directly wired to the electrical supply or they may have a four-pin (three pole plus ground) plug such as the NEMA 14-30R (Fig. 5.19d).

Flexes are available in a range of sizes and types and it is important that the correct type is used with an appliance. The main variations are (a) the number of conductors, (b) the thickness of the conductors and (c) the nature of the outer covering material. Flex may contain two, three or four conductors. An international colour code is used to identify these: the live wire is **brown,** the neutral wire **blue** and the earth wire **green/yellow.**

The maximum current-carrying capability of a conductor is determined by its cross-sectional area. A conductor of area 0.75 mm^2 can carry up to 6 A, 1.0 mm^2 up to 10 A and 1.5 mm^2 up to 15 A.

5.7 USE OF ELECTRICITY IN THE HOME

Before it can be used, electricity must be converted into some other form of energy, usually heat, light or mechanical energy. The use of electricity for lighting is discussed in more detail in Chapter 12.

5.7.1 Conversion of electricity into heat

In section 5.3.1 it was seen that when an electric current flows through a conductor, electrical energy is converted into heat. This principle is used to generate heat in ovens, water heaters and space heaters, where the heating element forms the electricity-carrying conductor. The amount of heat which is produced (P watts) depends upon the voltage V applied to the heating element and on the current I flowing through it:

$$P = V.I \ .$$

The current in turn is dependent on the resistance of the heating element (see section 5.1):

$$I = \frac{V}{R} \ .$$

For example, if a heating element has a resistance of 12 Ω and operates at 240 V, then the current will be given by:

$$I = \frac{240}{12} \text{ A}$$

$$20 \text{ A} \ .$$

Therefore the heating power of the element will be:

$$P = 240 \times 20 \text{ W}$$

$$\mathit{4800} \text{ W or } \mathit{4.8} \text{ kW} \ .$$

A domestic heating element consists of an electrical conductor which is covered by insulation. In heating elements used in immersion heaters (used in kettles and water heaters) the heating element is surrounded by a layer of magnesium oxide (an electrical insulator) which in turn is covered with a metal sheath to provide strength (see Fig, 5.20a). An alternative type of heating element used in radiant space heaters consists of a coiled heating element surrounded by a tube of silica glass. This type of heater responds very quickly when switched on or off. It is often called a 'quartz tube' heater (see Fig. 5.20b).

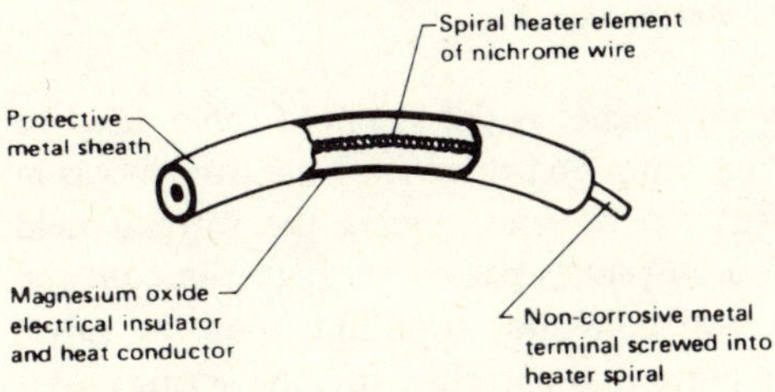

(a) sheathed heating element

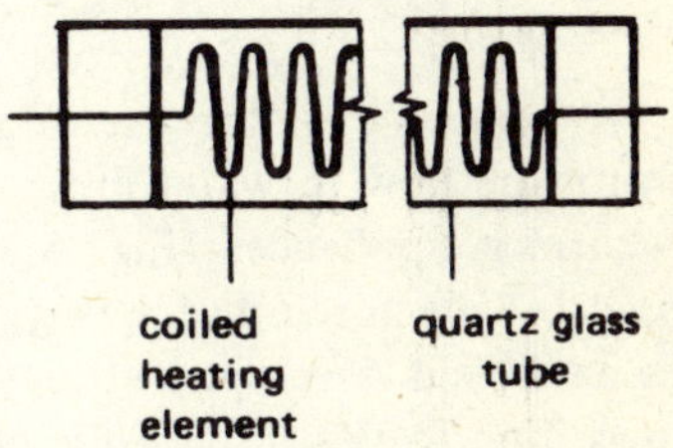

(b) quartz heating element

Fig. 5.20 – Electrical resistance heaters.

5.7.2 Conversion of electricity into mechanical energy

If an electric current is passed through a piece of conducting wire, a magnetic field develops around the wire, as shown in Fig. 5.21. If a piece of wire is wound

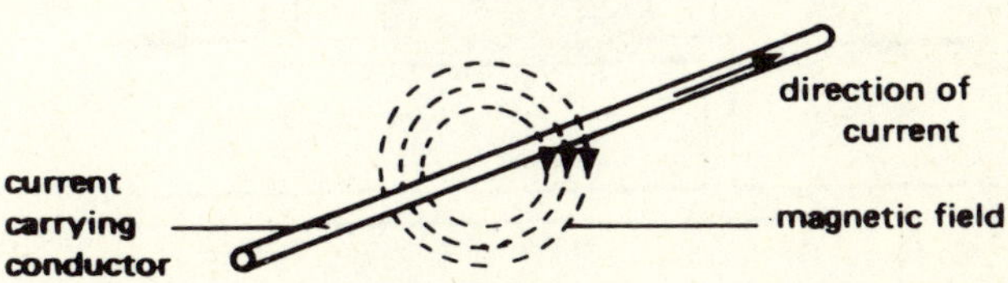

Fig. 5.21 – The magnetic field associated with flow of electricity.

around an iron bar, when an electric current flows through the wire all of these magnetic fields add together to produce a strong magnetic field (Fig. 5.22). If the direction of the electric current is reversed, so too will be the direction of the magnetic field. This magnet is referred to as an **electro-magnet** and is used in many electrical appliances and control devices.

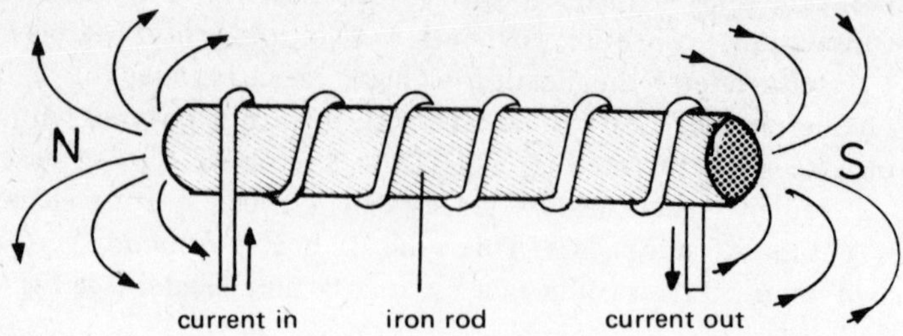

Fig. 5.22 – An electromagnet.

An example of the use of an electro-magnet is the solenoid valve, used to control the flow of water in washing machines and to control the flow of gas in gas-burning appliances (Fig. 5.23). With no current flowing the valve is held closed by the force of a spring. When a current is passed through the coil, the magnetic field developed attracts the iron core and, in pulling open the valve, allows fluid to flow. When the electric current is switched off, the spring closes the valve.

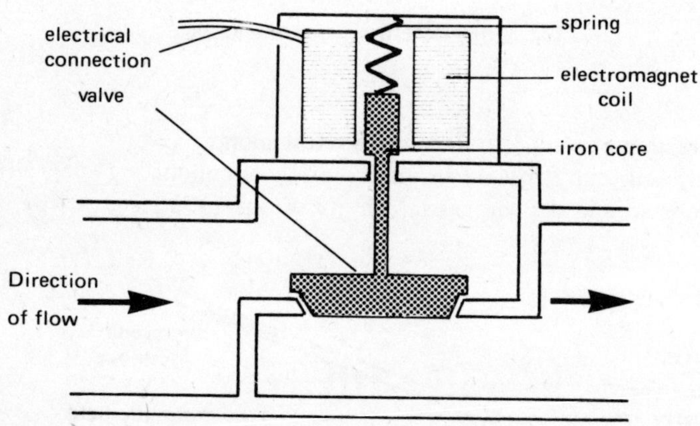

Fig. 5.23 – The solenoid valve.

The electromagnetic effect is also used in some electric motors. If a current is passed through a conductor which is in a magnetic field there will be a force on the conductor at right angles to the magnetic field, as shown in Fig. 5.24. If the conductor is in the form of a coil, mounted on a spindle and free to rotate, the force on the conductor will cause the coil to turn in the magnetic field. Motors used on alternating-current appliances are usually either **universal** motors or **induction** motors.

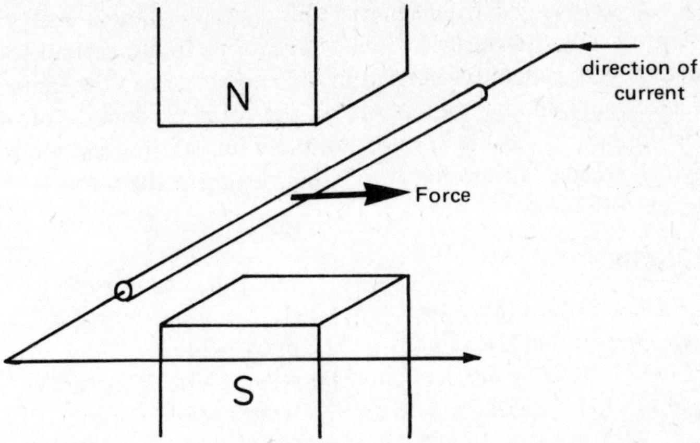

Fig. 5.24 – Movement of a conductor in a magnetic field.

The universal motor is so called because similar designs of motor can be used both with a.c. and d.c. In the universal motor, current is taken to the rotating coil (the **rotor**) through a **split commutator** against which press two **brushes.** The brushes are usually made of carbon and are kept in contact with the commutator by the pressure of a spring. During use the brushes gradually wear down and need periodic replacement. The rotor is surrounded by a magnetic field which is usually generated by an electromagnetic coil known as the **stator** (see Fig. 5.25).

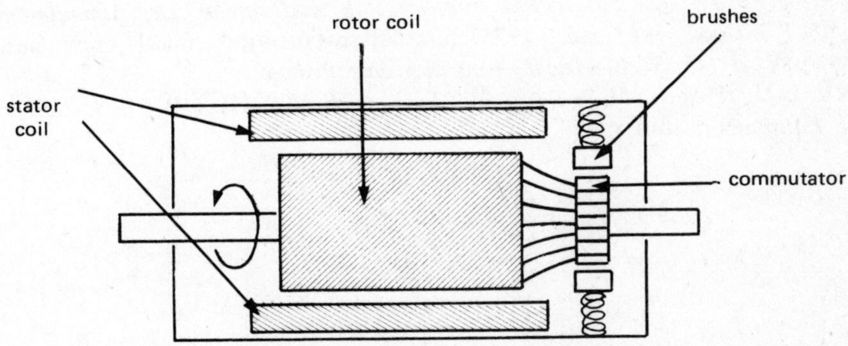

Fig. 5.25 – The universal electric motor.

There are a number of different designs of induction motor but they are all characterised by having no brushes or commutator. A common design of induction motor is known as a **squirrel-cage motor.** This has a rotor made of laminated sheets of iron with copper conductors running parallel to the shaft of the rotor.

The rotor is surrounded by magnetic coils. Because these are supplied with an alternating current the magnetic field appears to rotate around the rotor. The effect of this is to generate a current in the conductors on the squirrel cage. This flow of electricity in the conductors on the rotor produces a turning force on the rotor. Induction motors are not normally self-starting and require two coils, one of which receives electricity during the starting of the motor.

REFERENCES

Consumers Association (May 1976), p. 63–4.

Consumers Association (November 1979), p. 615–24.

Hall, F. (1977), *Building Services and Equipment*, **Vol. 2**, Longman.

Hickok, F. (1975), *Handbook of solar and wind energy*, Cahners.

McGuiness, W. J. & Stein, B. (1977), *Building Technology*, Wiley.

Overman, M. (1974), *Understanding Energy*, Lutterworth Press.

Reay, D. A. (1977), *Industrial energy conservation*, Pergamon.

Sumner, J. A. (1976), *Domestic heat pumps*, Prism.

BIBLIOGRAPHY

Cianchetti, C. J. & Jurewitz, J. L. (1975), *Studies in Electrical Utility Regulation*, Ballinger.

Electrical Association for Women (1971), *EAW Electrical Handbook*, English University Press.

Gaddis, B. (1976), *How to repair home kitchen appliances*, Tab Books.

I.E.C. (1979), *Plugs and sockets for domestic and similar use, Amendment N.1 to Document No. 83 (1975)*, International Electrotechnical Commission.

Ladley, R. H. (1969), *Electrical wiring diagrams*, Pitman.

Peet, L. J., Pickett, M. S., Arnold, M. G. & Wolf, L. H. (1975), *Household Equipment*, 7th Ed., Wiley.

CHAPTER 6

Water supplies and liquid waste disposal

6.1 INTRODUCTION

A supply of wholesome and palatable water has been historically an important consideration in the siting of settlements. The developments of civilisation have been paralleled by the development of water distribution systems, in domestic plumbing and in systems for the removal and treatment of waste water (Smith 1976). In modern society a supply of fresh safe water is regarded as an essential component of life.

In the UK current per capita consumption of water is between 220 and 350 litres/day and this is expected to rise by the year 2000 to 330 to 440 litres/day (IHVE 1974). Of the current consumption approximately one half (110 to 180 litres/day) is used in the home, principally for washing, bathing, WC flushing, washing eating utensils and laundry. There is a large seasonal variation in the demand for water. In the summer months people take more baths and showers and also water is used for garden watering, swimming pools etc. This poses a problem for the water supply companies as this period coincides with a season of the year when water supplies are not being replenished and when losses by evaporation from lakes and reservoirs are at a maximum.

6.2 SOURCES OF WATER

Water supplies are part of a natural cyclic process. Water is precipitated as rain to form streams, rivers, lakes and underground water sources, all of which finally flow into the sea. It evaporates from the exposed surface of seas, rivers and lakes. The evaporated water forms clouds from which it is precipitated as rain, thus completing the cycle (Fig. 6.1).

Water for domestic use can be extracted at a number of different points in this cycle, including rain water from roofs etc., surface water from streams, ponds and lakes; and groundwater from natural springs and wells. In addition to these natural sources of water, reservoirs are used to trap and store ground-

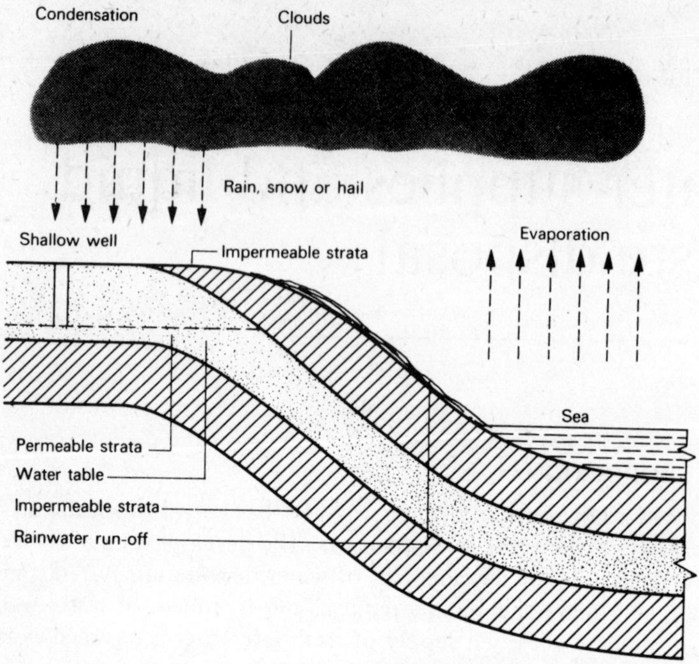

Fig. 6.1 – The Water Cycle (Hall, 1976).

water to allow for seasonal fluctuations in supply and demand. Although some houses have their own water supply from rain water, spring, stream or well, most receive their water from a piped water network. In the UK this is controlled by a regional water authority which is responsible for the supply of water, the treatment of sewerage and the control of waterways. Similarly in the USA, control is in the hands of regional and municipal water authorities. The mains distribution system has trunk mains from which street mains are taken. Often, in towns and cities, street mains are connected to two trunk mains so that supply can be maintained whilst one of the trunk mains is being serviced.

6.3 TREATMENT OF WATER SUPPLIES

In an industrial society water is re-used several times. Water, from a spring, stream or well is extracted and used and, with or without treatment discharged into a river. Lower down this river water is withdrawn for a variety of other uses (agricultural, domestic and industrial) and again returned to the river. Depending on the relative volumes of river water and returned wastewater, it is possible that the river will become polluted with chemicals and micro-organisms. Simultaneously it is possible that run-off water from agricultural land will be contaminated with residues of fertilisers and pesticides. Even natural unpolluted water supplies may contain materials which have been leached from underground soils and rocks and which cause discolouration and tainting of the water. For these reasons it is often necessary to process water supplies to provide water of an acceptable quality.

The quality of a water supply depends upon its use. If we exclude the specialised requirements of some industrial and laboratory processes which require ultra-pure water, the highest quality of water is needed for certain domestic applications: drinking, cooking and baking. This water needs to be free of harmful substances (chemicals and micro-organisms) and to have a good aesthetic quality (taste, odour and appearance). Water for other uses such as laundry, WC flushing and garden watering does not have to be of such high purity. Since current practice is to provide a house with a single supply of water for all of these purposes, this means that this water must all be to the standard required for drinking water. In the USA control of drinking water supplies is controlled by the National Safe Drinking Water Act (1974) which provides maximum levels for contaminants in drinking water (Clarke *et al.* 1977).

There are a number of different water treatments not all of which are necessary for all water supplies. Also water supplies vary and it may be necessary to introduce additional processes at certain times of the year. The principal processes are as follows:

(1) Screening: a sequence of coarse and fine screens and/or mesh drums to remove large particles.
(2) Flocculation and sedimentation: chemicals are added which coagulate and precipitate, removing as they do so suspended materials. The precipitate collects at the base of sedimentation tanks.
(3) Filtration: the water is passed under atmospheric or high pressure through sand filters to remove fine solids.
(4) Activated carbon: this removes taste and odour.
(5) Chemical treatments: disinfection (chlorine or bacteriocidal filter) fluoridation (to prevent tooth decay) softening (see section 6.5).

6.4 DOMESTIC WATER SUPPLIES

6.4.1 Connection of water main to building

Most houses receive their water from the street main (Fig. 6.2). On the pipe which connects the street main to the house, at a point outside the boundary of the property, is an on/off valve which marks the transition of responsibility from the water authority to the owner of the building.

An important property of a water supply is its pressure. This pressure is created by the difference in height between the surface of the water supply in a reservoir or storage tank and the tap outlet. If the water pressure is too low, the flow of water from the taps will be unsatisfactory and WC tanks, for example, will take a long time to fill. If the water pressure is too high there is a greater

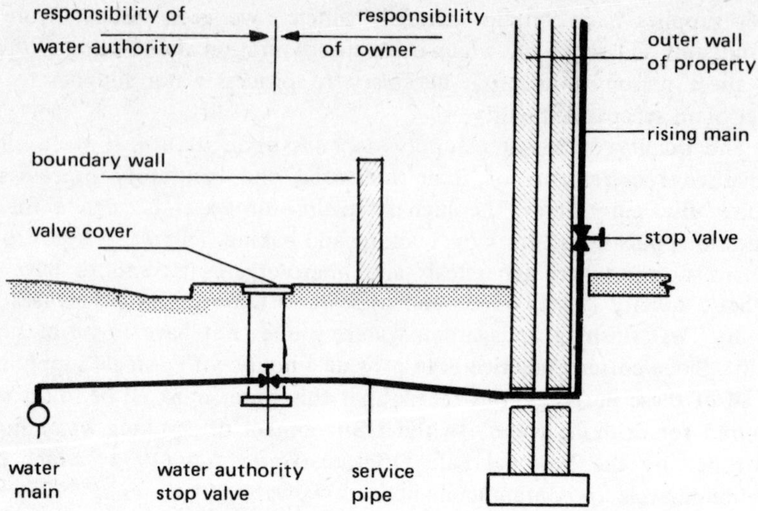

Fig. 6.2 – Connection of water main to domestic plumbing.

risk of the pipes leaking or bursting and the possibility of excessive noise in the pipes.

In the USA and most European countries, immediately following the stop valve is a water meter which is used to measure, and to charge for, water consumption. Domestic water users in the UK are not subject to the metering of supplies, water charges being at a fixed rate irrespective of consumption. The situation in the UK could change as the cost of supplying water increases, for example, the cost of extra reservoirs as consumption rises.

6.4.2 Domestic water distribution systems

Several types of water distribution systems are used within houses. In the *direct system* all sanitary fittings, taps (faucets) and the hot water storage tank are connected directly to the water main. This is the normal system in use in American and European homes (Fig. 6.3). An alternative system uses a cold water storage tank for hot water supply. In this system all sanitary fittings and taps are connected directly to the cold water supply as above, but there is a cold water storage tank feeding the hot water storage tank (Fig. 6.4). This system is permitted by some water authorities in the UK but most prefer the indirect system to be used in new installations. In the indirect system all the sanitary fittings, taps and the hot water storage tanks, with the exception of one cold water tap (usually the kitchen sink), are connected to a cold water storage tank. This is sited in the loft space to provide a good water pressure at the tap outlets (Fig. 6.5). The level of water in the storage tank is controlled by a ball valve.

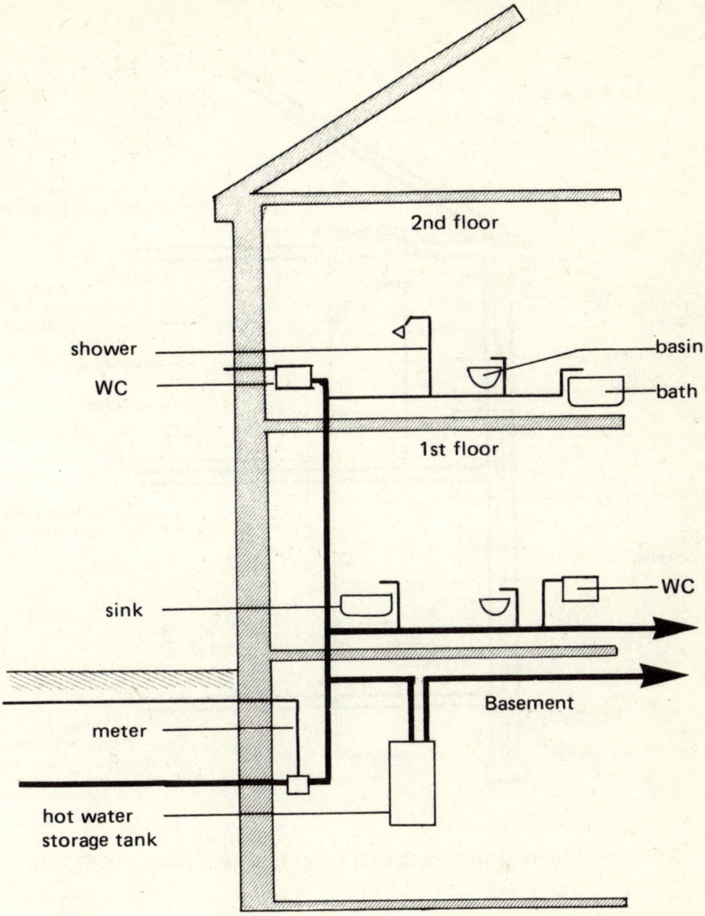

Fig. 6.3 – Direct cold water system.

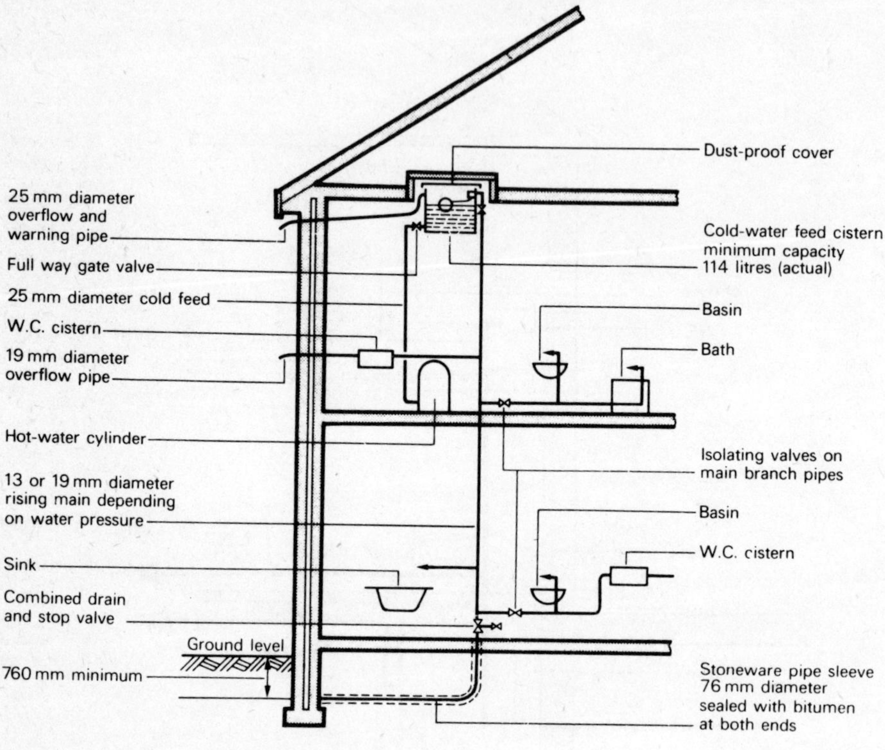

Fig. 6.4 – Direct system with cold water feed tank (Hall, 1976).

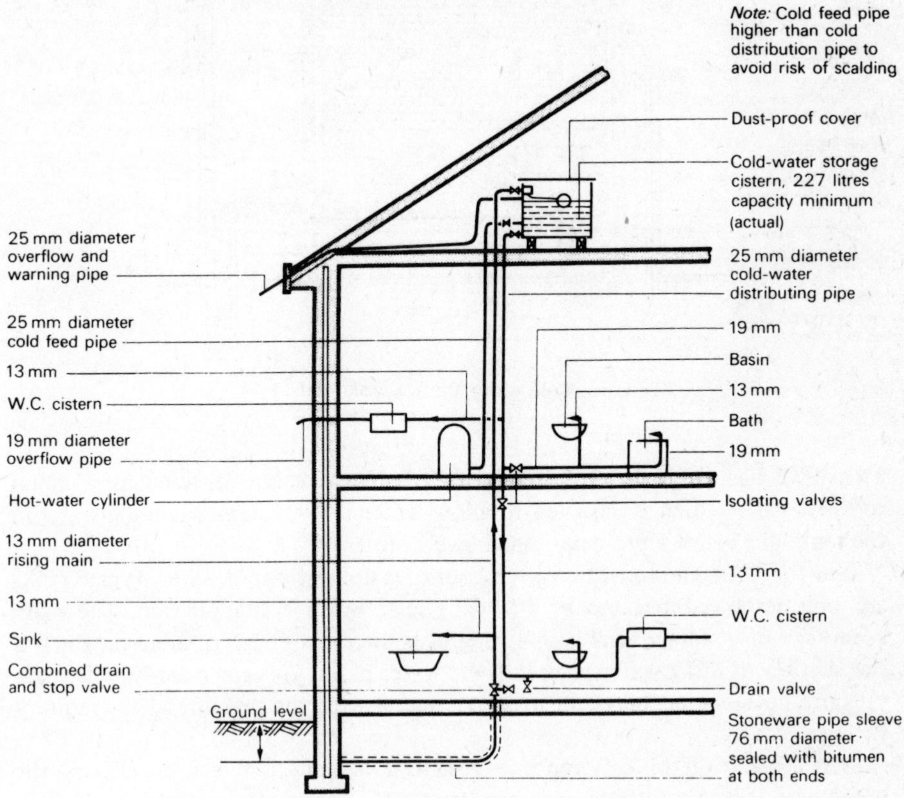

Fig. 6.5 – Indirect water system (Hall, 1976).

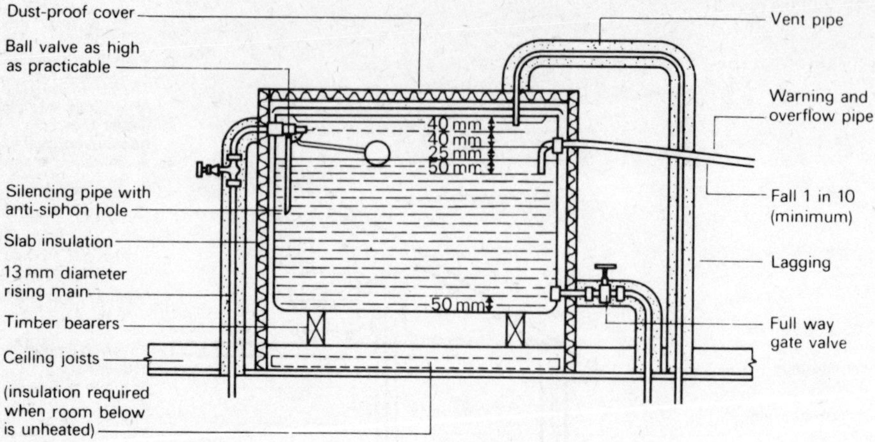

Fig. 6.6 – Cold water storage tank (Hall, 1976).

This valve has a metal or plastic floating ball which is connected by a metal rod to a valve which is designed to allow water into the tank when the water in the tank falls below a pre-determined level (Fig. 6.6).

In the USA and most European countries domestic cold water storage tanks are not permitted because of the danger of water contamination. The main advantage of a storage tank is that it reduces the risk of back siphonage, which is the suction of dirty water back into the water main. This can occur if a negative pressure develops in the water main compared with the pressure at the sanitary fitting.

As a result of this difference in design philosophy between the UK and the USA there is also a difference in the design of waste water systems to minimise the dangers of back siphonage.

The water distribution system is normally constructed from copper pipes of either 22 or 15 mm diameter. Polythene and PVC pipes are permitted for use with cold water supplies. The system has on/off valves at various points in the circuit to allow for the isolation of sections of pipework for maintenance. For a similar reason, drain valves are provided at low points in the pipework.

6.5 WATER SOFTENERS

With some water supplies which contain large quantities of calcium and magnesium salts it is often necessary or desirable to instal water softeners as part of the domestic plumbing. Without this process the hard water forms a scum with soap, does not give a lather and leads to the build up of scale in water heaters and boilers.

A water softener contains an ion exchange resin (zeolite) which is in the form of small beads. The surfaces of these beads have an excess of sodium ions. If a hard water is passed over this resin the calcium and magnesium ions from the water are exchanged with sodium ions on the beads, giving a soft water. After a period of time the resin becomes inactive because it becomes saturated with calcium and magnesium ions. It must then be *regenerated* by passing a solution of sodium chloride over it. During this regeneration there is an exchange of sodium ions from the solution with magnesium and chloride ions from the resin. At the completion of the regeneration process the resin is rinsed with water before it is ready for re-use as a water softener.

If small quantities of soft water are required, small portable softeners are available. Alternatively permanent water softeners can be connected into the rising main of the water supply. A clock is used to time the regeneration process so that it occurs when there is no demand for water, such as at certain hours of the night.

6.6 SANITARY FITTINGS

6.6.1 Water closets (WC's)

There are two parts to a water closet: a water tank and a pan. The two are designed to combine together to produce a strong flow of water which is sufficiently strong to clean the pan.

The tank is normally made of a ceramic such as vitreous china or a plastic material and it contains a ball valve to control the level of water in the tank. It also has a lever mechanism to discharge water into the pan as and when required.

The WC pan for domestic use is normally made of vitreous china. There are a number of different designs, the aim of the design being to produce a flow of water which gives the best possible cleansing action. The two main types are: (a) the *washdown* type, which relies chiefly on the force of water falling from the water tank to produce its cleansing action and (b) the *syphonic* type which, by producing a negative pressure in the air space behind the water trap, draws water out of the pan by a syphonic action.

6.6.2 Wash basins (Lavatories)

These are used for washing the hands and face in bathrooms, cloakrooms and bedrooms. They are normally manufactured out of vitreous china or acrylic plastic and can either be wall hung, mounted on a pedestal, or built into the top surface of a vanity unit. Normal fittings include separate hot and cold taps (faucets) and a drain. The drain is fitted with a stopper, which can be a hand plug or a pop-up waste fitting operated by a lever adjacent to the taps.

6.6.3 Baths

Baths are made of enamelled cast iron, enamelled pressed steel, acrylic plastic or fibre-glass. In some cases the bath is combined with a shower unit. The water supply to the bath may be from separate hot and cold taps or from a mixer valve. In the case of combined bath/shower units the tap has a valve for directing the flow of water either to the bath or shower head. As with wash basins, the drain may be either a hand plug or a pop-up plug operated by a lever adjacent to the taps.

6.6.4 Showers

A shower cubicle is made from glass or acrylic plastic screens or water-proof fabric, with a vitreous china or fibre-glass tray for collecting and draining the water. The shower head may be either a permanent wall fixture or connected to the water supply by a flexible hose. Water is supplied to the shower through a hot and cold mixer valve which is usually thermostatically controlled to provide a uniform water temperature and to minimise the risk of scalding. Alternatively water may be heated electrically from cold via a small wall mounted heater, with the rate of flow of water controlling the water temperature (see section 10.2.2).

6.6.5 Bidets

The bidet is not a common fitting in UK bathrooms but it is normal to find them in Europe and increasingly in North American homes. It was used to wash the genitals and is normally manufactured from vitreous china. A mixed supply of hot and cold water enters the bowl either from under the rim, a central spray on the base, or a combination of both. A stopper is used in the drain of the bidet and this is normally the pop-up type.

6.6.6 Kitchen sink

These are manufactured from stainless steel, enamelled pressed steel or enamelled cast iron. A number of different designs are in common use. In the UK they commonly incorporate a draining board either on one (single drainer) or both (double drainer) sides of the bowl. In the USA double bowl sinks are the most common, often with a waste disposal unit fitted in the base of one of the compartments (see section 16.4).

There may be separate hot and cold taps or mixer taps. In the UK some water authorities will not allow mixer taps where the cold water is direct from the mains and the hot water from a storage tank. In this situation mixer taps are available which keep the two flows of water separate until the point where they leave the tap. This prevents any danger of contamination of the cold water supply. In the USA kitchen sinks have a hose and spray in addition to fixed taps. Strainers are used in the drain fitting to prevent the entry of solid food particles into the drain-pipe.

6.7 SOIL AND WASTE REMOVAL

6.7.1 The requirements of a domestic drainage system

It is possible to classify the drainage from domestic appliances as either soil or waste. Soil appliances are those from water closets whereas all others are classified as waste appliances. Soil and waste drains should be designed to remove soil and waste from a building without any risk to the inhabitants and without any undesirable environmental effects such as noise and smell. There should also be no risk of contamination of mains water with contaminated waste.

6.7.2 Traps

All appliances which discharge waste or soil must have a water trap either as an integral part of the appliance, as in the case of a WC, or as a fitting directly below the appliance drain. In certain cases a group of appliances may have a single trap. A trap is a water seal which prevents gases from the drain entering the building through the open drain connection of an appliance. The required depth of water with the trap is dependent upon the type of drainage system used and there are a number of different designs of trap as shown in Fig. 6.7.

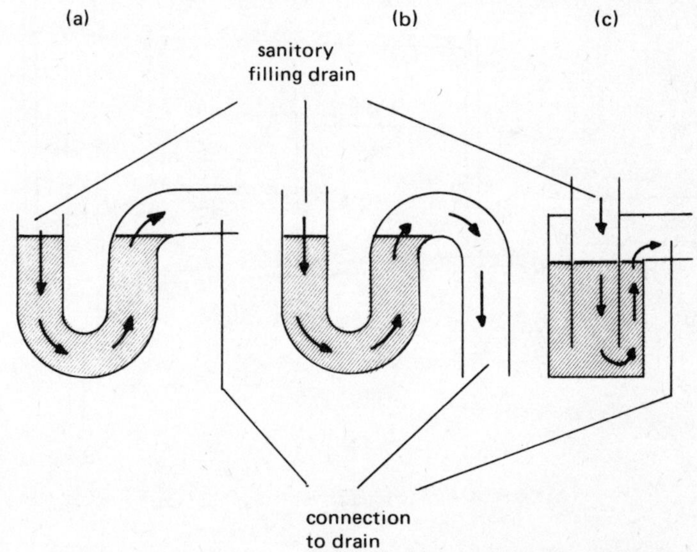

(a) P trap   (b) S trap   (c) Bottle trap

Fig. 6.7 – Appliance water traps.

6.7.3 Venting

If a negative pressure develops in the drain-pipe beyond a water trap, it is possible that water will be drawn out of the trap which is then rendered ineffective. In a similar manner, if a positive pressure develops in a drain-pipe it is possible for water to be forced out of a trap into the appliance. Careful design of the waste and soil drainage is required to prevent either of these situations arising. One way in which this can be done is to fit each appliance drain with an individual vent as shown in Fig. 6.8. At a level above the connection of the highest appliance drain, the vent pipe is connected to the drain-pipe, which is extended to roof level where it opens to atmospheric pressure.

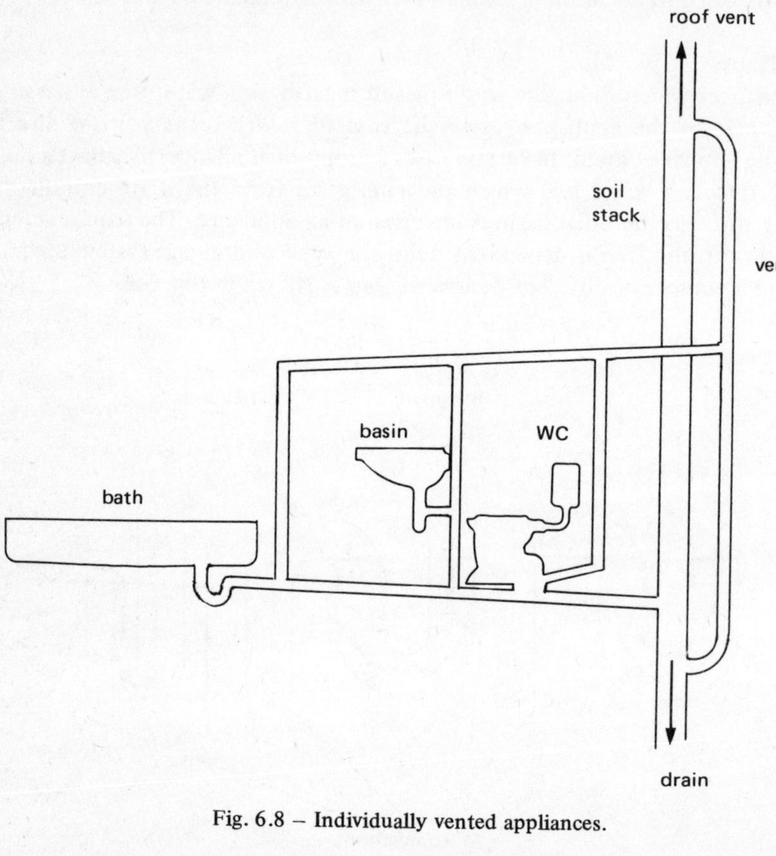

Fig. 6.8 – Individually vented appliances.

In some cases, particularly where appliance drains are short and where a group of appliances are connected in a cluster around a drain pipe which opens to the atmosphere at roof level, individual vents are not required. The system is used in new housing in the UK and it is allowed in the US by some authorities. A typical UK single stack drain is shown in Fig. 6.9. An alternative drainage method, which is found in older UK housing, is the two-pipe drainage system. In this method, a separate drain-pipe is used for soil and waste drains. These two pipes join at an inspection chamber close to the building.

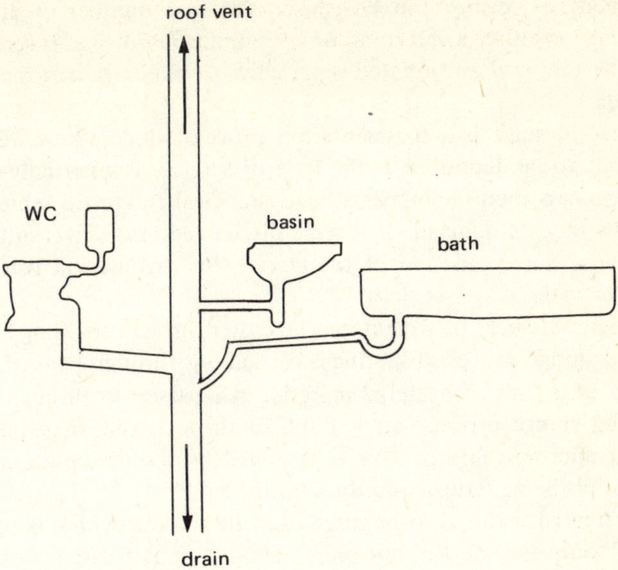

Fig. 6.9 – Single stack drainage system.

6.8 WASTE WATER TREATMENT

6.8.1 Private waste treatment

In some localities, usually rural areas, houses are not connected to a municipal sewerage network and must have their own sewage treatment on site. Two common methods of sewage treatment are the *cesspool*, which simply stores sewage until it is emptied by road tanker, and a *septic tank*, which processes the sewage.

A cesspool is a brick or concrete water-tight tank of about 20 m^3 capacity. It is sited at least 15 m from the house and away from any wells, springs or other sources of drinking water. A septic tank is much smaller than a cesspool as it stores water for periods of a few days only, during which it is digested by anaerobic bacteria. A sludge settles at the base of the tank and this must be removed periodically. The treated water, possibly with additional filtration, is then discharged into a ditch or stream.

6.8.2 Municipal sewage treatment

Waste water, soil and rainwater is collected and transported in a network of sewers. In some cases there are combined sewers which take all of these wastes. Alternatively, there may be separate sewers for dirty water, which needs treatment, and for surface (storm) water, which can be discharged directly into a river without any treatment.

Treatment of sewage can be carried out in a number of different ways. The first stage is either a screening of comminution process. If screening is used solids can be removed and treated separately. Comminution is the reduction in size of solids.

The second stage is a sedimentation process which allows 30–50% of suspended solids to be deposited at the base of a tank. The partially treated water from this process then undergoes a bacteriological treatment which is designed to reduce its oxygen demand. If it were discharged into a river in this condition it would cause deoxygenation of the water which would kill fish and produce foul smelling water.

The bacteriological treatment can be either through trickling filters or using an activated sludge process. Filtering is carried out through beds of stones which are covered in a film of bacterial material. Air passes up through the filter. In the activated sludge process, air is bubbled through the dirty liquid which is mixed with micro-organisms. This is followed by a sedimentation process, the deposited solids being returned to the aeration tank.

If the treated water is to be discharged into water which is to be used for recreational purposes or for hot water supply, it is treated with chlorine to destroy harmful bacteria.

REFERENCES

Clarke, J. W., Vussman, W. & Hammer, M. J. (1977), *Water Supply and Pollution Control,* IEP, 3rd Ed.

IHVE, (1974), *Domestic engineering services,* Institute of Heating and Ventilation Engineers.

Smith, N. (1976), *Man and Water,* Peter Davies.

BIBLIOGRAPHY

Fair, G. M., Geyer, J. C. & Okun, D. A. (1966), *Water and Wastewater Engineering,* Vol. 1, Wiley.

Hall, F. (1976), *Building Services and Equipment,* Longman.

McGuiness, W. and Stein, B. (1977), *Building Technology,* Wiley.

Tompson, T. A. (1972), *A guide to sanitary engineering services,* Macdonald and Evans.

Part II
ENVIRONMENTAL CONTROL OF THE HOME

CHAPTER 7

Comfort

7.1 INTRODUCTION

The objective of environmental control in the home is to create conditions of comfort for its occupants despite environmental changes which occur both within and external to the home. Comfort cannot be defined in absolute terms since it depends upon the individual, what he/she is doing and the type of clothing worn. Each individual has his/her own preference for conditions of optimum comfort and this varies, depending on whether the individual is reading a book, cooking a meal or doing housework. Normally we are more aware of, and therefore more likely to react to, conditions of discomfort rather than of comfort. Discomfort can be seen as an annoyance or distraction which is detected by one of the sense organs and which produces a response. The sensation can vary for a mild feeling of discomfort to extremes of pain and it can be seen as part of the human defence mechanism to move or protect itself from a hostile environment.

The human body is constantly monitoring its environment through thermal, visual, auditory, olfactory and tactile senses. Comfort does not represent absence of response from the senses. Indeed sensory deprivation can lead to unpleasant psychological effects. There is an optimal level of response from the sense organs which represents the condition of optimum comfort. Under these conditions mental and physical efficiency is at a maximum. Stress caused by conditions of discomfort leads to fatigue, irratability, loss of concentration and loss of efficiency.

In this and the following two chapters we will concentrate on thermal comfort. To some extent this is related to odour control since the control of humidity by ventilation and air conditioning removes odours from the atmosphere: this is considered in Chapter 11. Visual comfort, and its control through the provision of lighting, is the subject of Chapter 12.

7.2 CONTROL OF BODY TEMPERATURE

7.2.1 Physiological temperature control

The human body may be compared to an engine. Food is taken in and is utilised to perform work and, generally speaking, the more work that is performed by the body, the more food must be consumed. The performance of work is accompanied by the production of heat, the amount of heat being proportional to the work done. A person resting will generate between 20 and 40 W/m^2 of body surface, sitting 60–80 W/m^2 and standing 90 W/m^2. A person performing some physical activity will generate upwards of 100 W/m^2 (Brundrett 1973). If this heat were not lost from the body its temperature would rise. Conversely, if more heat were lost from the body than it generated internally, its temperature would fall. The body uses a number of different mechanisms to control its rate of heat loss thereby maintaining the temperature of internal organs, the brain and heart at 37°C. When the body loses heat at exactly the same rate as it is generated internally, it is said to be in *thermal equilibrium* and its temperature will be maintained at a constant value. A part of the brain known as the hypothalamus regulates the loss of heat from the body to control temperature.

When the rate of heat loss from the body is too low, blood is directed to the surface capillaries of the skin, raising its temperature and increasing the rate of heat loss. When the rate of heat loss from the body becomes excessive, it reacts by reducing the flow of blood through the skin. This diversion of blood from the capillary skin blood vessels to the inner mass of the body automatically results in a lowering of the temperature of the skin which reduces the rate of heat loss. The effect of this is to lower the temperature of the skin so that there is a temperature gradient from the core of the body, at 37°C, to the skin. Some parts of the skin, particularly the hand and foot fall to a much lower temperature than other parts of the skin (Brundrett 1973) as shown in Table 7.1.

Table 7.1
Skin temperatures of hand, foot and forehead

Part of body	Air temperature	
	20°C	15°C
Foot	25°C	22°C
Hand	30°C	28°C
Forehead	35°C	34°C

This mechanism works well providing that the difference between internal heat generation and heat loss is small. However the extent to which the skin temperature can be raised or lowered is small. If greater amounts of heat need to be lost from the body, pores in the skin are opened, allowing evaporation of water which produces a resultant cooling of the surface. This process is referred to as sweating. When the body is losing too much heat, the pores are closed so tightly that they stand out as 'goose-flesh'. Also, under conditions where too much heat is being lose, spasmodic muscular activity is used to generate heat, producing the sensation of shivering.

7.2.2 Heat transfer from the body

Heat is lost from the surface of the body by a number of different methods as shown in Fig. 7.1. Heat from the centre of the body is conveyed to the body surface by the flow of blood to the surface capillaries and by *conduction.*

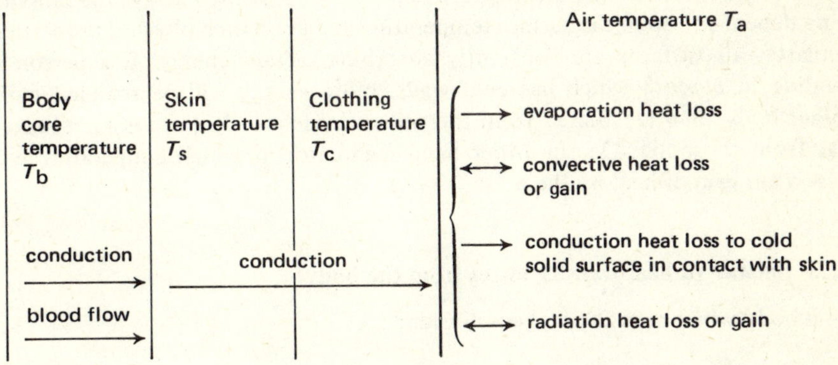

Fig. 7.1 – Heat transfer from the body.

Conduction is the flow of heat through a solid material caused by a temperature difference in the solid. In Fig. 7.1 the skin temperature, T_s, is lower than the body core temperature T_b and therefore heat flows from the body core to the skin. Similarly if the temperature of a layer of clothing is lower than the skin temperature, heat will flow from skin to clothing. However materials provide a *resistance* to this flow of heat, the degree of resistance depending upon the thickness of material (doubling the thickness will double the resistance) and upon the physical nature of the material. Some materials, such as metals, provide a low resistance to the flow of heat. This means that if a piece of cold metal is touched, a lot of heat will flow from our hand into the metal. It is therefore referred to as a *good conductor* of heat. Other materials, such as air, provide a high resistance to conductive heat transfer. Because of this, clothing which contains pockets of trapped air reduces the rate of heat loss from the body. It is therefore described as being a *poor conductor* of heat or it may also be considered to provide good *thermal* insulation.

Heat loss from the surface of a body may be either by **conduction, convection, evaporation** or **radiation.** *Conductive* heat loss occurs if a person stands on a cold floor particularly when the feet are uncovered. Similarly heat can be lost if a person sits on a cold surface.

Convection is the transfer of heat caused by movement of a fluid (gas or liquid). Cold air which is in contact with exposed skin or clothing is warmed by conduction and, as its temperature rises, it becomes less dense than the cold air and floats upwards. As it does so it is replaced by cold air which is, in turn, heated and the process continues. This is known as *natural* convection.

When a liquid is changed into a vapour it requires a source of thermal energy, known as *latent heat.* If water is evaporated from the skin surface the latent heat is extracted from the skin, producing a cooling effect.

Radiation is the transfer of heat from one surface to another. All surfaces within the domestic environement are constantly radiating energy, the amount being dependent upon the surface temperature and upon their physical properties. Similarly, all surfaces are constantly absorbing radiant energy. If a person is standing in a room which has cold walls, more energy will be radiated from his/her body than is radiated from the walls. Consequently there is a net loss of heat from the body. On the other hand a radiator at a high temperature will cause a net gain of heat by the body.

7.2.3 Balance of heat transfer losses from the body

For a body which is in thermal equilibrium:

$$M - W = E + R + C$$

where M is the metabolic rate

W is the energy expended usefully by the body doing work

E, R and C are evaporative, radiative and convective heat losses (Chrenko 1974).

Fig. 7.2 indicates the relative contribution of the three methods of heat transfer to the total body heat loss. At low temperatures almost no heat is lost by evaporation and this method of heat loss does not become significant until the combined radiation and convective heat losses decline. As the ambient temperature increases, it approaches that of the skin. This reduces the amount of heat transfer which can take place by convection and radiation. When the ambient temperature is higher than the skin temperature there is a gain of heat by the body from convection and radiation. This is offset by an increase in the amount of evaporative heat loss. Under normal interior conditions a person does not lose heat by conduction and the balance between radiation, convection and evaporation is approximately 2:2:1.

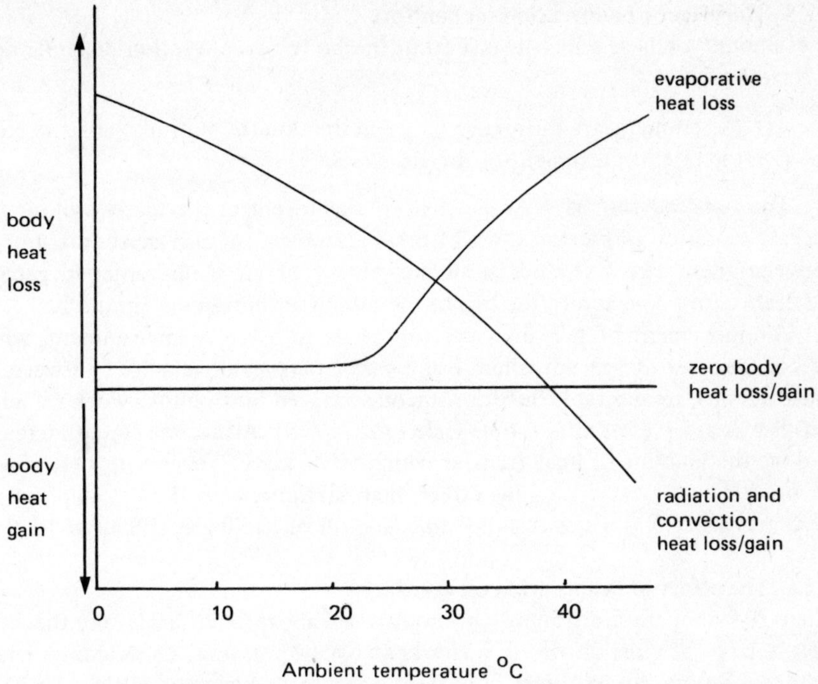

Fig. 7.2 – The effect of ambient temperature on body heat losses.

7.2.4 Effect of thermal conduction on comfort

As stated in section 7.2.3, the body does not lose much heat directly by conduction (caused by contact of the skin with a cold surface). For a person wearing shoes, these act as a layer of thermal insulation between a cold floor and the warm skin. Awareness of discomfort caused by standing on a cold floor depends partly on how long the person has been standing on the floor and also upon its thermal conductivity. If we compare two surfaces such as concrete and cork tile both at the same low temperature, the concrete will feel much colder than the cork tile because of its higher thermal conductivity.

Warm floors can also cause a feeling of discomfort. Floor temperatures in excess of 26°C are likely to cause discomfort in people who stand on the floor for long periods of time. It also causes fatigue and the sensation of tired feet.

A major influence of conductivity on comfort is through the use of clothing. This acts as a layer of insulation between the skin and its immediate environment. Clothing represents a conscious control of heat loss through the skin. As the environment becomes more hostile, we protect ourselves from it by the use of more efficient insulation (Humphreys 1970).

7.2.5 The effect of convection on comfort

The amount of heat which is lost from the body by convection depends upon two factors

(i) the temperature difference between the skin (or clothing) and the air
(ii) the rate of movement of the air.

The exposed skin is very sensitive to air movement. Perfectly still air produces a sensation of stuffiness whilst rapid air movement also causes discomfort. Between these two extremes some movement of air is desirable. In general, the more active a person is, the higher the rate of air movement should be.

Air movement of less than 0.1 m/s leads to a stuffy environment, which affects the level of concentration. For a seated person, air velocities between 0.1 and 0.15 m/s are desirable but for someone engaged in strenuous work velocities of 0.5 m/s are preferred (Croome-Gale *et al.* 1975). As air movement increases, so does the amount of heat transfer which takes place. This means that moving air produces a greater cooling effect than stationary air. For example air at 20°C moving at 0.5 m/s causes the same amount of cooling as still air at 14°C.

7.2.6 The effect of evaporation on comfort

When the temperature around the body rises above comfort levels, the body loses a large proportion of its waste heat by evaporation, or sweating. Under these conditions humidity is important because at high humidities the water does not evaporate but remains as a liquid on the surface of the skin. When this happens latent heat is not extracted from the body.

At low temperatures there is some loss of heat by moisture diffusion through the skin and by evaporation of moisture from the lining of the lungs, throat and mouth. At ambient temperatures above 34°C this, together with convection and radiation, causes insufficient cooling and the sweating mechanism is activated. At temperatures above 38°C practically all heat is lost by this method. In the zone of evaporative heat regulation, air movement greatly facilitates heat loss, provided that the air temperature is not above skin temperature. In these circumstances the body will be heated by convection and this can offset heat loss from the skin by evaporation.

7.2.7 The effect of radiation on comfort

The amount of radiant heat interchange depends upon the relative temperatures of the skin and walls and also upon the physical nature of the walls; some materials absorb a large proportion of the heat which falls on them, others reflect the heat. Normally skin absorbs most of the radiation which falls on it whereas some types of fabrics reflect a significant proportion of the incident radiation (Brundrett 1973).

In general, extremes of radiant heat loss and heat gain by a body produce feelings of discomfort, particularly if these are localised on a small part of the body surface. For this reason large areas of radiator surface at a low temperature are preferable to small high-temperature radiators.

Intense radiation from the body to a cold surface, such as a window, is often interpreted as a draught. This is because the heat loss is very localised since it only occurs on exposed skin which is facing the direction of the window.

7.3 CONTROL OF THE HOME ENVIRONMENT

7.3.1 Conditions for optimum comfort

Sensations of warmth or cold are purely subjective and depend on many factors such as age, sex, state of health, type of clothing and state of activity so that it is impossible to design conditions which will produce a reaction of comfort for all occupants. For example, although elderly people have similar comfort levels to young people, because they spend long periods of time inactively, they prefer slightly higher temperatures. Similarly, for the same kind of activity a woman may prefer a temperature up to 6°C higher than a man because of physiological and clothing differences.

As a further complication even for an individual involved in the same activity, the preferred temperature is not constant but depends upon the time of year. The body adjusts to long term changes in temperature but is more sensitive to sudden changes.

Because of these factors, general practice is not to specify a *point* of optimum thermal comfort but to define a range of conditions of a *comfort zone* in which, say, 70% of individuals will feel comfortable (see Fig. 7.3). The American Society of Heating and Ventilating Engineers produces charts giving comfort zones (ASHVE 1969).

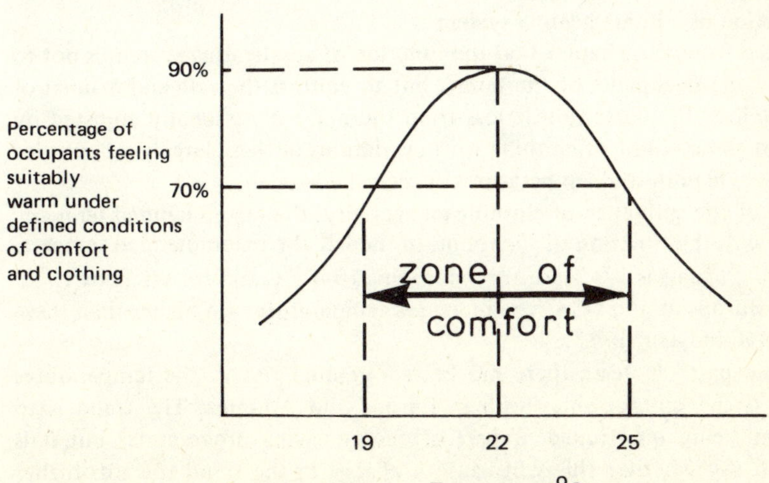

Fig. 7.3 – Conditions of optimum thermal comfort.

In addition to the average temperature, other factors contribute to feelings of comfort or discomfort.

(a) The proportions of the heat lost by radiation and convection must be similar. Neither high air temperatures with cool surroundings (low convection, high radiation) nor low air temperatures with intense radiation (high convection, low radiation) are acceptable for long periods of time. Under ideal conditions the average temperature of surrounding surfaces should be 1½-2°C above the air temperature, although this is difficult to achieve in practice. However a temperature which is no more than 1½-2°C below air temperature is acceptable. Insulation increases wall temperatures as do curtains over windows.

(b) Radiation should take place more or less uniformly in all directions. Intense radiation from one direction is distinctly unpleasant.

(c) Discomfort is experienced if the head is appreciably warmer than the feet. The temperature at head level should be 1-2°C below that at floor level.

(d) Air movement should be moderate and preferably variable in both speed and direction: 0.15 to 0.2 m/s is acceptable in a heated room in in winter; up to 0.40 m/s is tolerable in summer.

(e) Humidity should be about 50%. Below 30% the mucous membranes tend to dry whilst above 70% the skin may feel clammy and sweat evaporates less easily.

7.3.2 Function of a home heating system

It can be seen from this chapter that the function of any heating system is not to heat or cool the occupants of the home, but to control the rate and manner of bodily heat loss. In additon, heat loss from the body is personally adjusted by variations in the amount of clothing worn; variations in these largely account for the differences in preferred temperatures by men and women.

Because of the influence of clothing and activity, the recommended temperature varies with the function of the room. In the UK the recommended temperature for living rooms is 17-21°C, for bedrooms 10-15°C and for kitchens 15°C. In general, European and USA recommended temperatures are higher than these for the UK (Grandjean, 1973).

Over the past 50 years there has been a gradual rise in the temperatures considered to be comfortable, both in Europe and America. The trend is to some extent being questioned as part of energy saving programmes but it is too early to say whether this will cause a change in the trend towards higher temperatures. A one degree rise in room temperature can represent an increase in energy consumption of up to 10%.

REFERENCES

ASHVE (1969), *Heating ventilation and air conditioning guide,* American Society of Heating and Ventilating Engineers.

Brundrett, G. W. (1973), *An Introduction to thermal comfort,* Electricity Council Research Centre.

Chrenko, F. A. (1974), *Bedford's basic principles of ventilation and heating engineering,* 3rd Ed., Lewis.

Croome-Gale, D. J. & Roberts, B. M. (1975), *Air conditioning and ventilation of buildings,* Pergamon.

Grandjean, E. (1973), *Ergonomics of the home,* Taylor & Francis.

Humphreys, M. A. (1970), A simple theoretical derivation of thermal comfort conditions, *J. Inst. Heat. Vent. Engrs.,* **38**, p. 95.

BIBLIOGRAPHY

Allen, E (1980), *How buildings work,* Oxford University Press.

Chalkey, J. N. & Cater, H. R. (1968), *Thermal Environment,* Architectural Press.

McQuiston, F. C. & Parker, J. D. (1977), *Heating, Ventilation and Air Conditioning,* Wiley.

CHAPTER 8

The design of home-heating systems

8.1 INTRODUCTION

Of the energy consumed in a house, the largest proportion is used for space heating. The actual amount of energy used in this way depends on a large number of factors such as the local climate, the standard of construction of the home and the desired comfort levels of the occupants, but typically, 66% of total energy consumption is used for space heating in the UK (Agius 1979).

From this it can be seen that the design of a heating system is complex. The heat provided must be capable of maintaining a desired internal temperature (related to aspects of comfort as discussed in Chapter 7) despite changes in the climatic conditions outside the building. A heat balance can therefore be used to determine the heat losses and gains of the building and any difference between these must be provided by the heating system, if the desired internal temperature is to be attained.

8.2 HEAT BALANCE

As was the case with the human body (section 7.2.3), if a room temperature is to remain constant, the room must be in thermal equilibrium. When this occurs, the rate of heat gain must be equal to the rate of heat loss. If the former is greater than the latter the room temperature will rise, and vice versa. The factors contributing to heat losses and gains are shown in Table 8.1.

8.2.1 Personnel

It was shown in Chapter 7 that the human body is constantly loosing heat in order to achieve thermal equilibrium. Similarly any pets in a room will also be providing a source of heat to that room. In situations where there are many people in a confined space, this leads to overheating. Even in a house with the normal number of inhabitants there can be a significant contribution to the heat gains in this manner.

Table 8.1
Factors contributing to heat losses and gains in a house

Heat gains	Heat losses	
Occupants	Fabric losses –	roofs
Lighting		floors
Appliances		walls
Solar gain		windows
Heating	Ventilation losses –	draughts
		designed-ventilation
		flue losses
	Water losses	

8.2.2 Lighting
The conversion of electricity to light, particularly in the case of tungsten lamps (Chapter 12), results in the generation of heat. Also the light energy produced is eventually absorbed and converted into heat by the fabric of the house, except for that light which is lost through the windows. It is possible to assume that if a room has a total of 300 W of lighting, then this will result in 300 W of heat being gained by the room.

8.2.3 Appliances
Domestic appliances convert energy (chemical or electrical) into mechanical energy (as in a food mixer) or thermal energy (as in an oven). Eventually most of this energy ends up as heat gained by a room. Therefore, as in the case of lighting it is possible to assume that the energy consumed by appliances is providing an equivalent amount of heat to the building.

8.2.4 Solar gains
Solar energy enters a building mainly through the windows and the amount of energy absorbed in this way depends upon the hours of sunshine, the angle of the sun and on the direction and area of the windows. Even in winter significant amounts of heat can be gained in this way. Solar gain can be a problem in summer because of the discomfort caused by the resultant high temperatures. To prevent this happening in office blocks which have large areas of window, the windows are coated internally with a plastic film which reflects the long (invisible) infra red radiation, but allows the visible light to pass through. Unfortunately, this film will then reduce the useful solar gains in winter.

8.2.5 Fabric losses

Heat is lost from a building in a number of ways but a major factor is the conduction of heat through the fabric of the house – its walls, windows, roof and floor. These factors are discussed in detail in section 8.3.

8.2.6 Ventilation losses

Houses are not air-tight structures and therefore warm air escapes through gaps and cracks in the structure and is replaced by cold air from the outside. This is exacerbated by open doors and windows. A certain amount of ventilation is necessary to renew oxygen and remove carbon dioxide, odours and smoke (Chapter 11), but in most houses the ventilation which occurs naturally is more than enough to provide sufficient ventilation (Burton *et al.* 1979).

In the case of fossil-fuel burning appliances, large amounts of energy are lost in the form of hot-flue gases, which are the products of the combustion process (see section 2.7).

8.2.7 Water losses

Large amounts of heat are lost from a building when hot water from laundry, baths, showers and sinks is discharged into the drains.

8.3 CONTROL OF HEAT LOSS FROM A BUILDING

8.3.1 Fabric losses

Heat is lost from a building by the conduction of heat from the warm interior to the cold exterior. The magnitude of the heat loss by conduction depends on:

(i) the temperature difference between the inside (T_i) and the outside (T_o) of the building

(ii) the thermal conductivity (λ), surface area (A) and thickness (l) of the material through which the heat flows, as given by the following equation:

$$q = \frac{\lambda A}{l}(T_i - T_o) \text{ where } q \text{ is the rate of heat loss.}$$

When considering the heat loss through the building fabric, it is normal to combine the effect of material thermal conductivity and thickness together in a single value, known as 'the U value', that is,

$$q = U{\cdot}A\,(T_i - T_o) \text{ wher } U = \frac{\lambda}{l}\,.$$

The 'U value' is made up of three main components (Fig. 8.1):

(i) a boundary layer of air on the inside of the building

(ii) the layer(s) of material forming the building fabric

(iii) a boundary layer of air on the outside of the building.

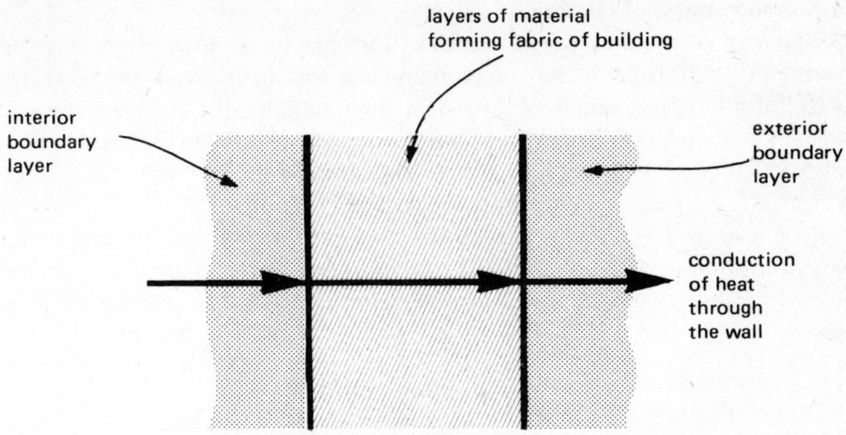

Fig. 8.1 – Factors contributing to the U value of a wall.

Since the thicknesses of the boundary layers are reduced in conditions of strong air currents, it is not surprising that U values are affected by winds. Because of this it is common practice to give a range of U values for a material, dependent on wind conditions. Tables of U values are available for normal UK (IHVE 1977) and USA (ASHRAE 1977).

If the U value is given in $W/m^2 \cdot {}^\circ C$, the area in m^2 and the temperatures in °C then the heat loss will be given in watts (joules/second). For example, for a single-glazed window the U value is given as 3.98 W/m·°C in sheltered conditions and 7.78 W/m^2 W/m·°C in severe winds. If the area of the window is 1 m^2 with an inside temperature of 20°C and an outside temperature of 5°C the heat loss can be calculated as follows:

Sheltered conditions

$$\begin{aligned} q &= U \cdot A\,(T_i - T_o) \\ &= 3.98 \times 1 \times (20-5)\ \mathrm{W} \\ &= 59.7\ \mathrm{W}\ . \end{aligned}$$

Severe winds

$$\begin{aligned} q &= U \cdot A\,(T_i - T_o) \\ &= 7.78 \times 1 \times (20-5)\ \mathrm{W} \\ &= 116.7\ \mathrm{W}\ . \end{aligned}$$

Heat losses, calculated as above, for different parts of the building fabric can be added together to give an estimate of the total heat loss from the building.

8.3.2 Effectiveness of insulation

The standard of thermal insulation varies from one house to another. A greater investment at the time of building a house can lead to reduced fuel costs over the life of a building. Similarly, for an existing building, the standard of insulation can be improved. This will require a capital investment but will lead to reduced fuel costs. It is useful to be able to determine the savings in energy from an investment in insulation as a basis for decision making.

One way of doing this is to calculate the *payback time* for the extra cost of insulation (McNair & Hitchin 1980), that is, how long it will take to recover the cost of insulation from the savings made in fuel costs. This may be estimated by:

$$\text{payback time (years)} = \frac{\text{cost of investment (£)}}{\text{annual fuel saving (£)}} .$$

This is a very simplistic method since, in reality, the calculation should take into account the cost of borrowing the money (or the interest accruing if the money were to be invested in some other way), and it should allow for fuel price rises over the payback time. However, it is a calculation which is relatively easy to apply, for example, to calculate the most economic thickness of insulation. The cost of insulation is directly proportional to its thickness as shown in Fig. 8.2.

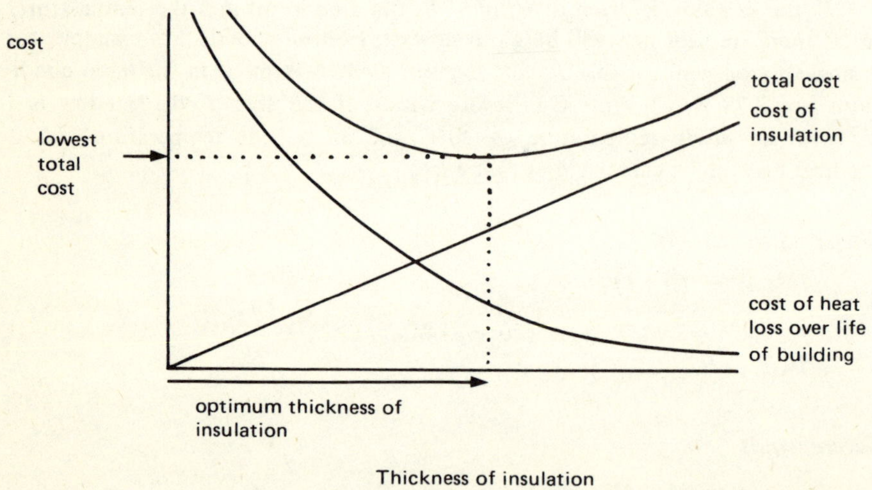

Fig. 8.2 – Graph showing the relationship between thickness of insulation and cost.

The cost of heat loss through the insulation decreases as the thickness of insulation increases but the amount of additional saving decreases, for thick layers of insulation. There is a point, as shown on Fig. 8.2 where the lowest total cost is achieved. Below this thickness of insulation money could be saved by increasing the thickness of insulation. With a greater thickness than the optimum, the additional energy savings are not justified by the higher capital cost.

In order to calculate the annual savings resulting from additional insulation it is necessary to calculate the reduction in heat loss over the heating season resulting from the additonal insulation. Using the relationship:

$$q = U \cdot A\,(T_i - T_o)$$

the amount of energy in Joules will be:

$$U \cdot A\,(T_i\ \text{(average)} - T_o\ \text{(average)}) \times \text{number of seconds in the heating season}\ .$$

For a whole building $U \cdot A$ is equal to the sum of individual values for each part of the building's fabric. Thus:

$$U \cdot A = U_w \cdot A_w + U_r \cdot A_r + U_f \cdot A_f + U_g \cdot A_g$$

where U_w and A_w refer to the walls

U_r and A_r to the roof

U_f and A_f to the floor

and U_g and A_g to the windows.

The length of the heating season depends on geographical location and altitude (Open University 1978). Similarly, average outside temperatures can be measured over a heating season or obtained from tables. The average internal temperature can also be calculated, based on comfort levels and heating patterns.

One method which can be used to combine the effect of heating season and average outside temperature is to use the *degree day*. For this assumptions are made about the average internal temperature (commonly 18.3°C) and the average external temperature (this again varies with geographical location and altitude (Diament 1971). For example, if the average external temperature were 4°C and the heating season 200 days long, this would be the equivalent of (200 × (18.3–4) degree days since:

$$\text{degree-days} = \text{length of heating season in days} \times (\text{average internal temperature} - \text{average external temperature}).$$

Degree-day figures are issued monthly by the UK Department of Energy since they can also be used to monitor fuel consumption variations.

8.3.3 Walls

Although the amount of energy lost through the walls of a house varies with the type of construction and the inside and outside temperatures, it normally accounts for a large proportion of the total heat loss (about one third).

In the UK the traditional brick or stone house was built with solid walls. Cavity walls were widely used after 1945 to prevent the passage of moisture from the outside wall of a building to the inside. They also had the effect however of decreasing the *U* value, thereby improving insulation. Additional improvements have been made by building the inner wall with aerated blocks.

The *U* value can be reduced even further by filling the cavity with a foamed material, such as urea formaldehyde, mineral wool or expanded polystyrene beads (see Table 8.2). The filling can either be applied at the building stage as sheet material or injected into the cavity of existing buildings. It reduces the convective heat transfer within the cavity. Care is needed at the application stage to prevent the introduction of cracks which can aid migration of water across the cavity.

Table 8.2

The effect of cavity filling on the *U* value of a wall

Construction	U value
Solid wall + 10 mm plaster	1.8–2.0
Cavity wall + 10 mm plaster	1.3–1.5
Cavity wall (filled) + 10 mm plaster	0.6–0.8

Insulation can also be added to the outside or inside of solid walls, thus decreasing their *U* value. When insulation is added to the outside it must be protected by a waterproof layer, such as wooden weather boarding.

8.3.4 Windows

The amount of heat loss which occurs through the windows of a building depends on the area of the windows. A typical figure for UK housing would be about one tenth of the total heat loss (Department of Energy 1979).

Windows lose heat by both conduction through the glass and also by radiation. Additionally, they can lead to sensations of discomfort by creating a cold downflow of air close to the window and by localised radiation from exposed areas of skin to the cold window surface. The presence of a cold surface within a warm moist atmosphere can also lead to condensation on the surface of the glass. The window frame, as well as the glass itself, can be a cause of heat loss by air leakage, particularly where these are made of metal.

The heat loss from windows can be reduced in a number of ways. The use of heavy curtains, shutters or blinds can reduce the heat loss by 30%. Metal window frames are available with a plastic thermal barrier between the inside and the outside of the building. However the most common method for reducing heat loss through windows is by double glazing.

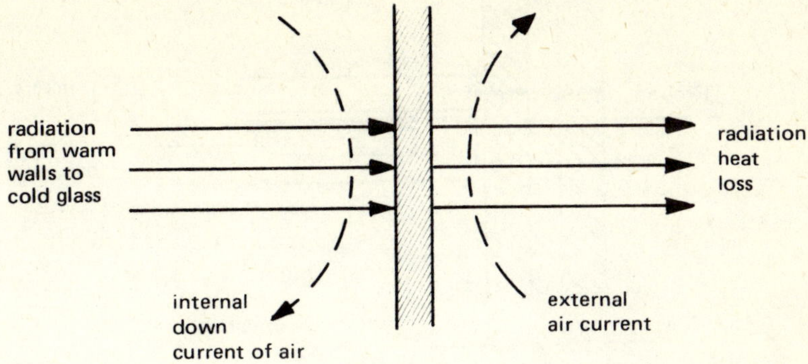

Fig. 8.3 – Heat losses associated with windows.

Double glazing consists of two layers of glass separated by an air space, which to be effective must be at least 20 mm. The effect of the air space is to reduce the amount of heat lost by conduction, whilst still allowing solar heat gains. There are two main methods of double glazing:

(1) Sealed units: two panels of glass sealed together with dry air between them to prevent condensation problems.
(2) Secondary windows: a second pane of glass fixed inside the normal pane. Ventilation of the air space is important to prevent condensation problems (Diament 1971).

A metalised film may be incorporated into double glazing which reduces radiative heat loss without reducing the transparency of the glass.

8.3.5 Floors

About 15% of the heat loss from a house is through the ground floor. Suspended floors consist of floorboards on wooden joists with a ventilated air space between the floor and the ground. Improvements in *U* value can be made by sealing cracks through the floorboards and by using heavy carpets and underlays. A layer of insulation below the floorboards can make a significant reduction in the *U* value. This is easy to apply at the construction stage but more difficult for an existing building.

Solid floors consist of a layer of concrete on hardcore with a thin damp proof membrane. Heat is lost by conduction through a solid floor, with a large proportion of heat escaping through a narrow strip close to the outer walls. This can be reduced by the installation at the construction stage of vertical and horizontal insulation at positions A and B on Fig. 8.4.

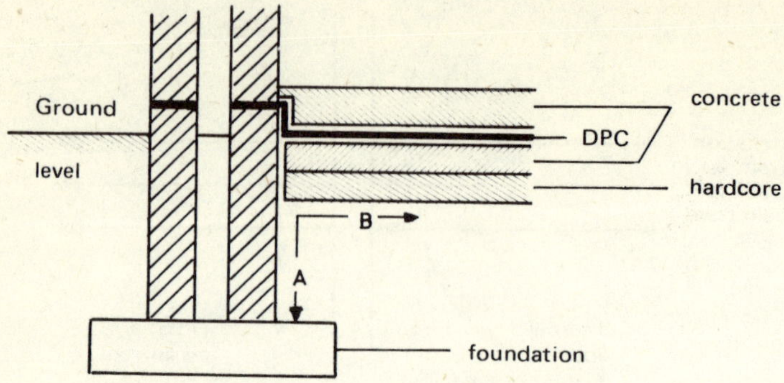

Fig. 8.4 – Heat loss through a solid floor.

8.3.6 Roofs

Approximately one quarter of the heat from a typical family house is lost through its roof. For a pitched roof U values given in published tables relate to the ceiling area and not to the external surface area of the roof. Current standards of construction give roofs with a U value of 0.6 $W/m^2 \cdot {}^\circ C$ but older properties may have U values as high as 2.4 $W/m^2 \cdot {}^\circ C$. These older buildings may be improved by the application of insulation between the roof joists or rafters. A layer of 100 mm of insulation will produce a significant reduction in heat loss.

8.3.7 Ventilation

Recommendations are given by a number of authorities on the level of ventilation required (ASHRAE 1977). This is usually stated as the number of air changes per hour. For a home it varies from 1 to 1½ for dining room, living areas and bedrooms to 2–3 for bathrooms and kitchens. The heat loss associated with this ventilation can be calculated from a knowledge of

(1) the volume of the room V (m^3)
(2) the number of air changes C (h^{-1})
(3) the inside temperature T_i (°C)
(4) the outside temperature T_o (°C)
(5) the specific heat capacity of air c($J/m^3 \cdot {}^\circ C$).

If a room has volume V and there are C air changes per hour, then in one second the volume of fresh air to be heated (v) will be:

$$v = \frac{V \cdot C}{3600} \, m^3/s \ .$$

The amount of heat required to heat this volume of air is given by:

$$q = v \cdot c \, (T_i - T_o) \, W \ .$$

As an example if a room has a volume of 125 m^3 with an air change rate of 1.5 per hour, the volume of air to be heated per second is:

$$\frac{125 \times 1.5}{3600} = 0.052 \text{ m}^3/\text{s} .$$

If the specific heat capacity of air is 1200 J/m$^3 \cdot$ °C, the inside temperature is 18°C and the outside temperature is 4°C, the heating system would need to provide:

$0.052 \times 1200 \times (18-4) = 874$ W to maintain the temperature of air within the room at 18°C.

Ventilation over and above that needed can be reduced by the use of draught excluders around windows and doors (Consumers Association 1972), by sealing cracks between the floorboards and by sealing unused fireplaces (Townsend and Colesby 1975).

8.4 CALCULATING THE HEATING REQUIREMENTS FOR A ROOM

As discussed in section 8.2, a heating system needs to be able to supply the difference between the heat losses and the heat gains for a room, in order to maintain its temperature at the desired comfort level. Using techniques discussed in the earlier parts of this chapter, assumptions must be made about the outside and inside temperatures and estimates made of heat gains and heat losses. The difference between these is then equal to the designed heat needs of that room.

As an example, consider the heating requirement for a living room with a designed internal temperature of 20°C to be maintained when the outside temperature falls to −5°C. Assume that there are three occupants each providing 100 W (Chalkey & Cater 1968) with an additional heat gain of 200 W from lighting and appliances. The room has a total of 20 m^2 of external walling of U value 1.6 W/m$^2 \cdot$ °C, 2 m^2 of windows of U value 2.5 W/m$^2 \cdot$ °C and 20 m^2 of floor of U value 1.25 W/m$^2 \cdot$ °C. The volume of the room is 60 m^3, with one air change/hour.

Heat gains:	Occupants		300 W
	Lighting and appliances		200 W
	Total heating gains		**500 W**
Heat losses:	External walls	$20 \times 1.6 \times 25$	800 W
	Windows	$2 \times 2.5 \times 25$	125 W
	Floor	$20 \times 1.25 \times 25$	625 W
	Ventilation	$\frac{60 \times 1}{3600} \times 1200 \times 25$	500 W
	Total heat losses		**2050 W**

Heating requirement of room $= 2050 - 500$ W
$= \mathbf{1550}$ **W**

In this calculation it has been assumed that there is no heat loss or gain from other parts of the house. This is normally the case where all the rooms are at a similar temperature. Where this is not so it may be necessary to calculate heat gains from adjacent rooms which are warmer and heat losses from rooms which are colder. Once a figure has been calculated for the heat requirements of a room it can be used to choose a suitable room heater, as discussed in the next chapter. Where there is a central heating system, individual room needs can be added to give the total heat requirement of the central heat source.

REFERENCES

Agius, P. J. V. (Ed.) (1979), *A warmer house at lower cost,* Watt Committee on Energy.

ASHRAE (1977), *Fundamentals,* Am. Soc. of Heating, Refrigeration and Air Conditioning Engineering.

Burton *et al.* (1979), Reducing energy demand in existing houses, in *A warmer house at lower cost,* Watt Committee on Energy.

Chalkey, J. N. & Cater, H. R. (1968), *Thermal Environment,* Architectural Press.

Consumers Association (1972), *Which,* **Nov.**, p. 185.

Dept. of Energy (1979), *Energy conservation,* Energy Paper No. 33, HMSO.

Diament, R. M. E. (1971), *The Internal Environment of Dwellings.*

IHVE (1977), Guide No. 3, Thermal and other properties of Buildings, Inst. of Heating and Vent. Eng.

McNair, H. P. & Hitchin, E. R. (1980), *Gas Engineering and Management,* **Nov/Dec.**, p. 447–451.

Open University (1978), *Energy in the home*.

Townsend, P. & Colesby, J. (1975), *Keeping warm for half the cost,* Prism Press.

BIBLIOGRAPHY

Allen, E. (1980), *How Buildings Work,* OUP.

Burberry, P. (1970), *Environment and Services,* Batsford.

Watt Committee 1978, Report No. 3.

CHAPTER 9

Space heating equipment

9.1 INTRODUCTION

Having decided on the amount of heat required to maintain the temperature of a room under the chosen design conditions (external air temperature, ventilation etc.) decisions can then be made about the manner in which this heat is supplied to the room so that the occupants will be confortable (see Chapter 7).

Each room may have its own heater(s) which can be used independently and each of which generates its own heat. Alternatively the generation of heat can take place centrally with the heat then distributed throughout the house. This is commonly referred to as a 'central-heating system'. Additionally decisions must be made on the balance of radiative and convective heat transfer. The importance of this was discussed in Chapter 7. Heaters are often classified as radiant or convection but usually they involve a mixture of both types of heat transfer.

Convection heaters rely on the upward movement of warm air (caused by a decrease in its density) to transfer heat from its source within a room to the room's environment. *Radiant heaters* generate radiant energy which at low temperatures cannot be seen. Sources of radiant heat at higher temperatures produce a visible glow of red, orange or yellow light. Radiant heaters provide a more instantaneous effect than convectors. Extreme radiant heat can cause discomfort because the heating is localised to those parts of the body which can be 'seen' by the source of radiant heat.

A further factor which must be taken into account is the siting of the heater in a room. As shown in Fig. 9.1, a heater can either reduce or accentuate the downflow of cold air associated with a window when there is a low outside temperature. As described in section 8.3.4 a window is often associated with a downflow of cold air. A heater sited under a window will counteract this flow of air.

9.2 ROOM HEATERS

A room heater generates and distributes heat to its surrounding environment.

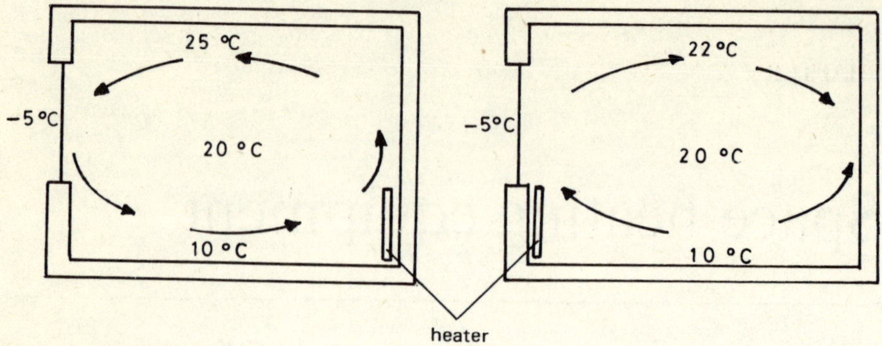

Fig. 9.1 – Effects of different sitings of a convection heater in a room.

The heating effect is localised to all, or part of, the room in which the heater is sited. The heat generation may be from a primary fuel (coal, wood, gas or oil) or from electrical-resistance heaters.

9.2.1 Radiant heaters

9.2.1.1 *Solid-fuel open fires*

A solid-fuel open fire must be designed so that the air necessary to maintain combustion of the appropriate fuel is supplied (section 4.7). This is much more critical for coke and smokeless fuels than it is for bituminous coals and wood. Without control of the air supply the efficiency of an open fire is low (25% for coal, 35% for coke) and the ventilation is excessive (4–5 changes/hour – section 8.3.7). Since most of the heating effect is by radiation, the combustion characteristics and the radiation output of the fuel are important (Fig. 4.4).

In some situations an electric fan is mounted in the front of the fire to provide a controlled flow of air through the fire. Alternatively, air may be supplied from outside the room, using an underfloor air supply. This is either brought into the appliance through a duct in a solid floor or else is drawn from the space beneath a suspended floor.

9.2.1.2 *Electrical radiant heaters*

The conversion of electrical energy to heat is discussed in section 5.7.1. In the case of electrical radiant heaters, the heating elements, which may be metal sheathed or in a quartz (silica) tube, are mounted in front of a reflector which directs heat forward into the room. The efficiency of an electrical radiant heater is 100%. Radiant elements are often incorporated into heaters which produce a simulated log-burning effect. Elements can also be incorporated in ceilings and floors to provide large areas of low temperature radiating surface.

9.2.1.3 *Gas radiant fires*

Radiant-type gas fires are normally fixed at hearth level using a non-aerated burner (see Section 2.6.5). They do not have a flue connection. They are only 40–50% efficient with 90% of the heat transferred by radiation. They have now largely been replaced by convection or radiant-convection heaters (Sections 9.2.2 and 9.2.3).

Gas fires are available which produce a coal or log effect with open incandescent (yellow) flames burning between simulated logs. The efficiency of these fires is low (less than 30%) and about one half of the heat transfer is by radiation. They are normally used as a replacement for inset solid-fuel fires to create a visual effect together with some heating.

9.2.1.4 *Oil radiant fires*

Radiant fires are available which burn kerosine, as described in section 3.6.2. As with the electrical radiant fire, the heat is thrown forward off a polished metal reflector (Fig. 3.4).

9.2.2 Convection heaters

Convection heaters transfer the major part of their heat to air which then circulates around the room by natural or forced convection.

Natural convection is the circulation produced by warm air which has been in contact with a heater rising and, when it looses the heat to the environment, falling again. This process produces a natural circulation of air within a room. In *forced convection* the air moves by the action of a mechanical device such as an electric fan.

9.2.2.1 *Electric convection heaters*

The simplest form of electric convection heater consists of a metal cabinet with air openings at the base and also at the top or front. Inside are a number of electric heating elements. The output may be controlled either by switching on a different number of heaters or by using a thermostat (section 9.4.1). Cold air enters through the base, is heated and passes out through the top of the heater (Fig. 9.2). The efficiency is 100%.

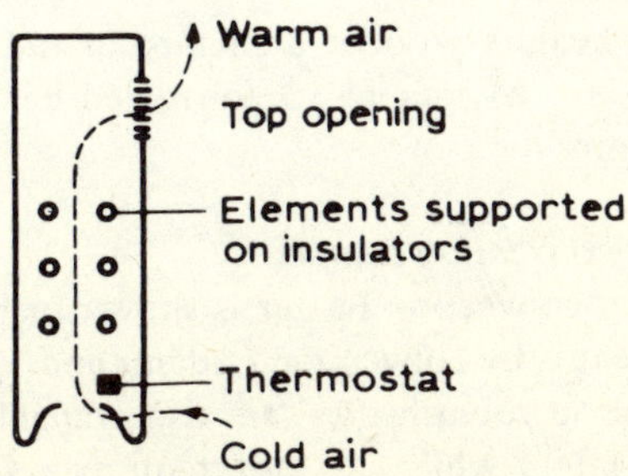

Fig. 9.2 – A convection heater.

A fan heater is similar to a convection heater except that air movement is caused by a small electric fan. There are separate controls for air speed and heat output. For a given heat output this type is much more compact than a convection heater.

9.2.2.2 *Gas convection heaters*

Gas convection heaters are available with conventional or balanced flues (section 2.7). In the case of small output heaters they may not require a flue. Fig. 9.3 shows a gas convection heater with a balanced flue. Air from the room is heated by contact with the hot metal of the heat exchanger which is heated by the combustion products.

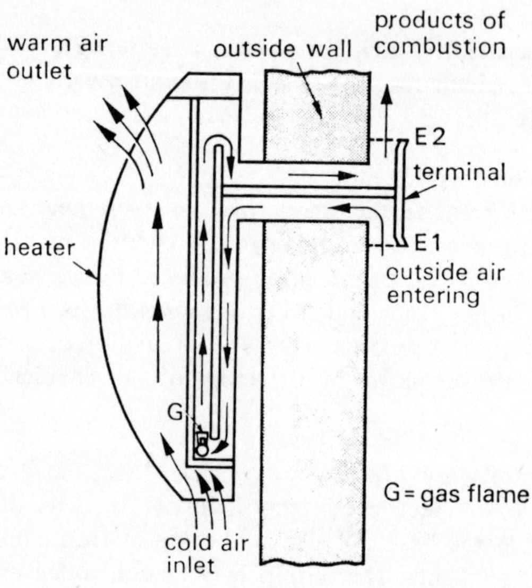

Fig. 9.3 – Balanced flue gas convection heater.

9.2.3 Radiant-convection heaters

Radiant-convection heaters produce a balance of radiant and convected heat. In some cases the balance of radiant to convected heat can be controlled by the occupants of the room.

9.2.3.1 *Gas radiant-convection heaters*

A natural gas radiant-convection heater is shown in Fig. 9.4. Between one-third to one-half of the heat is by convection (Adams and Vaisey 1970).

Radiant-convection room heaters are also available which utilise LPG. These are free-standing heaters which do not require a flue and have therefore an efficiency of over 90% (Consumers Association 1979).

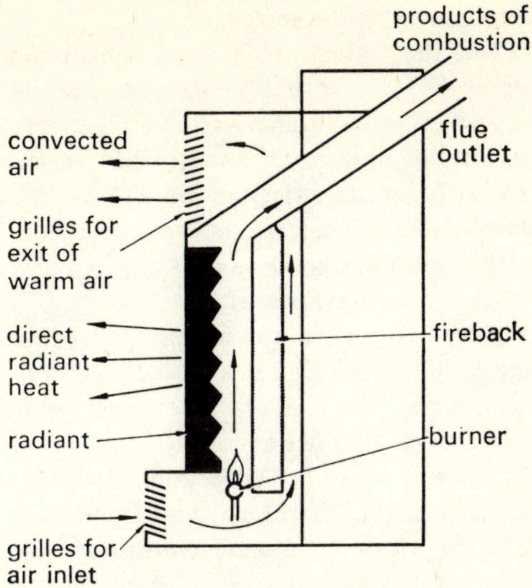

Fig. 9.4 – Gas radiant-convection heater.

9.2.3.2 *Solid-fuel radiant-convection heaters*

There are two types of solid-fuel appliances which provide a balance of convective and radiative heat transfer: the convection open fire and the closed room heater. *The convection open fire* is similar to a conventional open fire with the addition of an air space allowing cold air to circulate over hot fire bricks at the back of the fire. There must be no direct contact between this air and the air and combustion products in the flue. Approximately half of the heat transfer is by convection and the efficiency is 33% for coal and 44% for coke. Open wood-burning stoves are also available which give a balance of radiant and convected heat. These are usually free standing rather than inset in a fire place.

A *room heater* is a closed fire usually made from cast iron. Many types have a glass panel in the front to provide both direct radiant heat and the visual effect of a solid-fuel fire. They may be either built into a fireplace or free standing. Thermostatic control is normally used. This regulates the volume of air entering the appliance for combustion. Convected heat is obtained from an air space surrounding the fire. Compared with open fires they provide more-controlled, longer burning which gives longer periods without refuelling; more constant heat output; more flexibility in combustion rate, including low-rate overnight burning; and ease of ash removal. The efficiency is about 50% for coal and 75% for coke. Wood-burning closed room heaters are a traditional method of heating rooms, particularly in Scandinavia. They are made of cast iron with control over the amount of air entering for combustion. Because they are normally free standing they have a high efficiency. Some models can be operated either open or closed.

9.2.3.3 *Electric radiant-convection heaters*
One method of obtaining the required balance of radiant and convected heat is to combine both types of heater into a single unit. The user can then choose between radiant heat (which gives a faster warm-up time from cold) and convected heat (which gives more uniform heating). The convection elements are normally operated by a thermostat. Heaters are also available in the form of skirting board mounted tubular heaters and wall-mounted oil-filled radiators. The air adjacent to the warm surface heats up and is convected through the room. Heat is also radiated from the hot surface.

Storage heaters were introduced to increase the demand for electricity between midnight and 8.00 a.m. During this period electricity is converted into heat which is stored in a specially manufactured brick which has a high heat capacity. During this period electricity consumption is measured using the 'offpeak rate' section of the White meter (section 5.4.1). Heat is retained in the storage heater by surrounding the bricks with layers of insulating material, which allows the heat to be released at a slow, controlled rate.

There are three different types of storage heater. In the simplest, heat is conducted from the bricks, through the layers of insulation to the outside walls of the cabinet. It is transferred into the room by radiation and convection. The disadvantage with this type is that the only control on the heat output is the amount of heat stored the previous night. The second type of storage heater contains spaces which permit the passage of air between the bricks and out, through a damper, into the room as shown in Fig. 9.5(a). Control of the heat

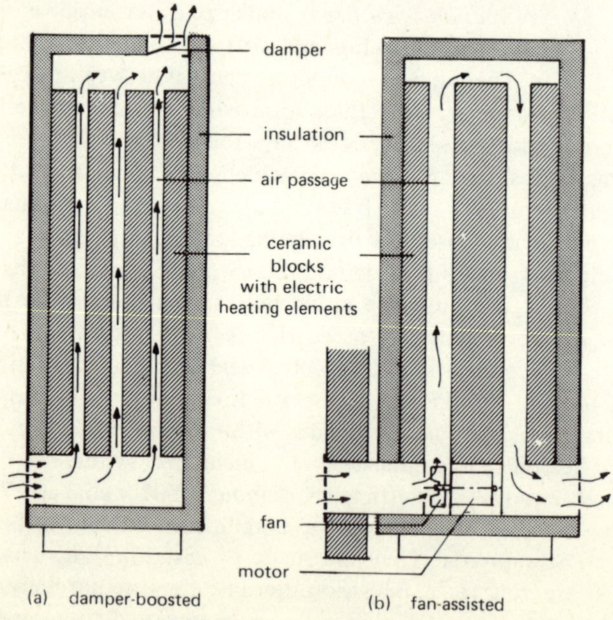

Fig. 9.5 – Electric storage heater.

output is possible by adjustment of the extent to which the damper is open. An extension of this principle is the fan-assisted storage heater shown in Fig. 9.5(b). Air circulation through the heater is controlled by a fan which can be used to boost the heat output at particular periods of the day.

One problem with storage radiators is that the heat output gradually declines throughout the day as the temperature in the core of the heater falls. To some extent this does not match the normal requirement for a slightly higher temperature in the evening than during the day, because of the difference between daytime and evening activities (section 7.2.1). The peak output of a storage heater occurs around midday, with the temperature declining in the evening (see Fig. 9.6). They also give out heat over a 24-hour period even if there are no occupants in the house.

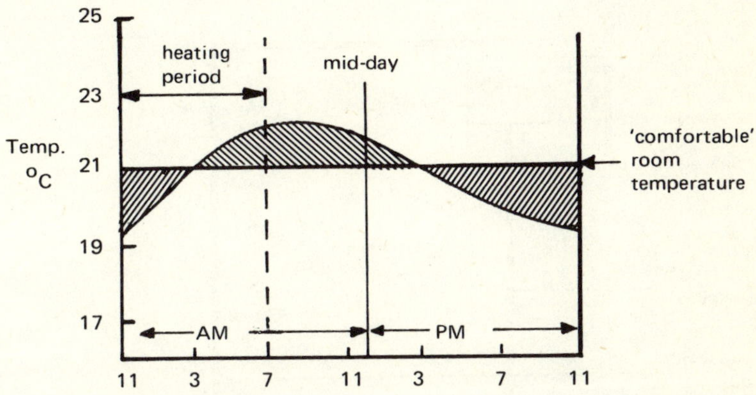

Fig. 9.6 – Temperature changes in a room heated by a storage heater.

9.2.3.4 *Oil radiant-convection heaters*

Wick type paraffin heaters produce a balance of radiant and convection heat transfer (section 3.6.1 and Fig. 3.3). Also, radiant-convection heaters of the type shown in Fig. 9.4 are available for use with kerosine and domestic fuel oil. The combustion of liquid fuels in flueless appliances is discussed in section 3.4.

9.3 CENTRAL HEATING SYSTEMS

A central heating system consists of four main parts:

(1) a heat generator which converts either chemical (wood, coal, oil or gas) or electrical energy into heat;
(2) a heat distribution system which uses a fluid to transport the heat to different parts of the house; in domestic central heating systems the fluid is normally either air or water;
(3) space heaters which release the heat at a number of points throughout the building;

(4) a control system which is designed to provide the desired level and balance of heat throughout the house, at the required time.

9.3.1 Water-circulation central heating

In this type of heating system hot water is circulated from a boiler around a pipe network to heating appliances situated around the house and back to the boiler. Additionally, the water boiler provides a source of hot water for domestic use (Chapter 10).

9.3.1.1 *The Water Boiler*

Most domestic central heating systems operate at a low pressure which is maintained by having a feed/expansion tank at some height above the water boiler, as shown in Fig. 9.7.

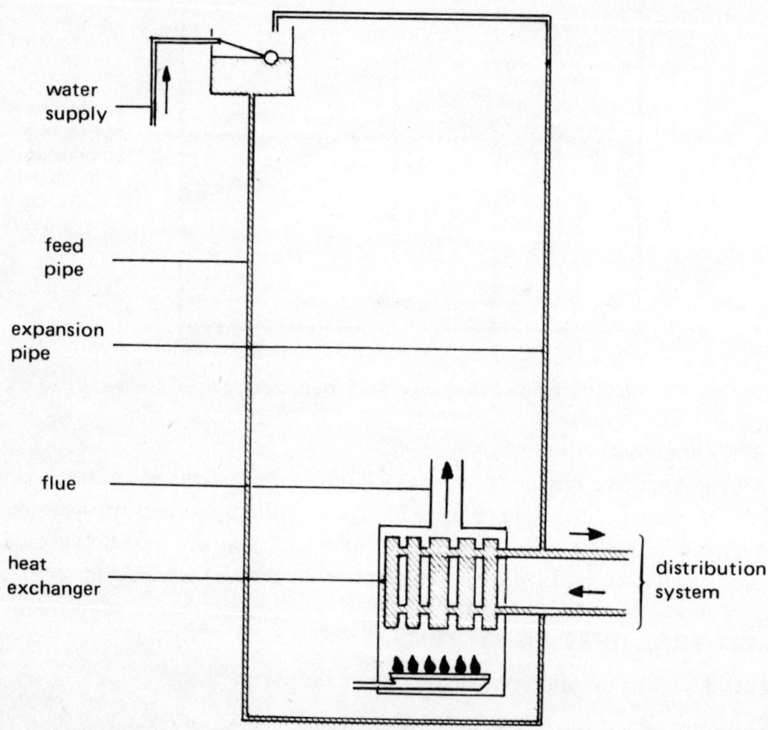

Fig. 9.7 – Central heating water boiler.

Free standing water boilers are available for use with all fossil fuels and wood. Some can be converted easily from one fuel to another, for example, from oil to gas. These boilers normally have cast-iron water vessels which hold a large volume of water. Whilst at their full output they have a high efficiency

(70–80%), when they are required to produce small volumes of hot water, such as for domestic use in summer, their efficiency falls to 40–50%.

In the case of gas and oil, small-capacity boilers are available which overcome this problem. Water circulates through the boiler in a finned pipe which holds only a few litres. Even when the output of these boilers is down to one tenth of the maximum they still operate at 80% efficiency (Pickup & Miles 1978).

Low-capacity back boilers are also used with room heaters as shown in Fig. 9.8. Externally they have the appearance of a conventional gas, oil or solid fuel room heater but they have a water boiler installed at the back. In the case of solid-fuel room heaters, the heat from the burning fuel is used to heat both the room and water in the boiler. Gas and oil room heater/back boiler combinations have separate burners and controls for the two functions.

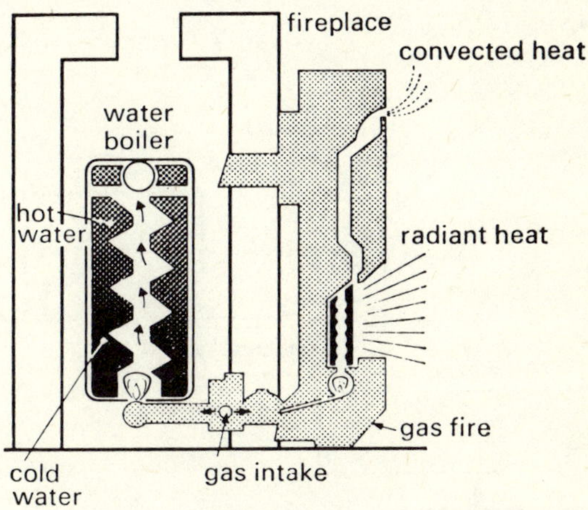

Fig. 9.8 – Gas-fired back boiler.

In some situations it is an advantage to maintain the water at a higher pressure using a *sealed system*. This avoids the need for a feed tank and it also means that the boiler can be mounted at a high point in the building. The expansion tank is sealed, but has a flexible diaphragm which allows for the expansion of water within the pipework. Normally these systems operate at a higher water temperature (up to 110°C) without any boiling noise and, because of the exclusion of air, with less risk of corrosion. Only periodic 'topping up' of the water is needed and this can be done either manually or automatically.

9.3.1.2 *The distribution pipework*

Water may be circulated around the distribution system either by *gravity* (natural convection) or by a pump (forced convection). *Gravity systems* are quiet and do not need an electrical supply for the pump but require large diameter pipes (25 to 30 mm) and bulky radiators to obtain the necessary water velocity. For this reason most new installations use pumped circulation using either *small-bore* (12 mm diameter) pipes or *mini-bore* (6, 8 or 10 mm diameter) pipes.

A *small-bore system* is similar to the ring-main electrical circuit (Fig. 5.12). In a small house there may be a single pipe loop supplying all radiators, but in other cases two or more separate loops may be used. There are two different methods used for connecting heating appliances to this pipe loop, as shown in Fig. 9.9. In the single-pipe system the flow and return connections of each

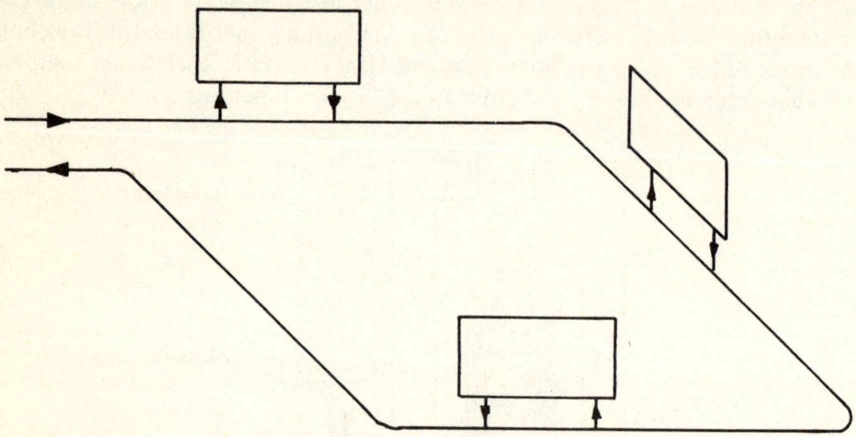

(a) one-pipe system.

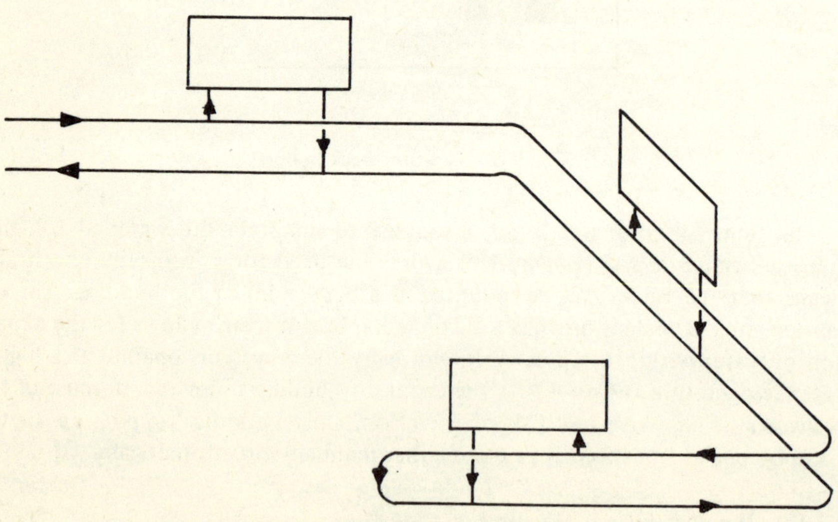

(b) two-pipe system.

Fig. 9.9 – Small-bore pipe distribution system.

appliance are connected to a single pipe. The two-pipe system has separate pipes for flow and return. This gives more uniform temperatures for all heaters since, with the one-pipe system, cooled water from the first heater is discharged into the hot water flowing to the second heater and so on. This means that the last heater in the circuit is running off much colder water than the first. The two-pipe system is more expensive to install because more pipe is required.

In the case of the mini-bore central heating system (Fig. 9.10), a radial distribution system is used, similar to electrical radial wiring (Fig. 5.11a). Each heater has a separate flow and return pipe which is fed by a distributor. The distributor has a number of flow and return connections and is centrally sited to minimise the lengths of the radial pipes. These radial pipes are made from thin, flexible copper tubes, usually 8 mm in diameter and, depending on the size of the installation, there may be one, or a number of distributors. Mini-bore systems use a higher water velocity (1.5 m/s) than small-bore systems (0.9 m/s).

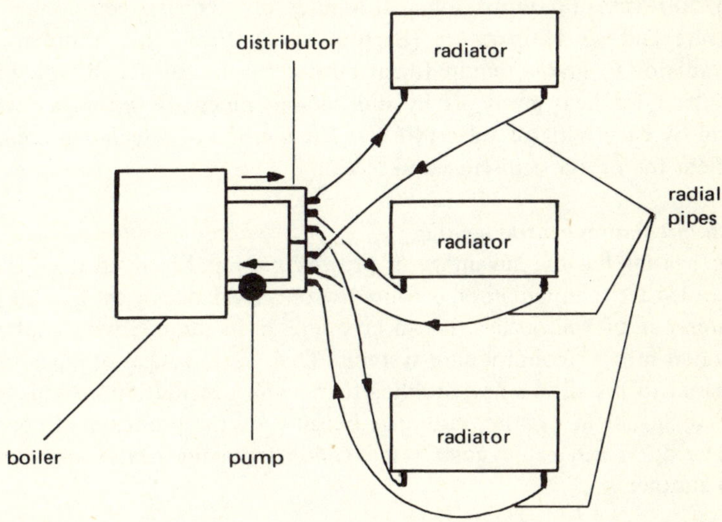

Fig. 9.10 – Mini-bore distribution system.

9.3.1.3 *Central heating space heaters*

There are a number of different methods of transferring the heat from the circulating hot water to a room. The most common method is by the use of *radiators.* Early gravity-feed radiators were made of cast iron. They have now been replaced by pressed-steel panel radiators which are more acceptable from an aesthetic viewpoint although slightly less efficient. Although described as radiators, only a small proportion of their heat is transferred by radiation, the bulk of the heating effect being by convection. Radiators normally have a valve at each end, one an on/off (wheelhead) valve and the other a pre-set (lockshield) valve which is used to balance the heat output of all the radiators. In the case of mini-bore systems valves are available providing both flow and return connections to a single end of the radiator.

Fan convector heaters are sometimes used in hot-water circulation heating systems. These consist of a finned heat exchanger over which air is blown by an electrical fan. The main advantage of this type of heater is that it is more compact than the conventional radiator and provides more rapid heating from cold.

Natural convection skirting board (baseboard) heaters are also available. These are expensive but give a better distribution of heat to a large room than the more localised radiator. In some cases water may also be circulated through underfloor or ceiling pipework to provide large areas of low-temperature radiant heat (floor less than 27°C, ceiling 45–50°C).

The size of radiator required is normally assessed by means of data on the heat output, as supplied by the manufacturer. Single panel radiators provide between 500 and 600 W/m^2 for a difference of 50% between average water temperature and air temperature (Barton 1970). From this information the size of radiator to give a required heat output can be calculated. Allowance is usually made for heat given off by uninsulated pipework within a room (40 W/m) and by each radiator valve (40 W). The method of calculating the heating requirement for a room is discussed in section 8.4.

9.3.2 Air circulation central heating

Warm-air heating has the advantage of providing a rapid heat up from cold and it is less obtrusive than radiators, because the only evidence of the heating system is the presence of wall grilles. It can be operated in summer with cool air and incorporated into air conditioning systems. There is no danger of pipes freezing. It is cheaper to install in a new building than water distribution systems but it is difficult to install in existing buildings because of the problems of concealing ducts. The ducts can cause noise transmission from the heater and from one room to another.

9.3.2.1 *The air heater*

Combustion products to air (direct) heat exchanger. This operates with oil or gas fired burners. A metal heat exchanger is heated by hot combustion products. Cool filtered air is passed over the heat exchanger and in the process is heated. The warmed air is circulated by a large volume low-velocity fan. This system does not provide domestic hot water (Fig. 9.11a).

Water to air (indirect) heat exchanger. In this method the circulation air is drawn over a heat exchanger, which is heated by hot water from a conventional boiler, with the advantage that domestic hot water can also be provided (Fig. 9.11b).

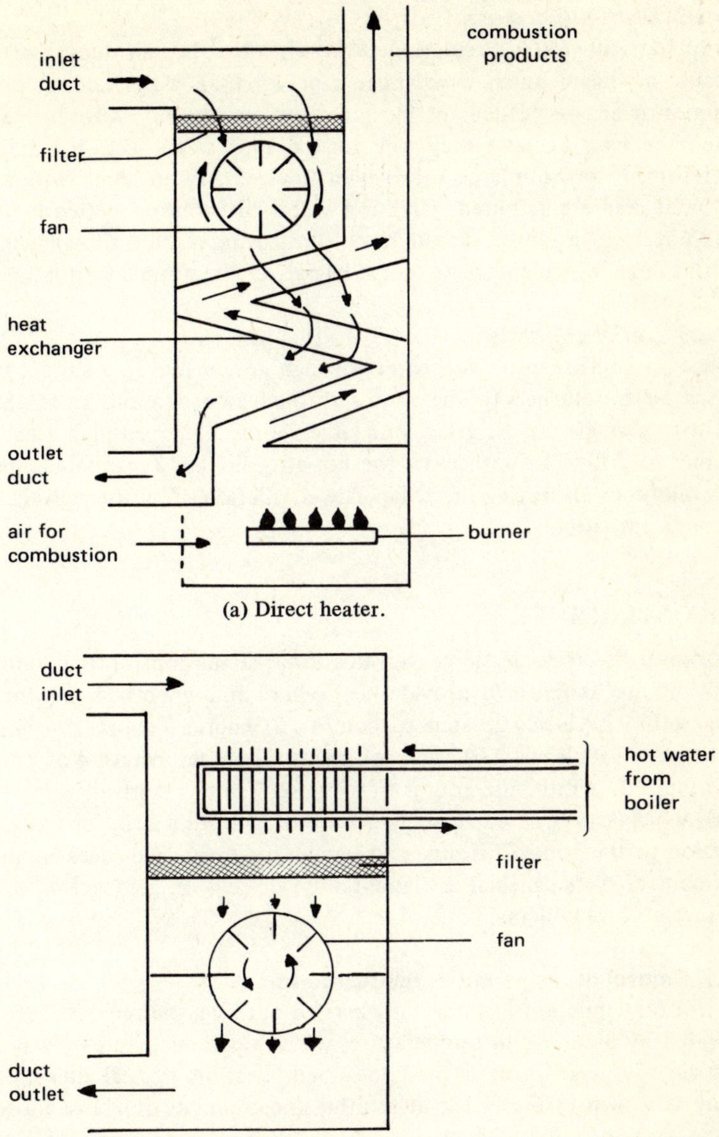

(a) Direct heater.

(b) Indirect heater.

Fig. 9.11 – Central air heaters.

Electric-storage heat exchanger. This works on the same principle as the fan storage heater (Fig. 9.5b) but with a larger heat capacity to provide heat for the whole house.

9.3.2.2 *Distribution ducts*

Compared with water circulation central heating, hot-air ducts are large. This is because air has a much lower heat capacity than water and therefore greater volumes of air are needed for the same thermal output. Also, high air velocities cannot be used because they can cause draughts (see section 7.2.5) and noise. Therefore, to provide large volumes of low-velocity air, ducts with a large cross-sectional area are required. Unlike a water distribution network, the air duct does not form a closed circuit since, on leaving a duct, the air will be mixed with fresh air which enters the house as part of the normal ventilation.

9.3.2.3 *Outlet grilles*

Cold air is drawn into the heater through grilles and, in a similar manner, the heated air is returned to the rooms through one or more grilles. Small rooms will have a single hot-air grille, but large rooms and open plan areas will have a number of grilles. In both cases the hot-air grille has a manual damper to adjust the volume of air leaving the grille. The circulation of air through the room leads to good ventilation.

9.4 CONTROLS

Most controls on domestic central heating systems control time, temperature or both and are designed to provide heat where and when it is required. Since air temperature is closely related to comfort (Chapter 7), a *thermostat* is used to control comfort level. A thermostat measures air temperature at one or a number of points within the rooms of a house. Time is controlled by an electrical clock which can be used to sequence events within a daily or weekly cycle. In addition to the control of time and temperature, safety devices are incorporated into control systems, such as flame-failure devices on gas (section 2.9) and fuel-oil (section 3.7) boilers.

9.4.1 Control of temperature: the thermostat

Control of temperature is normally carried out by a thermostat. This is a switch, activated by a change in temperature, which operates in such a way as to maintain the temperature at a predetermined level. A typical thermostat control circuit is shown in Fig. 9.12, illustrating the elements of a closed-loop feedback control system.

The measuring element detects a change in the condition of a *controlled variable* which, in the case of a thermostat, is the temperature of air or water. The measuring element converts a change in temperature to a change in some other physical property. In many cases thermal expansion is used to convert a change in temperature into mechanical movement, which can then be used to open or close a gas valve or switch on and off an electrical heater, a solenoid, or a motorised valve (Fig. 5.23).

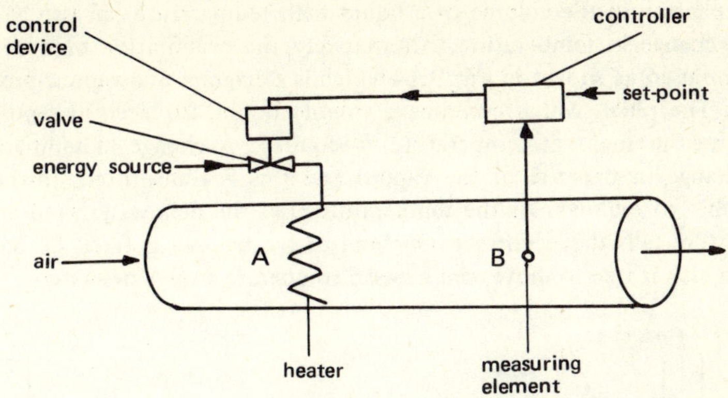

Fig. 9.12 – A closed-loop temperature control system.

The expansion of metals can be used as the measuring element of a thermostat. Normally the difference in the coefficient of thermal expansion of two metals is used, as in the bi-metallic strip (Fig. 2.14 and 2.15) and in the rod-and-tube thermostat (Fig. 9.13). Brass expands more than steel for a given change in temperature and therefore the relative lengths of the brass tube and steel rod vary with temperature. This causes the end of the steel rod to move into and out of the brass tube as the temperature rises and falls.

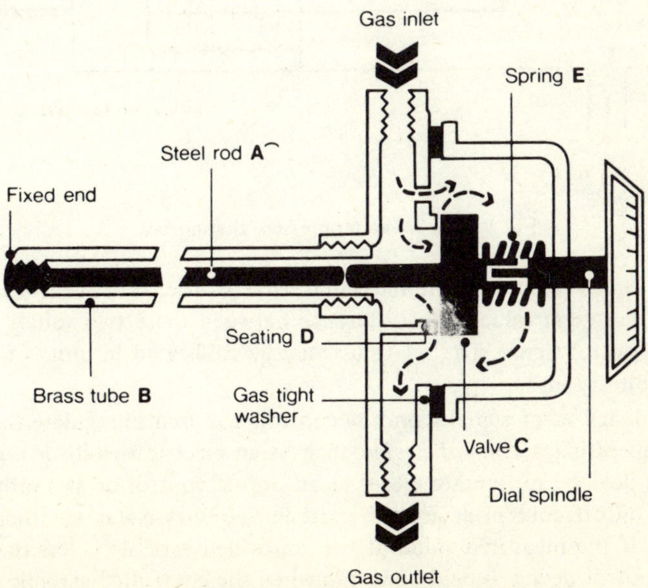

Fig. 9.13 – Rod-and-tube thermostat.

The expansion in the volume of a liquid with temperature can also be used to sense a change in temperature. Alternatively the evaporation of a volatile liquid can be used as shown in Fig. 9.14 which is a diagram of a vapour-pressure thermostat. The phial, which contains a volatile liquid, is placed in a position where the measurement of temperature is required. A change in temperature causes a change in pressure of the vapour and this is transmitted, through a capillary tube, to bellows. As the temperature rises the bellows expand and as the temperature falls they contract. One end of the bellows is fixed so that the other end, which is free to move, can be used to operate a valve or switch.

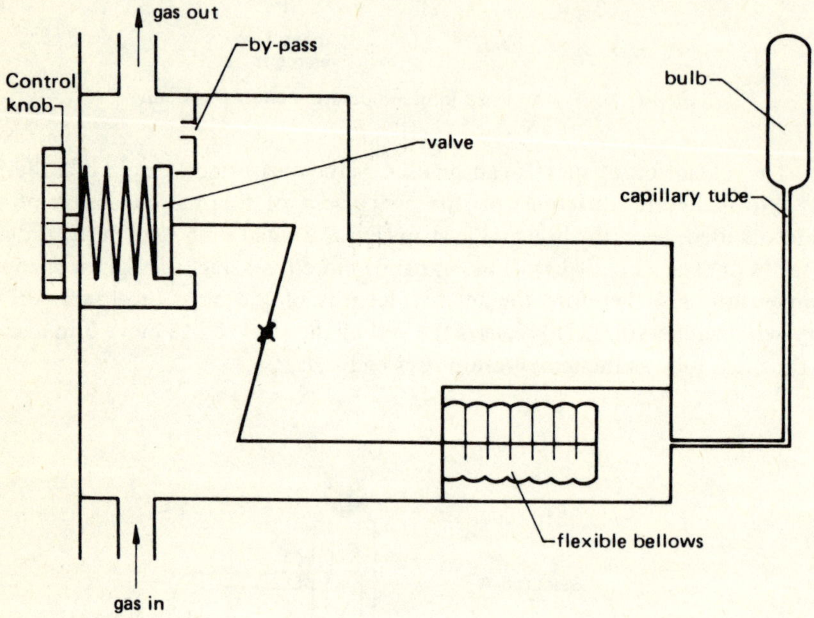

Fig. 9.14 – A vapour-pressure thermostat.

The *controller* compares the measured value of the controlled variable with the pre-selected (set) value. Any difference between these two values produces a *feedback* control signal. The set value may be calibrated in units of temperature or in arbitrary units.

The feedback error signal from the controller is used to regulate the flow of energy by operating a *control device* such as an electric switch or a gas valve. This control device can operate either as an on/off control or as a proportional control. An on/off control device can exist in one of two states: either fully on or fully off. If the measured value of the controlled variable is less than the set value, the control device is held open and when the controlled variable is greater than the set value the control device is closed. In the case of proportional control the control device can be held in any position between fully closed and fully opened, depending on the size of the difference between the controlled variable and the set value.

One of the problems with feedback control is the effect of time lags in the system. Reference to Fig. 9.12 shows that there is a time delay between when the air at point A is heated to its set value and when it arrives at the measuring element (B). By the time this occurs the air at A will have been overheated. Also, with an on/off control, there is a difference between the temperature which opens the valve and the temperature which closes it. The effect of this, together with the time lags, results in the control producing a cyclic variation in temperature on either side of the set point, as shown in Fig. 9.15.

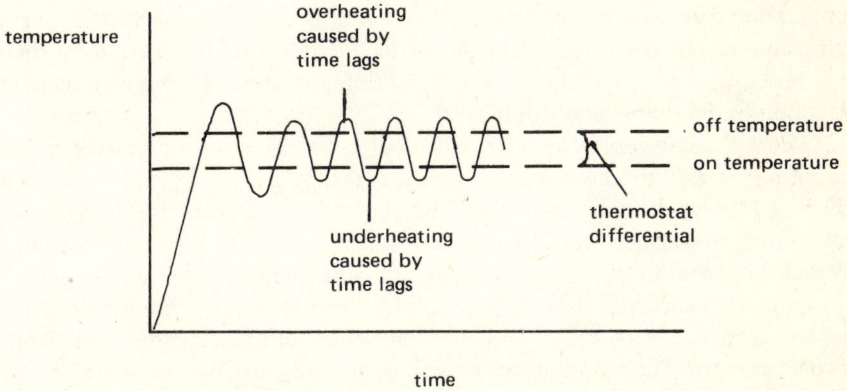

Fig. 9.15 – Temperature cycling caused by thermostat differential and lags.

9.4.2 Boiler thermostats

Water boilers have adjustable thermostats to regulate the temperature of water within the boiler. These can be rod or vapour-pressure thermostats. In the case of solid-fuel boilers the thermostat regulates the position of an air damper which controls supply of air for the combustion process. However the control action takes a long time and therefore there are larger lags. Gas boilers use a gas valve which can be an on/off valve (with re-ignition by a pilot flame or an automatic spark ignition device) or a proportional (modulated) valve. In oil central heating systems the thermostat operates an electrical switch which starts and stops the oil pump.

9.4.3 Thermostatic radiator valves

Normally radiators have two valves one of which is an on/off valve and the other a pre-set valve to control the flow rate of water through the radiator. One of these valves can be replaced with a thermostatic valve. This measures the temperature in the room and uses a liquid-filled bellows to control the flow of water through the radiator. Control of room temperature is possible using a calibrated valve head which also holds the temperature sensor. In some situations a remote sensor is used, particularly where the valve-head sensor would be inaccurate because of obstruction by curtains or localised air draughts.

9.4.4 Room thermostats

Room thermostats operate from a gas-filled bellows or a bi-metallic strip. Simple room thermostats are of the on/off type. Normally there are large time lags because the room thermostat is not sited very close to the radiator. Room thermostats are available which contain a mechanism for reducing the lag in control. This device is known as an *accelerator* (Haselden 1979). This is a small electrical heater used with a bimetal thermostat, which is activated when the air temperature is below its set value. This results in the air close to the bimetal strip being marginally hotter than the air in the room and the thermostat therefore anticipates the air temperature reaching its set point. In this way there is a reduction in the thermostat differential.

Where a single room thermostat is used to control a central-heating system, the siting of this thermostat is very important. It should be sited in a room which is frequently used and should be about 1–$1\frac{1}{2}$ m above floor level, away from direct sunlight, electrical appliances and draughts. There should be no heating appliances in the room which are not controlled by the room thermostat.

In certain circumstances a zone-control system gives a more satisfactory performance. Fig. 9.16 shows a dual-zone control for a water-circulation central heating system. The distribution circuit is divided into two separate circuits, one for the ground floor and a second for the upper floor. Each circuit has a room thermostat which operates a motorised valve. This means that each zone

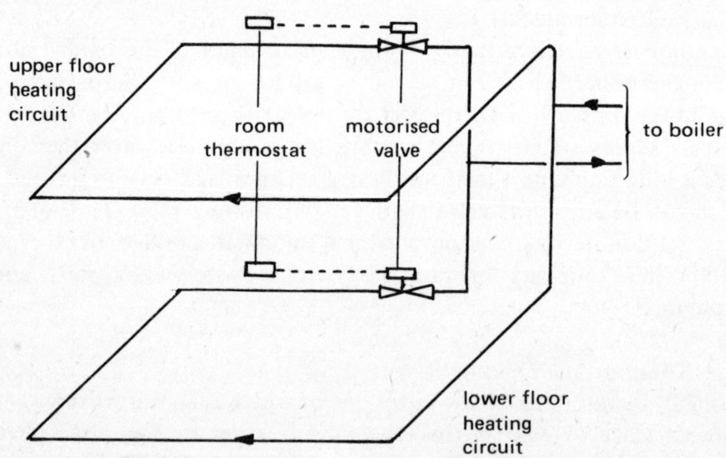

Fig. 9.16 – Dual-zone control central heating.

can be set to a different temperature and/or operated during different times of the day by using a programmer for each circuit.

9.4.5 Programmers

A programmer is a time clock which controls the heating cycle during a 24-hour (or in some cases seven day) period. It controls heating and domestic hot water supplies to match the needs of occupants. Programmers vary in sophistication and flexibility. They usually have tappets which can be set manually by the occupants to suit their life style. Examples of the type of control available from a programmer are as follows (HW = hot water, CH = central heating):

(1) Simultaneous control of central heating and hot water: Off/HW & CH.
(2) Hot water alone or hot water and central heating: Off/HW/HW & CH. On once, off once during 24 hours.
(3) As (2) but on twice, off twice during 24 hours.
(4) Hot water and central heating separately: Off/HW/CH/HW & CH. On twice, off twice during 24 hours.
(5) As (4) but day divided into 12 time periods, each of which can be programmed.

Because a house takes time to heat up and cool down it is necessary to anticipate the needs of the occupants. Heating should come on half to one hour before it is needed. Programmers are also available with two thermostats, one for day-time use and the other for providing a low level of background heat during the night.

9.4.6 Water-circulation control system

Control systems of this type vary from one installation to another but two typical examples are given.

(1) the programmer controls the electric circulation pump and the gas solenoid valve or oil pump based on a room thermostat. The boiler thermostat controls the temperature of the circulating water (Fig. 9.17).
(2) The system provides separate control of a pumped water supply both to the central heating and the domestic hot water. One three-way valve may be used to divert the flow of water to either the central heating, hot water or both. Alternatively two separate motorised valves may be used, one on the hot-water supply and one on the central heating (Fig. 9.18). Priority may be given to hot water or to central heating.

9.4.7 Warm-air circulation control system

In a typical installation a programmer controls the operation of a gas burner and air-circulation fan. A room thermostat controls the demand for heat during an 'on' period of the programmer. The gas heater also has a high limit thermostat

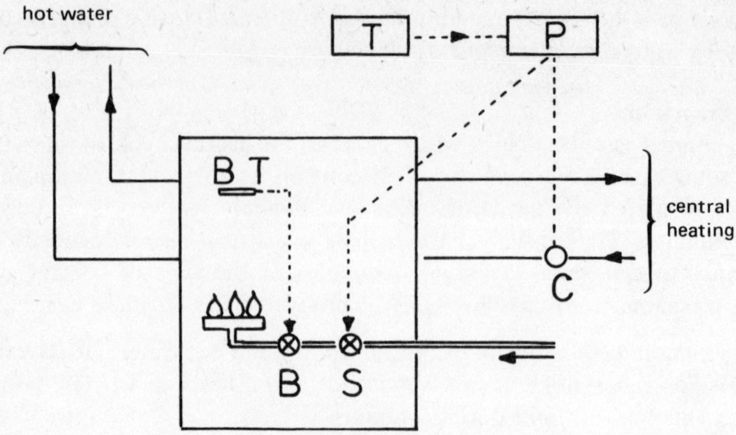

T room thermostat
BT boiler thermostat
C water circulator
P programmer
B boiler thermostat valve
S gas solenoid or oil pump

Fig. 9.17 – Programmed central heating; gravity hot water.

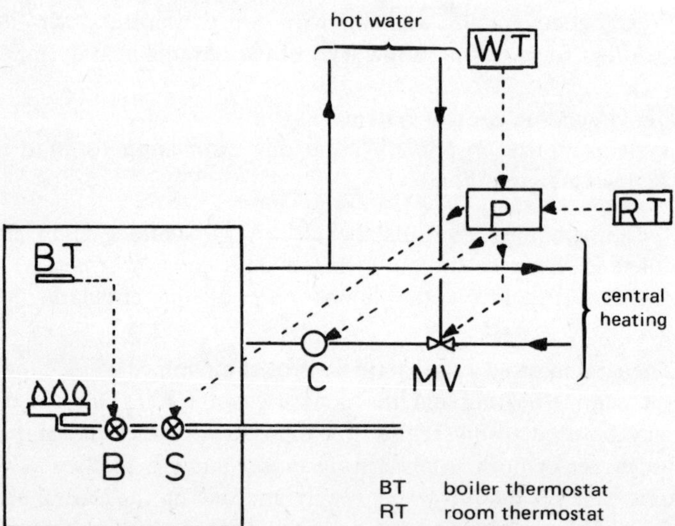

BT boiler thermostat
RT room thermostat
WT water thermostat
C circulator
P programmer
B boiler thermostat valve
S gas solenoid or oil valve
MV motorised valve (2-way)

Fig. 9.18 – Programmed central heating: pumped hot water and heating.

which switches off the burner if the air temperature close to the burner rises excessively. There is also a low-limit thermostat which is coupled to the fan control to prevent circulation of cold air when the gas burner is initially ignited and which keeps the fan running (after the burners are turned off) until the residual heat within the appliance is dissipated (see Fig. 9.19).

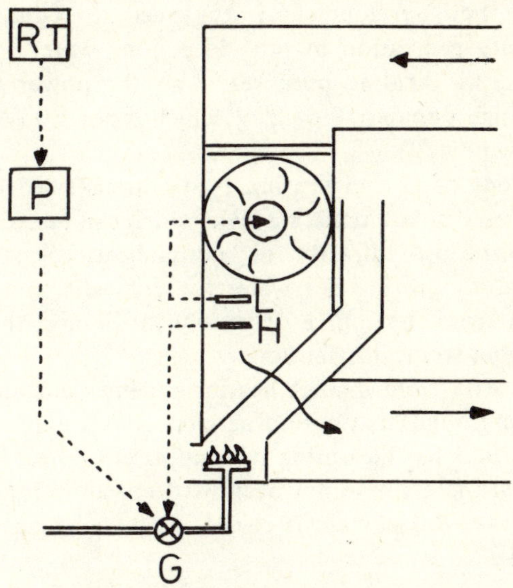

L low level thermostat
H high level thermostat
RT room thermostat
P programmer
G gas solenoid

Fig. 9.19 – Programmed warm-air central heating.

9.5 DISTRICT HEATING SCHEMES

District heating can be defined as the centralised generation of heat and its distribution to more than one building. In Scandinavia 85% of new houses are linked to district heating schemes and for Germany the figure is 35%. Similarly in the USA and USSR district heating schemes are being used in a large number of installations.

There are several types of district heating schemes but they can be generally classified into one of two types: one where the fuel used to generate heat is for the sole purpose of district heating; and the second type where the heat is waste cooling water from fossil-fuel or nuclear electric power stations. The latter method is often referred to as a *Combined Heat and Power Scheme* (CHP).

For district heating schemes using fuel directly, the fuel utilisation can be much more efficient than is the case when each household has its own boiler. Other advantages include greater convenience, lower maintenance cost and the ability to use cheaper fuels such as heavy heating oil and even domestic refuse (Field 1975).

CHP schemes are common in the USSR, USA and Europe. One of the problems restricting their development is the need to reduce the operating efficiency of electricity generation to provide cooling water at a high enough temperature for domestic heating purposes. Also the power station must be relatively close to a high population density, which is not always advisable from an environmental view point.

With both methods of district heating, heat is transferred from the source of generation to the user by a circulating fluid which can be either hot water or steam. This fluid can be used directly for central heating systems but a heat exchanger is required to provide hot water for domestic use. The steam/hot water are distributed from the source of supply to the user through insulated pipes in a similar manner to gas distribution.

Problems which arise from district heating schemes include corrosion and leaks within the underground distribution network and the difficulty of finding a cheap accurate method for measuring consumption of heat – *a heat meter*. In schemes where heat meters have not been used consumption has been found to be very high (Courtney & Jackman, 1976).

9.6 HEAT PUMPS

Heat pumps utilise the mechanical refrigeration cycle, which is described in Chapter 15. The most extensive use of heat pumps is in the USA where reverse-cycle heat pumps are used for cooling (in air conditioning systems) in summer and heating in winter. These units are large and are designed primarily as cooling units. There is a need in the UK and Europe for small domestic heat pumps designed specifically for winter heating.

Most units which have been investigated are of the air-to-air type, that is, they extract heat from the outside air and heat up air inside the building as a part of warm-air central heating. Air-to-water heat pumps are available but the water temperature is rather low for use in radiators. Other sources of heat suggested include the ground, solar energy and domestic heat wastes (Reay & MacMichael 1979).

Using air as the heat source, coefficients of performance (COP) of 2.8 to 3.0 can be obtained for an outside temperature of 5°C. As the outside air temperature falls the COP falls and some form of back-up heating is required (Blundell 1976). A comparison of gas warm-air central heating and an electrically driven heat pump indicated that the heat pump did not justify the higher cost of electricity (Heap 1976).

An alternative to the use of electrical mechanical heat pumps is the use of the absorption refrigeration cycle (section 15.3.3). These could provide an efficiency of 113% for space heating compared with 75% for a gas boiler (Jesson & Johnston 1979). This means that the heat output from the fuel is higher than its calorific value.

9.7 SOLAR SPACE HEATING

There is a poor match between the availability of solar energy and the demands for domestic space heating. However solar energy can form a significant heat input to space heating in many areas of the world, particularly as the standards of insulation of housing are improved. Solar water-heating appliances are described in Chapter 10. Solar air heaters are also available for the provision of natural or forced-air circulation space heating (Coltan 1978).

If solar energy is to provide a significant proportion of domestic space heating, some form of heat storage is required to even out the daily fluctuations in the availability of energy. The simplest method of doing this is to store the heat in a passive material such as water, stone or concrete. The latter can be a part of the fabric of the building as in the 'Trombe' Wall (Foley 1976). Research into other means of storing solar heat, using the energy associated with phase changes, and chemical reactions is underway.

9.8 DOMESTIC HEATING IN THE FUTURE

As the standard of insulation of homes becomes better the proportion of domestic energy consumption attributable to space heating will decline. Space heating is about two-thirds of domestic energy consumption. In addition to improvements in the standard of insulation, more sophisticated energy use within buildings, and techniques for recovering waste heat in ventilated air and hot water are available. These latter two techniques are costly and at the moment only cost effective for commercial and industrial uses. However, as the cost of fuel increases, the cost of this equipment will become economically viable even for the domestic consumer. With these changes, it is likely that more energy will be consumed for water heating rather than space heating.

REFERENCES

Adams, A. J. & Vaisey, G. A. (1970), Evolution of the modern gas fire, *IGE,* **10** p. 797.

Barton, J. J. (1970), *Small-bore heating and hot water supply,* Newnes/Butterworth.

Blundell, C. J. (1976), Domestic heat pump design for the British climate, *Heating and Ventilating Engineer,* **June**, p. 6.

Consumers Association (1979), October, p. 592.

Coltan, N. (1978), *Solar heated houses,* Tutor Press.

Courtney, R. G. & Jackman, P. J. (1976), A study of three district heating schemes, *Heating and Ventilating Engineer,* **March**, p. 6–14.

Field, A. A. (1975), District heating and total energy, *Energy World,* **Feb.**, p. 11.

Foley, G. (1976), *The Energy Question,* Penguin.

Haselden, F. G. (1979), Technical aspects of room thermostats, *Heating and Ventilation News,* **January**.

Heap, R. D. (1976), A domestic heating comparison between a heat pump and a gas warm air heater, *Heating and Ventilating Engineer,* **November**, p. 14.

Jesson, P. F. & Johnston, I. W. (1979), Some thoughts on domestic gas fired heat pumps, *Gas Engineering and Management,* **November**, p. 553.

Pickup, G. A. and Miles A. J. (1978), The performance of domestic wet heating systems in contemporary and future housing, *Gas Engineering and Management,* **June**, p. 188.

Reay, D. A. & MacMichael, D. B. A. (1979), *Heat pumps – design and application,* Pergamon.

BIBLIOGRAPHY

Barton, J. J. (1970), *Domestic Heating and Hot Water Supply,* Applied Science,

Bradbury, K. (1973), *Solid Fuel in the Home,* Women's Solid Fuel Council.

Consumers Association (1978), *Saving Money on Heating Costs,* **August**, p. 446.

Consumers Association (1981), *Central Heating,* **July**, p. 417–22.

Electrical Association for Women (1971), *The EAW Electrical Handbook,* English University Press.

Haines, R. W. (1977), *Control Systems for Heating Ventilation and Air Conditioning,* Van Nostrand.

Hayes, D. & Kennedy, P. (1970), Methods of controlling domestic central heating systems, *Heating and Ventilation News,* **June**, p. 645.

Heap, R. D. (1980), *Heat Pumps,* E & F Spon.

Miles, V. C. (1965), *Thermostatic Control,* Newnes Butterworth.

Plant, E. M. (1971), *Gas and its Domestic Uses,* Women's Gas Federation.

Cowan, H. J. (1980), *Solar Energy Applications in the Design of Buildings,* Applied Science.

CHAPTER 10

Water heating

10.1 INTRODUCTION

Of the 220 to 350 litres of water consumed per person per day (section 6.1) a significant proportion is required in the form of hot water. Although usage varies widely from person to person and depends on the facilities available in a house, current consumption levels are of the order of 100–150 litres/person/day (Smith & Julian 1976), of which the largest proportion is used for personal hygiene.

For domestic use, water is stored at 70°C if the water is soft and at 60°C in hard water districts which do not have water softening facilities (see section 6.5). For some applications such as dishwashing and laundry, water is needed at this temperature but for washing and bathing the preferred temperature is 35–45°C. To achieve this the hot water must be diluted with cold water. Hot water flow rates and typical consumption figures for various domestic activities are given in Table 10.1.

Table 10.1
Hot water requirements for domestic activities

Use	Flow rate (litres/s)	Typical consumption of hot water at 60°C (litres)
Bath	0.3 -0.4	70
Shower	0.10–0.12	10–15
Handwashing	0.10–0.12	5
Dishwashing	0.2 -0.3	5
Clothes washing	0.3	25–30

A number of different methods are available for the supply of domestic hot water and the method chosen in any situation will depend on a number of factors, such as fuel availability, the size of house and the number of occupants.

Hot water may be produced *instantaneously*, that is the water is heated when there is a demand for hot water as indicated by a flow of water through the heater (section 10.2), or the water may be held in a *storage tank* so that water is available to match demand (section 10.3). Alternatively, the supply system may be *centralised* with a single appliance responsible for the generation of hot water which is then distributed to the hot-water outlets (section 10.4). Hot water may be generated at a number of *localised* heaters each responsible for a single outlet.

10.2 INSTANTANEOUS WATER HEATERS

In its simplest form an instantaneous water heater consists of a heat exchanger through which cold water flows directly to the outlet, gaining heat as it does so, as shown in Fig. 10.1.

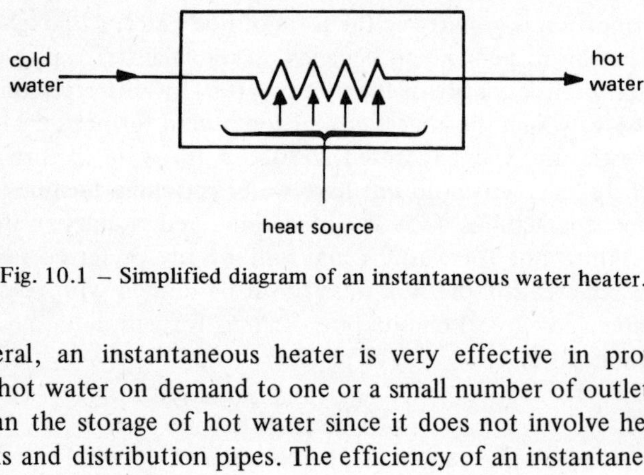

Fig. 10.1 – Simplified diagram of an instantaneous water heater.

In general, an instantaneous heater is very effective in providing small volumes of hot water on demand to one or a small number of outlets. It is more efficient than the storage of hot water since it does not involve heat loss from storage tanks and distribution pipes. The efficiency of an instantaneous gas-fired water heater is 70 to 75%.

10.2.1 Gas instantaneous water heaters

Gas instantaneous water heaters are available as single or multiple outlet units. A single-outlet heater is fixed over a sink or bath and supplies water to that outlet only. A multipoint unit supplies water to a number of outlets. Both types work on the same principle, as illustrated in Fig. 10.2. The essential part is a gas valve which is operated by the flow of water through the heater, in this way ensuring that when a tap is opened gas is supplied to the burner. A number of different methods are used to sense the flow of water through the heater and convert this into movement of a rod which is used to actuate the gas valve (Taylor 1973). Of these, many use the change in pressure which occurs when water flows through an orifice in the water inlet pipe, as shown in Fig. 10.3. When there is no water flow, the pressure on both sides of the restrictor is the

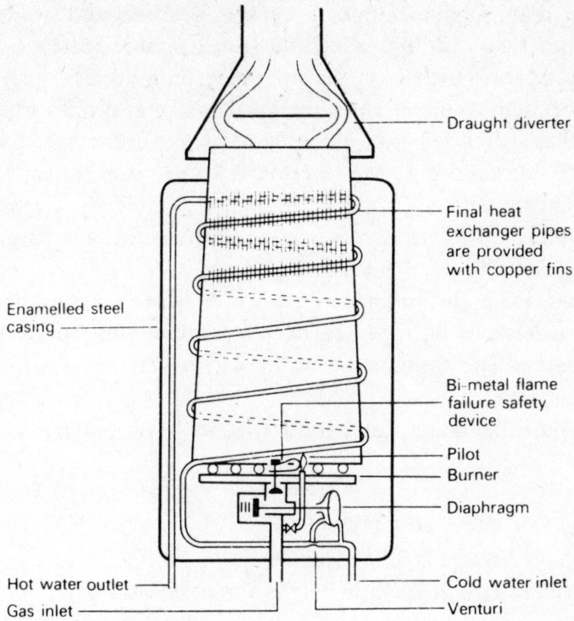

Fig. 10.2 – Single outlet instantaneous water heater (Hall, 1976).

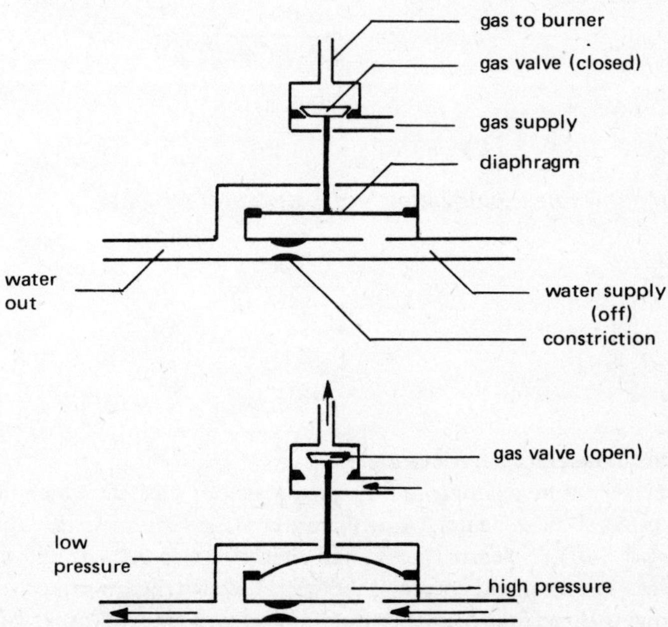

Fig. 10.3 – Water-flow actuated gas valve.

same and the diaphragm occupies a central position and holds the gas valve closed. As water flows through the orifice, in response to the opening of a tap outlet, the pressure is higher on the inlet side than on the outlet side and this causes the diaphragm to move, opening the gas valve as it does so.

The gas flow is ignited by a pilot flame (see section 2.8.1) which has a bi-metallic strip flame failure device (section 2.9.1) to prevent gas flow if the pilot is not alight. On modern appliances the pilot is ignited by a piezo-spark igniter (Fig. 2.12). Single outlet heaters do not require a flue but larger heaters have either a conventional or a balanced flue (section 2.7).

The output from instantaneous gas water heaters varies from 6 kW for a small heater which can be installed over a kitchen sink to 23 kW for a large multipoint heater. The volume of water which this can heat is calculated as follows.

Assume that the water temperature is to be raised by 45°C from 15°C to 60°C.

6 kW heater 6 kW = 6000 J/s

The specific heat of water is 4200 J/kg·°C
Quantity of heat = mass × specific heat × temperature rise

(J/s) (kg/s) (J/kg.°C)

For 1 second 6000 = mass/s × 4200 × 45

$$\text{mass/s} = \frac{6000}{4200 \times 45}$$

$$= 0.032 \text{ kg/s}$$

1 kg of water = 1 litre

Therefore, flow rate is 0.032 litres/s of water at 60°C.

23 kW heater

$$\text{mass/s} = \frac{23\,000}{4200 \times 45}$$

$$= 0.12 \text{ kg/s}$$

$$= \textit{0.12 litres/s of water at 60°C}$$

10.2.2 Instantaneous electric heaters

The most common example of an instantaneous electric water heater is the wall-mounted shower heater. As cold water passes through the appliance it is heated by a 5–7 kW electric heater and then flows directly to the shower head. As with gas instantaneous heaters, the heater is switched on only when there is water flowing through the unit. In the simplest unit, control over water temperature is by adjustment of the flow rate of water but more sophisticated units have variable power controls and/or thermostats. All units have thermal cut-out controls to prevent overheating of the water. The efficiency of these units is very high but the limitation to their use is caused by the relatively low water flow rate. For example, if water from a shower is heated to 40°C from an inlet temperature of 15°C, the maximum flow can be calculated as follows.

For a 7 kW heater
Using the method of calculation illustrated in the previous section:

$$\text{mass/s} = \frac{7000}{4200 \times 25}$$

$$= 0.07 \text{ kg/s} = 0.07 \text{ litres/s}$$

The maximum flowrate is 0.07 litres/s at 40°C

From this it can be seen that large volumes of water cannot be heated instantaneously because of the high power level which would be required.

10.3 STORAGE HEATERS

10.3.1 Combined storage tank/heater units

These units consist of an insulated storage tank (made of copper, galvanised iron or steel) with a built-in electrical heater and thermostat. The capacity of these units varies from 6 to 70 litres for free-outlet heaters (see Fig. 10.4), which

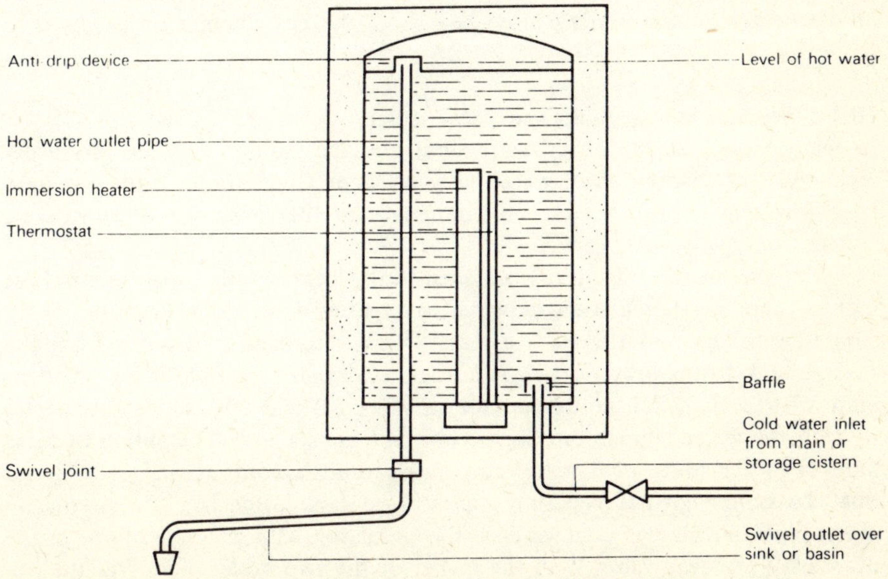

Fig. 10.4 – Free-outlet storage heater (Hall, 1976).

provide water for a single outlet to between 100 and 240 litres for a centralised storage tank providing hot water to all of the taps in a house, Some of the electrical storage heaters have two horizontal heating elements, one above the other, each with its own switch and thermostat. The advantage of this arrangement over a single heater is that, when small volumes of hot water are required, only the top heater need be used, giving greater economy. In all cases the cold water inlet is at the base of the unit and hot water is drawn off from the tap. This reduces the mixing of hot water with cold water and hence leads to a greater overall efficiency and hotter water at the taps.

Since hot water expands on heating, and also because dissolved air is released from water when it is heated, it is necessary to have some means of releasing pressure from the tank. In the USA the tank is fitted with a pressure release valve whereas in the UK the storage tank has a vent/expansion pipe which overflows into the cold water feed tank (see section 10.4). Tanks may contain an anode rod made of magnesium. This prevents corrosion of the tank and its fittings. Some tanks have a glass lining for the same reason.

Effective insulation of the storage tank is very important since an uninsulated tank can lose heat at the rate of 500 W/m^2. In some cases an insulated jacket is incorporated during manufacture of the tank, but sometimes the householder must provide an insulated jacket around the storage tank. A 5 cm layer of fibreglass can reduce the heat loss by 90%.

A gas storage water heater is similar to the electric version with the addition of a burner in the base of the unit and a central flue. A baffle in the flue slows down the flow of combustion products and in this way increases the efficiency of the heater by transferring more heat from the hot combustion products to the water.

10.3.2 Separate storage tanks and water heater

In many situations the water heater unit and the storage tank are physically separated with circulation of water between them. One example is the *gas circulator* which is a water heating unit designed for the conversion of an existing storage tank.

A circulator (see Fig. 10.5) is a small water heater with a high rate of heat input. There is a circulation of water, by natural convection (section 7.2.2), from the heater into the storage tank with a return of cold water from the storage tank to the heater. The gas burners have piezo ignition or a permanent pilot with a flame-failure device (see sections 2.8 and 2.9) and a thermostat to keep the water temperature in the unit at a pre-set level. The output of these units is of the order of 3–4 kW and they require a conventional or balanced flue. To obtain greater economy the storage tank often has two alternative outlet pipes, which feed cold water to the circulator, with selection of one or the other by a two-way valve. With the higher of the two outlet pipes (A), the volume of stored hot water is reduced from 140 litres to 50 litres. This is used when

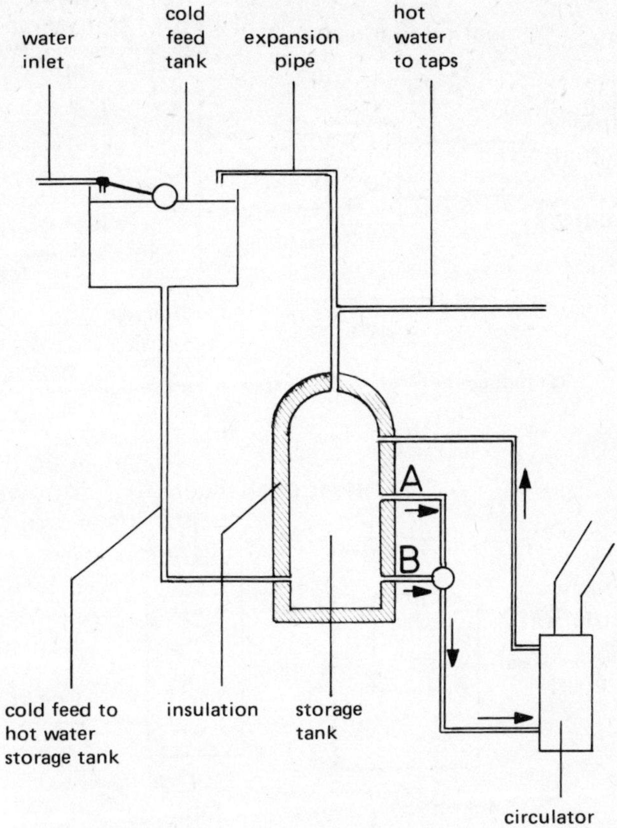

Fig. 10.5 – Hot water gas circulator.

only small volumes of hot water are required. When greater volumes are required, such as for bathing, the valve can be rotated to allow the circulator to heat the full tank of water (B).

A second example of the separation of the water heater unit from the storage tank is where space heating is by water circulation through radiators (section 9.3.1) and where a single boiler is to be used for both space and water heating. To prevent scale build up in the boiler and to separate the flow of water for domestic use from that for space heating it is necessary to have a heat exchanger between the water boiler and the water in the storage tank. This heat exchanger will transfer heat between the primary hot-water system (used for space heating) and the secondary hot-water system (used for domestic hot water). This heat exchanger may be situated:

(a) in the boiler (Fig. 10.6a).

(b) in the storage tank (Fig. 10.6b).

(c) externally to both boiler and storage tank (Fig. 10.6c).

(a)

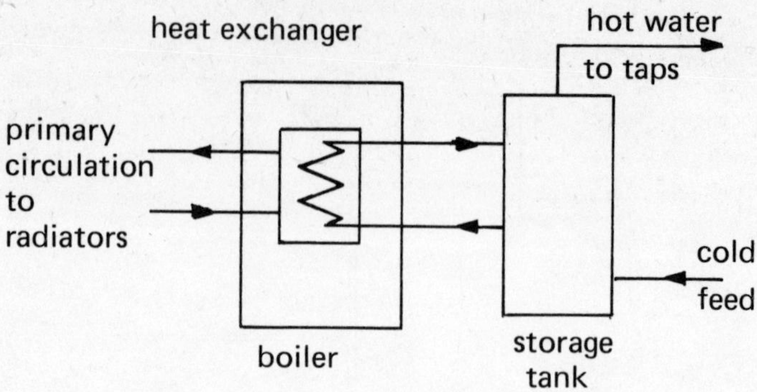

(a) Indirect hot water with heat exchanger in boiler.

(b)

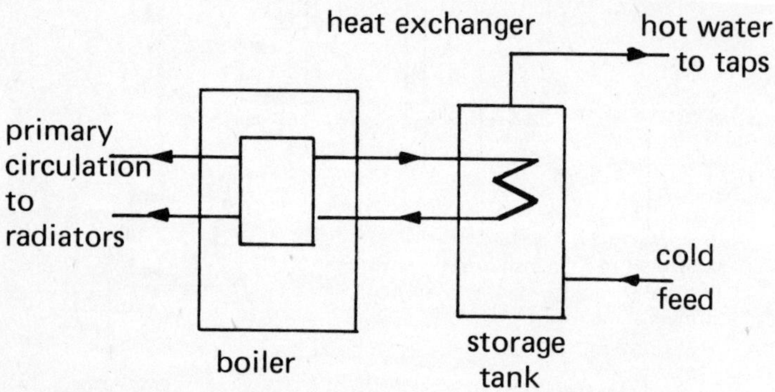

(b) Indirect hot water with heat exchanger in storage tank.

(c)

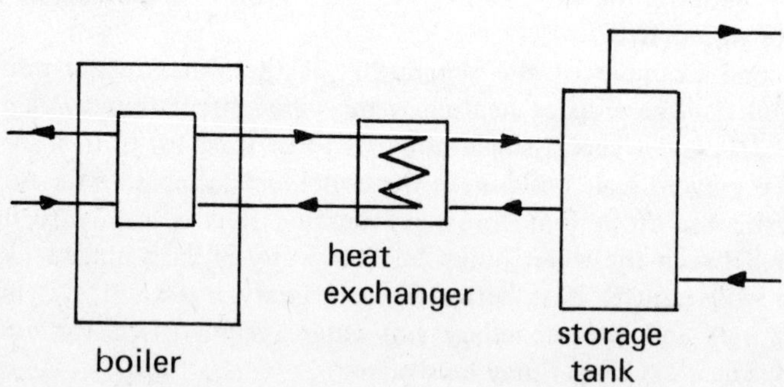

(c) Indirect hot water with external heat exchanger.

Fig. 10.6 – Single boiler used for space and water heating.

10.3.3 Solar water heaters

In its most basic form a solar water heater consists of a solar panel and a hot-water storage tank with a circulation of water between the two. A solar panel consists of pipes in or on a black metal absorber panel. This is mounted beneath a glass cover and has an insulated backing (Fig. 10.7). It is usually fastened to a South facing roof at an angle of 30° to 50° to the horizontal and where it is

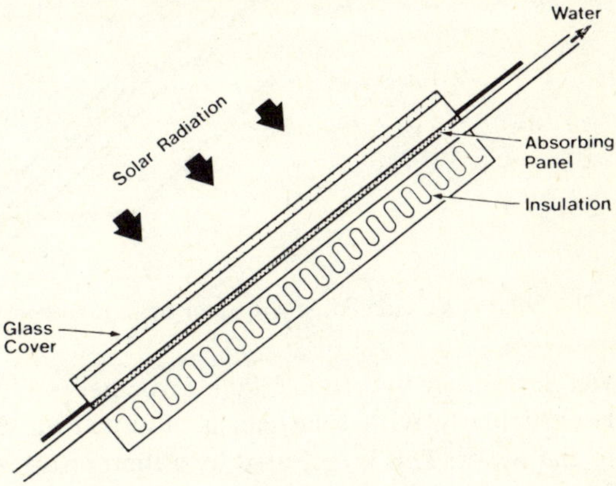

Fig. 10.7 – Construction of a simple solar panel.

not in the shade of trees or other buildings. The amount of heat which can be obtained is directly related to the collector area and this must also be related to the incident solar radiation available in a particular geographical location. One of the advantages of solar water heating when compared with solar space heating is that hot water is required throughout the summer months, when there is a high level of solar radiation. The most cost-effective use of solar energy is to design a system which will provide most of the hot-water requirement in the summer months and to use a back-up source of heat in the winter.

As an example, if a household has a need for 400 litres of water per day, this is equivalent to 67.2 MJ or 18 kWh/day. If during the summer months the incident solar radiation is 5 kWh/m^2 per day, with an efficiency of 40% the energy available is 2 kWh/m^2 per day. From this it can be seen that to supply all of the hot water needed in the summer, 9 m^2 of collector area would be required. If in the winter months the incident radiation falls to 0.5 kWh/m^2 the solar heating system (with 9 m^2 of collector area) will only supply one tenth of the required supply of hot water.

However to supply all of the winter hot water requirement would need a vast area (90 m^2) of solar collector and in addition a large proportion of this heat would be wasted in the summer.

The solar panel may be connected to the storage tank as shown in Fig. 10.8 which uses natural convection. The disadvantage with this is that the solar panel

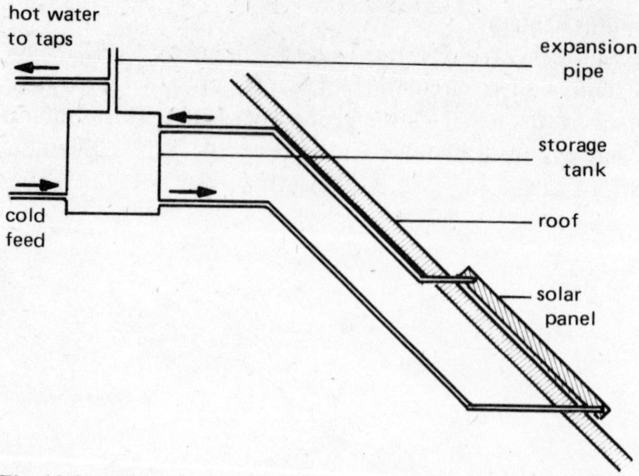

Fig. 10.8 – Solar panel circulating hot water by natural convection.

must be at a lower level than the storage tank (Wozniak 1979) and it is often more convenient, particularly with solar panels mounted on the roof, to use a pump to circulate the water. This is activated by a thermostat on the solar panel which switches the pump on and off depending on the temperature of the water. To minimise corrosion it is better to use an indirect storage tank. This has the additional advantage that anti-freeze and anti-corrosion chemicals may be used in the primary circulation between the panel and the storage tank. The storage tank may contain a second heat exchanger for supplying alternative heat from a central heating boiler or it may have an electric immersion heater (see Fig. 10.9).

In more complex solar water systems there may be two storage tanks, with a solar heating tank feeding a conventional storage tank. This means the conventional tank pre-heats water for the conventional storage tank. If the temperature of the water in the solar heating tank is sufficiently high it can supply the hot water distribution directly, by-passing the conventional storage tank (Wozniak 1979).

10.4 HOT WATER DISTRIBUTION SYSTEMS

Water from the storage tank or multipoint heater is distributed by a network of distribution pipes to sinks, baths etc. In order to minimise heat loss from these pipes it is important that they are kept as short as possible. For this reason the siting of the storage tank is very important. Also, hot water distribution pipes should be lagged to reduce the temperature drop in the water between the storage tank and the sink.

Two different methods are used for distributing hot water. The first, which is used in small houses and where the lengths of pipe runs are relatively short, consists of a radial distribution system (see Fig. 10.10). The second method is to

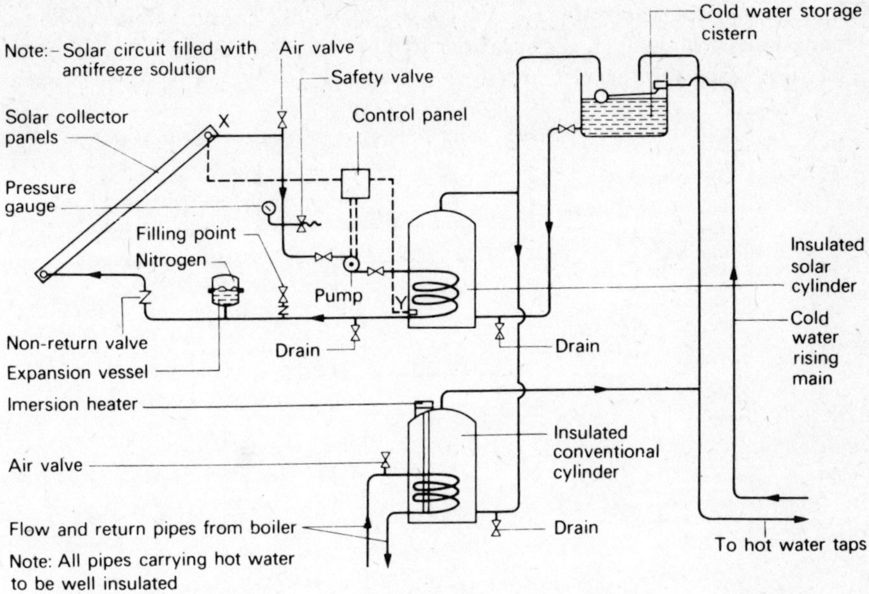

Fig. 10.9 – Solar panel with pumped circulation and indirect tank (Hall, 1980).

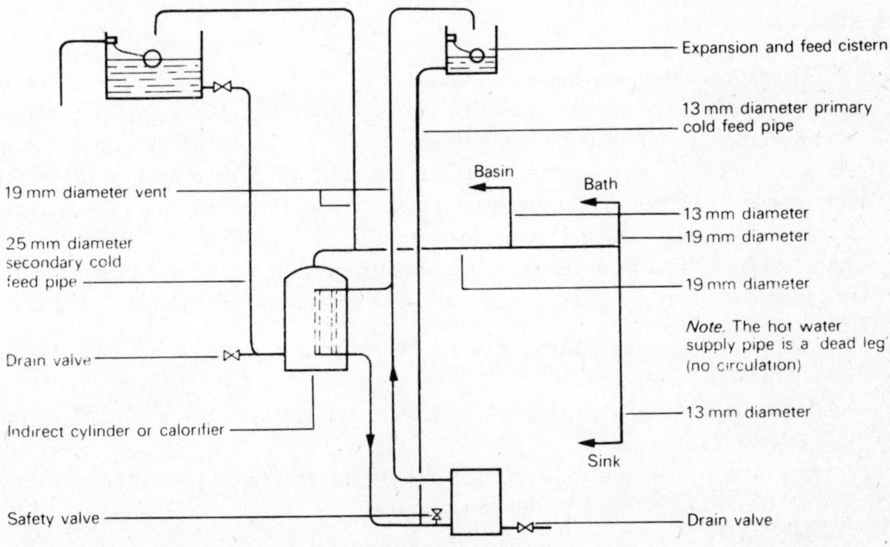

Fig. 10.10 – Typical UK hot water distribution (Indirect) (Hall, 1976).

circulate the hot water in a continuous loop, to which short lengths of pipe are connected to individual outlets (Fig. 10.11). These methods are analogous to electrical radial and ring circuits (section 5.4).

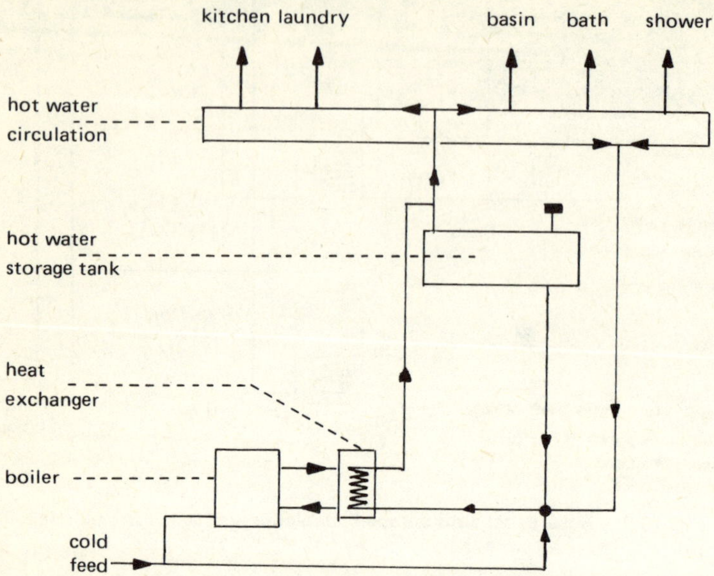

Fig. 10.11 – Typical USA hot water distribution.

As with cold-water systems (Chapter 6) there are major differences in plumbing practices between the USA and the UK. Fig. 10.11 shows a typical American hot water distribution system. Water in the entire system is under pressure from the street main and it is this pressure which drives water out of the taps. Each tap has an expansion air chamber to absorb energy when the taps are closed and to take up the pressure from the expanding water (Allen 1980). Check valves and vacuum breakers are used to prevent the reverse flow of water back into the water main. There are three distinct natural circulation flows:

(a) from the boiler, through the external heat exchanger and back to the boiler
(b) from the external heat exchanger to the storage tank and back to the heat exchanger
(c) from the storage tank, around the closed loop (with tap-off connections to taps), and back to the storage tank.

To contrast with this, a typical UK hot water distribution system is shown in Fig. 10.10. The primary water circulation is the water circulating between the boiler and the inner heat exchanger of the hot water storage tank. Pressure in this circuit is provided by the expansion/feed cistern. In the case of a combined space heating/water heating system this tank also controls the pressure of the water which is distributed to the radiators. The secondary circuit provides a feed of cold water to the storage tank and to the hot water taps. Pressure relief in this circuit is provided by a vent pipe into the cold water storage tank. In soft water districts and where central heating is not supplied a direct system may be used. If this method is used a separate primary feed and expansion tank is not required and a direct cylinder, without an internal heat exchanger, can be used.

REFERENCES

Allen, E. (1980), *How Buildings Work,* Oxford University Press.

Smith, P. R. & Julian, W. C. (1976), *Building Services,* Applied Science.

Taylor, B. (1973), *le Fevre's Domestic Utilisation of Gas,* Ernest Benn.

Wozniak, S. J. (1979), *Solar heating systems for the UK,* HMSO.

BIBLIOGRAPHY

Anderton, B. (1976), *The Solar Home Book,* Cheshire Books.

Barton, J. J. (1970), *Domestic Heating and Hot Water Supply,* Applied Science.

Brinkworth, B. J. (1978), *Active Collection and Use of Solar Energy in Ambient Energy and Building Design,* Randell, J. E. (Ed.), The Construction Press.

Coltan, N. (1978), *Solar Heated Houses,* Tutor Press.

Hall, F. (1976), *Building Services and Equipment,* **Vol. 1,** Longman.

Hall, F. (1980), *Heating, Ventilating and Air Conditioning,* Construction Press.

McGuiness, W. J. & Stein, B. (1977), *Building Technology,* Wiley/Ripka.

Sadaby, P. (1978), *The Solar House,* Newnes-Butterworth.

Which (1977), Solar Heating, **August**, 448–453.

Vale, B. & R. (1975), *The Autonomous House,* Thames and Hudson.

CHAPTER 11

Ventilation and air conditioning

11.1 INTRODUCTION

This chapter is concerned with the control of the air supply within a house and, in particular, the supply of oxygen for respiration and combustion, the control and removal of airborne contaminants and with the control of air temperature and moisture content.

The main contaminants of air within a house are:

(a) carbon dioxide from respiration of the occupants,
(b) carbon dioxide and combustion products from flueless appliances,
(c) water vapour (humidity) from cooking, respiration and perspiration,
(d) perspiration odours from the occupants,
(e) cooking odours,
(f) tobacco smoke,
(g) dust.

In addition to the control of the levels of these contaminants, a supply of air for fossil-fuel burning appliances which do not have their own ducted-air supply may be required.

11.2 VENTILATION

As shown in Fig. 11.1 the ventilation system of a house is concerned with the inlet of fresh air into the house and the removal of airborne contaminants from it. Of these contaminants, the best evaluation of the adequacy of a ventilation system is the absence of body odours, dust and fumes in the atmosphere. In practice these are detected before any physiological effects resulting from the build up of carbon dioxide are apparent. Excessive ventilation is as undesirable as insufficient ventilation because it can lead to low room temperatures, draughts and high heating bills. In a typical house, with normal levels of ventilation, heat losses caused by ventilation can amount to one third of the total heat loss. In a well insulated house this can rise to one half (Gale 1979). The heat losses associated with air changes are discussed in section 8.3.7.

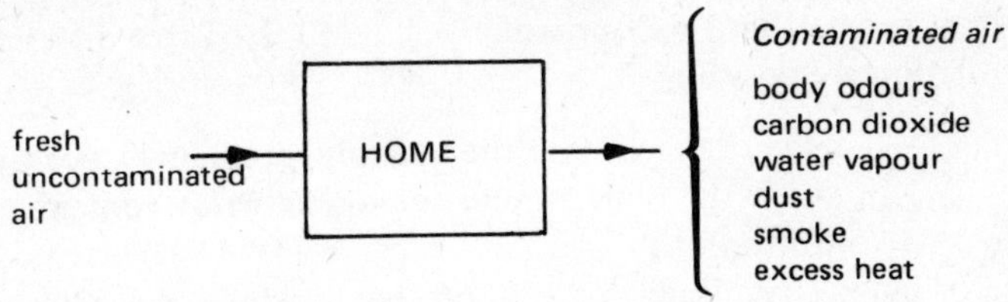

Fig. 11.1 – The ventilation system of a house.

11.2.1 Natural ventilation

Natural ventilation is made up of two components: *inadvertent* infiltration of air through gaps, cracks and non-closeable ventilators in the structure of the building, and *optional* ventilation through open windows and doors. In both cases the ventilation is caused by natural convection currents (see section 7.2.2) resulting either from the *stack effect* or from the *wind effect*.

The stack effect is the name given to ventilation caused by changes in the *density* of air at different temperatures. Air inside a building is warmer (and therefore less dense) than the air outside a building. This causes cooler air to enter a building at the lower floor level and warmer air to leave the building at the upper floor level (Fig. 11.2a). The stack effect is very noticeable in the case of warm chimneys and flues. An open coal-burning appliance can lead to four or five air changes per hour.

The wind effect is caused by differences in *pressure* between the windward and leeward sides of a house. A positive pressure on the windward side of a house results in air entering the building on that side. A negative pressure on the leeward side draws air out of the building (Fig. 11.2b).

In practice these two mechanisms of ventilation work together as shown in Fig. 11.2c. The control of natural ventilation is difficult because it is influenced by temperature variations and by wind velocity and direction. For this reason there is a trend towards much better-sealed houses with controlled ventilation provided by mechanical means (Gale 1979).

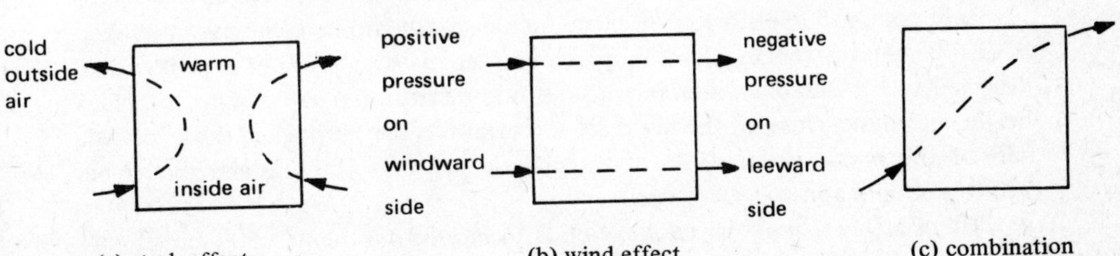

Fig. 11.2 – Natural ventilation.

11.2.2 Forced ventilation

Forced ventilation is provided by electrically-driven fans which produce *forced convection*. For domestic use the fan is usually of the propellor or centrifugal type and may be sited in a wall or window (Fig. 11.3). Three different types of forced ventilation are possible:

(1) Inlet fan — a fan draws air into the building, providing a positive pressure inside the building and the discharge of contaminated air by infiltration.

(2) Extractor fan — a fan draws air out of the building, providing a negative pressure inside the building, and the ingress of fresh air is by infiltration.

(3) Combination — air is drawn into building by an inlet fan and the contaminated air removed by an extractor fan.

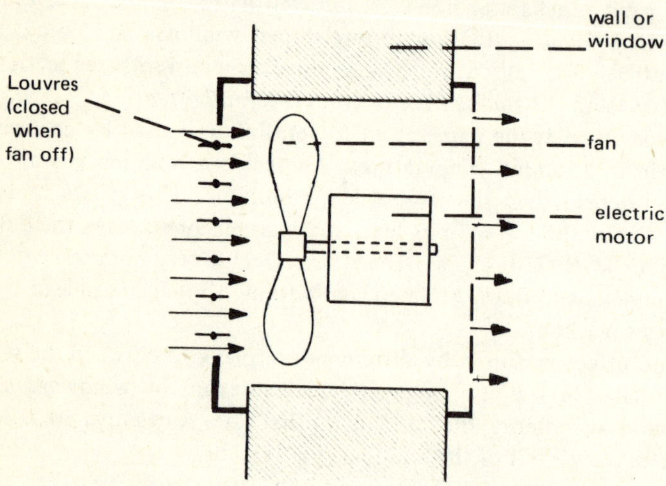

Fig. 11.3 – Wall or window mounted extractor fan.

The advantage of methods (2) and (3) is that the extractor fan can be placed close to a source of contamination such as a cooker or WC. In the case of methods (1) and (3) the inlet air can be filtered to provide dust-free air.

One common application of extractor fans in the home is the use of cooker hoods. These are rectangular canopies mounted over the hob of a cooker and designed to extract the water vapour and odours from the cooking process. The height and dimensions of the hood are important as they influence the effectiveness of the hood. It should be wider and deeper than the hob and should be 550 to 750 mm above it.

There are two types of cooker hoods. A *ducted hood* extracts the contaminated air from the kitchen to the outside (see Fig. 11.4). The *ductless hood* purifies the air and returns it to the kitchen.

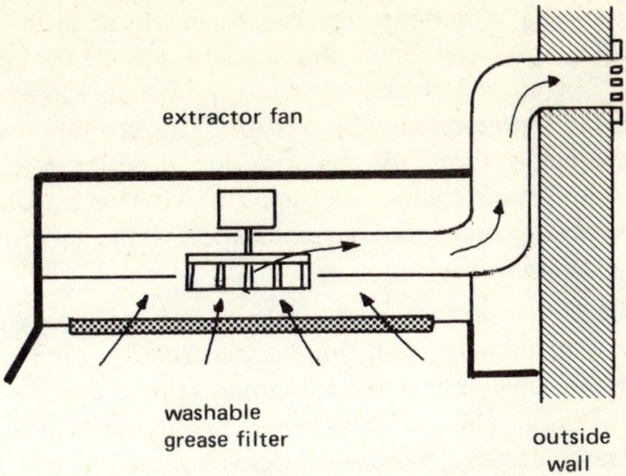

Fig. 11.4 – Cross-section of ducted cooker hood.

The ducted hood consists of a metal canopy, a grease filter, a centrifugal fan and a duct which transfers the air to the outside environment. Grease and dust are trapped in the metal mesh filter which can be washed periodically. A ductless hood filters the air, first though a metal mesh filter (to remove dust and grease) and then through an activated charcoal filter (to remove odours) which must be replaced periodically. The ductless hood does not extract excessive heat and humidity from the kitchen, whereas the ducted type does do so. The ducted hood must be placed on an outside wall and requires a hole to be made in the wall. The ductless hood, on the other hand, can be placed on an outside wall and needs only to be fastened to it.

11.3 HUMIDIFICATION

11.3.1 Relative humidity

If a dish of water is left standing in a warm room the water will eventually 'disappear'. It actually *evaporates,* that is, water molecules leave the liquid water and are dispersed into the air in the room as water vapour. This process also requires latent heat. In this case it comes from the warm air in the room. Boiling is also evaporation but in this case the evaporation is so rapid that it takes place not only at the liquid surface but inside the bulk of the liquid itself. The bubbles observed during boiling are bubbles of vapour formed inside the liquid, next to a hot surface, for example, next to the element in a kettle or the bottom of a pan.

The molecules of vapour in the air exert a pressure: the *vapour pressure*. If the water is in a corked bottle then the air above the water contains molecules of water vapour. There is a continuous exchange of molecules between the liquid water and the air. Molecules leave the water and enter the air space (evaporation), whilst others return to the water (condensation). Eventually however an equilibrium situation is reached when the air in the bottle can hold no more molecules of water vapour: the number entering the air just equals the number leaving. The air is now *saturated* and the pressure exerted by the water vapour is called the *saturation vapour pressure.* This saturation vapour pressure of a particular liquid is always the same at any given temperature. It is equal to the atmospheric pressure when the liquid boils. This explains the effect of pressure on the boiling point. Examples illustrating the importance of vapour pressure include the following:

(i) Volatile liquids, for example, cleaning solvents, exert a higher saturation vapour pressure than non-volatile liquids. 'Dry cleaning' is only dry in the sense that a volatile liquid is used and this evaporates off easily in the machine without resort to the use of heat. 'Wet cleaning' with water involves the use of less-volatile water and hence drying is more difficult.

(ii) Drying clothes in a small, non-ventilated room may cause excessive condensation. The water evaporates from the clothes but eventually the air in the room becomes saturated with water vapour. If this saturated air comes into contact with a cold surface, such as a window, its temperature is lowered and some of the water vapour condenses on the cold surface. This may be prevented by ventilating the room, i.e. allowing saturated air (damp air) to escape and replacing it with dry air.

(iii) Clothes do not dry quickly outside on a damp, still day because the atmosphere contains a lot of water vapour and the air next to them soon becomes saturated. The ideal day is a sunny, breezy day, because the heat from the sun provides much of the latent heat for evaporation (less needs to come from the air) and the moist air near the clothes is continuously blown away and is replaced by dry air.

Warm air can hold more water vapour than cold air, that is, the saturation vapour pressure is higher at higher temperatures. If warm air containing some water vapour is cooled therefore, it eventually becomes saturated with the water vapour. On further cooling the excess water vapour will condense on any available surface. This is the reason that when warm moist air comes in contact with a cold window condensation occurs. The temperature at which the water vapour condenses on the cold surface is called the *dew point*. This is used as a measure of the amount of water vapour in the air.

An alternative method of expressing the water-vapour content of air is to use *relative humidity* (RH). This is a measure of the mass of water vapour contained in a given volume of air compared with how much water vapour the air could contain if it were fully saturated.

$$\text{RH} = \frac{\text{mass of water vapour in a given volume of air}}{\begin{array}{c}\text{mass of water vapour in the same volume of}\\ \text{saturated air at the same temperature}\end{array}} \times 100\%$$

As an example, if air at 15°C contains 8.4 g/m^3 of water vapour and saturated air at 15°C contains 14 g/m^3 of water vapour, its RH will be:

$$\frac{8.4}{14} \times 100 = 60\% \ .$$

In winter, when cold air is taken from outside and heated, its relative humidity is decreased and therefore it may need humidifying (section 11.3.2). In summer if warm moist air is cooled its relative humidity may become excessive, in which case it can be dehumidified during air conditioning (section 11.4).

The increasing use of central heating has lead to problems of dry atmospheres in houses and this has been exacerbated by the trend for householders to have higher average temperatures in their homes now compared with twenty years ago. This dry atmosphere leads to feelings of discomfort because of drying of the mucous membrane of the ears, nose and throat. The dry atmosphere also causes static electricity, particularly when found in conjunction with carpets and furnishings made of artificial fibres. Additionally the dry air can cause damage to furniture and to paintwork because wood shrinks on drying. This causes loose joints and cracking of paint on wooden window frames etc. High levels of RH can also lead to feelings of discomfort. The body temperature cannot be controlled by evaporation and a person becomes overheated and drowsy. High humidities can also cause condensation problems.

11.3.2 Humidification

As discussed in section 7.3.1, the RH in a room should be about 50%. In winter, when cold air is taken in from outside and heated, its RH may be as low as 20%. For a room of volume 300 m^3 with two air changes per hour and with an air temperature of 20°C and an RH of 20%, the air will require the addition of almost 1 litre of water per hour to maintain its humidity at 50%. This amount of water cannot normally be maintained by evaporation of water from plants or even by trays of water placed in a room and must be added to the atmosphere by a *humidifier*. There are a number of different designs of humidifiers (Leisinger 1970) but the two most common types for use in the home are the *steam evaporation humidifier* and the *fan induced evaporation humidifier*.

A steam evaporation humidifier is shown in Fig. 11.5. A container of water (7 to 10 litres) is heated by an electrical element (100–500 W) to a temperature of just below boiling. This temperature is controlled by a thermostat. In addition to this the electrical supply is isolated if water in the container falls below a pre-determined minimum level. This control operates via a sensor which detects the weight of water in the container. There is a slow release of steam from the container through slots in the lid. One major advantage of this unit for use in the home is that it is silent in operation.

In the fan induced evaporation humidifier a fan draws air into the humidifier and blows this air over a wick saturated with water. A number of different

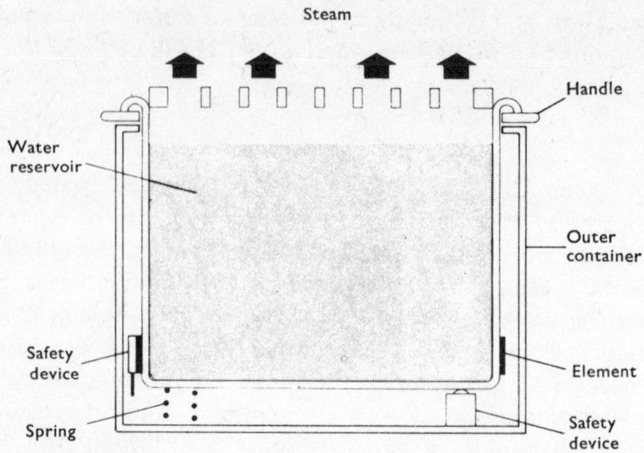

Fig. 11.5 – Steam evaporation humidifier.

methods are used to ensure that the unit remains saturated with water (Leisinger 1970) such as static wicks, and revolving filter belts. The water is supplied from a reservoir tank (25–30 litres) and since a water heater is not required only 40–50 W are required for the electrical fan. This type of humidifier is also suitable for mounting in the duct of a warm air circulation central heating system, in this way providing whole house humidification. For water circulation heating systems radiator humidifiers are available. These are made to hang on to the top of the radiator and use the natural convection currents induced by the radiator to absorb moisture from a wick, which is supplied with water from a reservoir.

11.4 AIR CONDITIONING

The term 'air conditioning' is used to mean two different things. In its most general sense it refers to the control of the air supply system for a house, maintaining the supply of clean, odourless air at a desired temperature and humidity level. A more limited use of the term air conditioning refers to the control of the air supply during the summer months, and is concerned with the supply of chilled and dehumidified air.

Referring to the first definition, an air-conditioning system is concerned with the provision of a comfortable indoor climate and involves:

(a) removing excess heat in summer,
(b) providing heat in winter,
(c) humidification of air in winter,
(d) dehumidification of air in summer,
(e) removing dust particles from air,
(f) removing odours and smoke from air,
(g) maintaining the correct balance of CO_2 and O_2.

From this it can be seen that air conditioning, using this definition, involves all of the topics covered in Chapters 9 and 11 so far, with the addition of the provision of chilled, dehumidified air in summer.

Air chilling and dehumidification makes use of the mechanical refrigeration cycle (Chapter 15). Referring to Fig. 11.6, it can be seen that heat is absorbed from the internal environment of the building by contact with the cold surface of the evaporator coil. This cooling of the air causes it to fall below its dew point (section 11.3.1). Because of this water vapour condenses on the cold evaporator surface, in this way reducing the humidity of the air. Control of the RH is achieved by controlling the temperature to which the air is chilled, since the lower the temperature of the air, the lower is its saturated vapour pressure. Following chilling of the air to remove water vapour, it is sometimes necessary to re-heat the air.

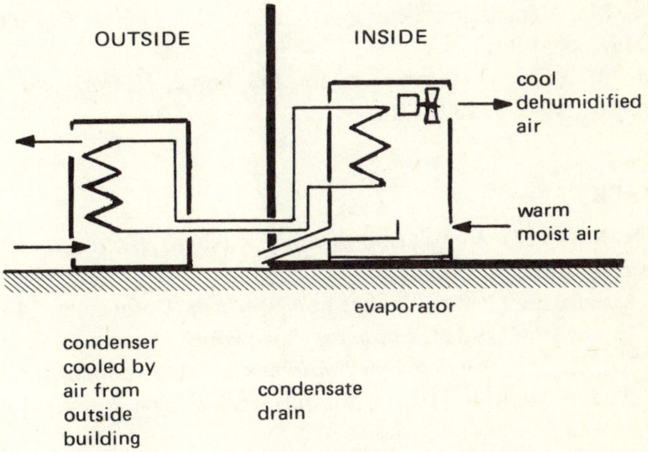

Fig. 11.6 – Air chilling and dehumidification.

Heat which is extracted from the air at the evaporator is released to the outside environment at the condenser. Air is drawn into the condenser unit by a fan and passes over the hot surface of the condenser, from which it gains heat. The warmed air is then returned to the environment.

Appliances for providing air chilling and dehumidification may be individual room units or whole-house systems. Individual room units are of two types. *A self-contained room unit* is mounted through a wall or window with the condenser and evaporator in the same unit. *A split unit* has the condenser and evaporator in separate units, connected by pipes containing the recirculating refrigerant (Chapter 15). The advantage of the split unit is that the room chiller does not have to be against an outside wall or window. Also, since the compressor is outside the building, it is quieter in operation than the self-contained unit. Centralised chilling and dehumidification sytems use metal ducting, similar to that used for warm air heating (section 9.3.2), to transfer the chilled air around the building and to return warm stale air. For this reason it is common practice to combine heating and cooling systems together to provide full air conditioning from a single unit.

An alternative way of providing summer cooling and winter heating from a single unit is to utilise a reverse cycle heat pump. In this unit the direction of the refrigerant can be reversed using a valve giving the following alternatives:

Season	*Inside coil*	*Outside coil*	*Effect*
Summer	Evaporator	Condenser	Air cooling
Winter	Condenser	Evaporator	Air heating

REFERENCES

Gale, R. (1979), Ventilation heat loss outside, *Gas Engineering and Management,* **Nov.**, p. 563–572.

Leisinger, F. M. (1970), Humidifiers for the home, *Heating and Ventilation Engineering,* **Nov.**, p. 232–236.

BIBLIOGRAPHY

Billington, N. S. (1972), Ventilation of home kitchens, *Heating and Ventilation Engineering,* **April**, p. 512–518.

Consumers Association (1979), Cooker hoods, *Which,* **October**, p. 543–546.

Consumers Association (1981), Extractor Fans, *Which,* **May**, p. 292–300.

Hall, F. (1976), *Building Services and Equipment,* **Vol. 1**, Longman.

McGuiness, W. J. & Stein, B. (1977), *Building Technology,* Wiley.

CHAPTER 12

Artificial lighting

12.1 INTRODUCTION

Lighting within the home is supplied by both natural daylight and by artificial lighting. During daylight hours, sufficient light is available in the home from windows for all but the most exacting tasks. During periods of dull weather, at dawn and dusk and when detailed tasks are being performed, the available daylight must be supplemented by artificial light. At night all the lighting requirements must come from artificial sources. This chapter is concerned with the generation and uses of artificial light sources and with the design of lighting requirements to match the needs of a household.

12.1.1 The physical nature of light

Light is a type of energy known as electromagnetic radiation (Milson & Kirk 1980). An electromagnetic wave has a magnetic field and an electrical field and, as the wave travels through space, these fields vary in magnitude in the same way as alternating current electricity (see section 5.3.2). The relationship between the electrical and magnetic fields is shown in Fig. 12.1. As the wave travels through space the two fields are at right angles to each other.

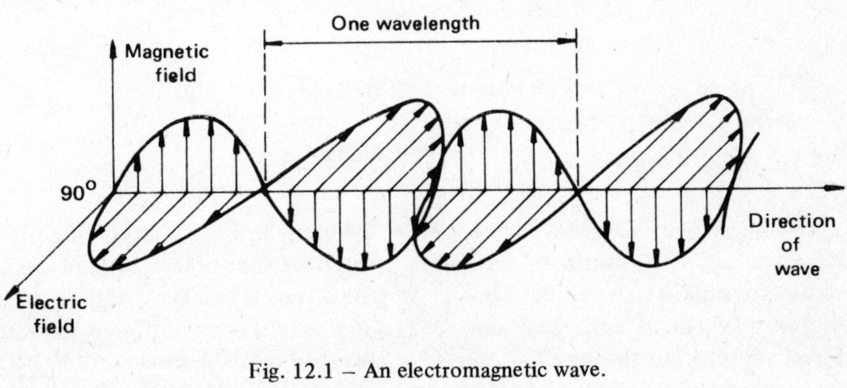

Fig. 12.1 – An electromagnetic wave.

There are many different kinds of electromagnetic waves including those used for radio and television transmissions, microwaves and radiant heat. These waves have varying wavelength and frequency as shown in Fig. 12.2.

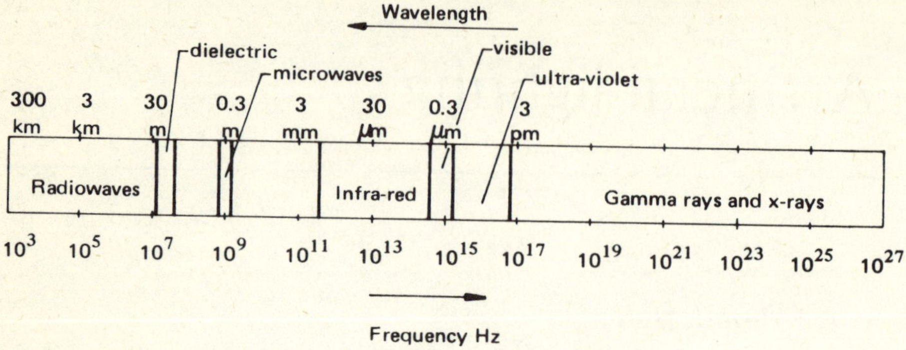

Fig. 12.2 – The electromagnetic spectrum.

Only a small portion of the electromagnetic spectrum can be detected by the human eye and this is known as the visible spectrum. The eye is most sensitive to light of wavelengths 500 to 600 nm (1 nm $\equiv 10^{-9}$ m), but the visible spectrum covers the wavelengths 380 to 760 nm, as shown in Fig. 12.3. The eye interprets these different wavelengths as particular colours, with red and orange colours between 570 and 760 nm, green and yellow between 520 and 570 nm and blues and purples between 380 and 520 nm (these follow the colours of the rainbow). For more details of the physiology of vision see Nuckolls 1976, Rees 1970 and Grandjean 1973.

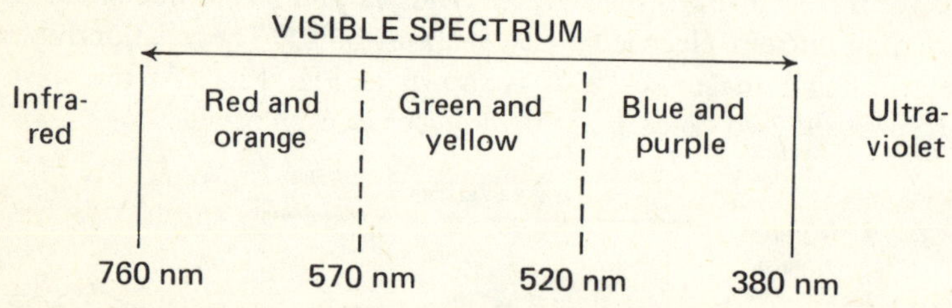

Fig. 12.3 – Visible radiation.

When we see an object, a number of different processes are taking place, each of which contributes to the image created of that object. A light source produces visible electromagnetic radiation which travels as a wave until it strikes an object. When the light strikes an object, it may either be reflected from the object, transmitted through the object, or absorbed by it, as shown in Fig. 12.4.

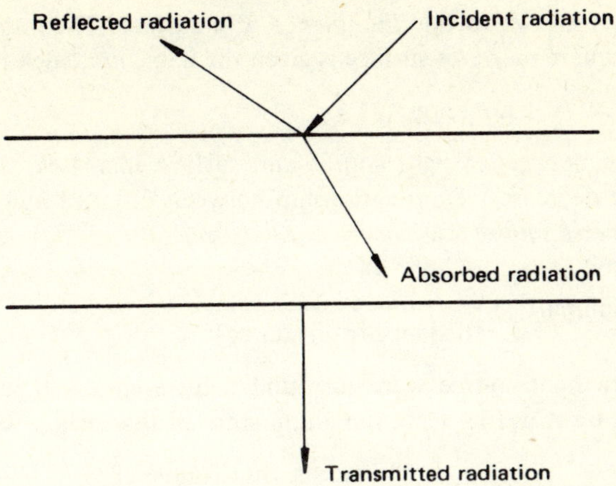

Fig. 12.4 – Interaction of light with an object.

In practice it is likely that a combination of these effects will take place. Light which is reflected from, or transmitted through, an object then passes to the eye (see Fig. 12.5).

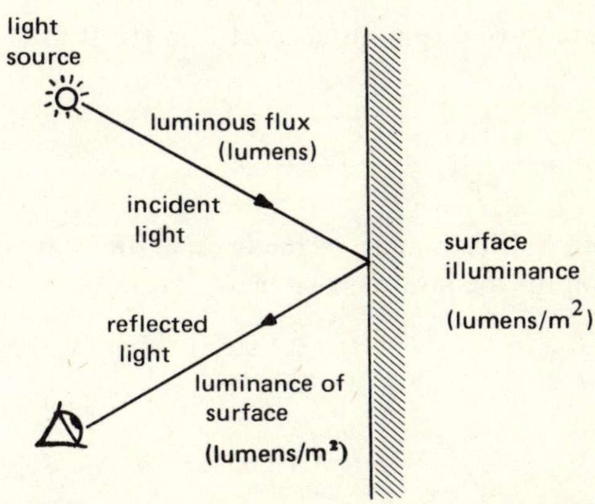

Fig. 12.5 – Physical relationships between light source and surface.

The intensity of light emitted from a lamp or bulb in a given direction is known as the *luminous intensity*. The SI unit of luminous intensity is the *candela* (cd). The *total* amount of light leaving a light source is referred to as the *luminous flux* which is measured in lumens (lm). For example, a 100 W tungsten filament bulb (see section 12.2.1) produces a luminous flux of approximately 1200 lumens. This unit can also be used as a measure of the light received by a surface, which is known as the *illuminance* of a surface. The number of lumens falling on one square metre of surface is given the name *lux*, such that

$$1 \text{ lux} = 1 \text{ lumen/m}^2$$

As the distance between a light source and surface increases, the illuminance of that surface decreases. The relationship between distance and illuminance is given by the *inverse square law*:

$$\text{illuminance} \propto \frac{1}{(\text{distance from source})^2} .$$

If a 100 lumen light source were surrounded by a spherical surface 1 metre away, as shown by A in Fig. 12.6, the illuminance of the surface would be:

$$\text{illuminance (lux)} = \frac{\text{luminous flux (lumens)}}{\text{surface area} \quad (\text{m}^2)}$$

$$= \frac{100}{4\pi r^2} .$$

Since $r = 1$ m, the illuminance is $\frac{100}{4\pi 1^2} = 7.96$ lux.

If the surface were moved to a distance of 2 m (to B in Fig. 12.6), the illuminance would fall to:

$$\text{surface B} \quad \frac{100}{4\pi 2^2} = 1.99 \text{ lux} .$$

It can be seen from this simple example that the only variable is $\frac{1}{r^2}$, which explains the reason for the inverse square law.

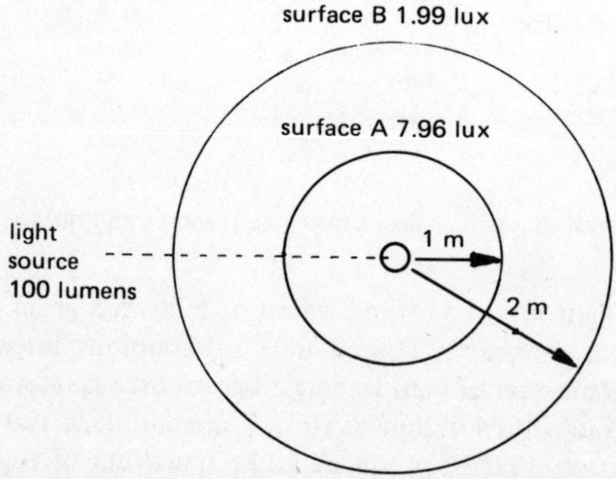

Fig. 12.6 – The effect of distance on illuminance.

The amount of light reflected by a surface is known as its *reflectance.* A reflectance of 0.4 indicates that 40% of the incident light is reflected back from the surface. In general, light surfaces have a high reflectance and dark surfaces a low reflectance. The light given off by a surface is known as its *luminance* and is measured in lumens/m^2. The relationships between the various terms defined in this section is illustrated in Fig. 12.5.

Light sources may be direct, in which case all of the light travels in parallel straight lines or it may be diffused, as shown in Fig. 12.7. Direct light sources produce very distinct shadows whilst diffused light sources produce weak shadows. In practical light sources, a combination of direct and diffused light is produced, but the balance of the two can be controlled to produce the desired lighting effect.

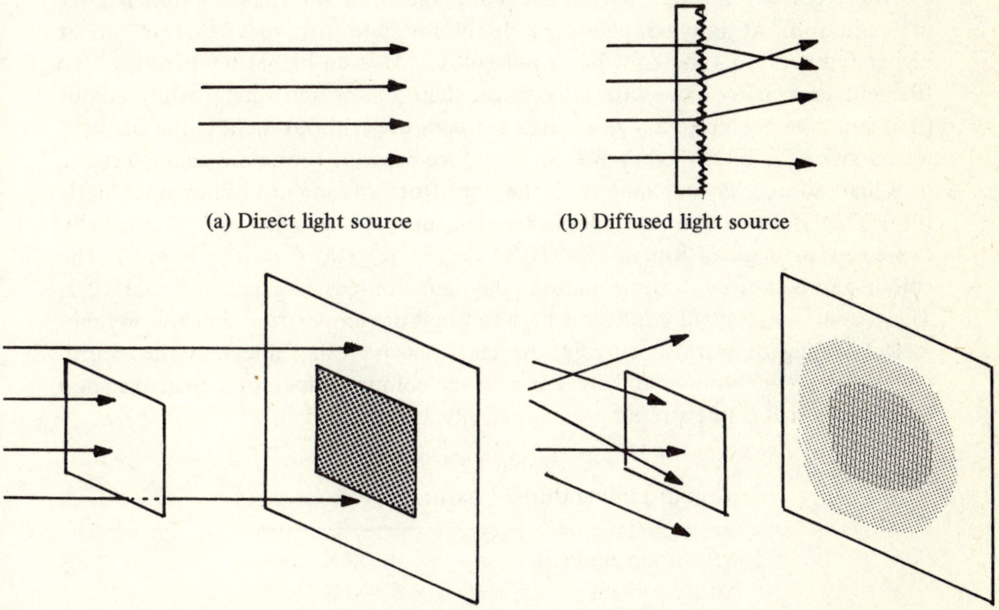

(a) Direct light source (b) Diffused light source

(c) Strong shadows from direct light source (d) Weak shadows from diffused light source

Fig. 12.7 – Direct and diffused light sources.

With a direct source the contrast between the illuminance of light and shade may be as high as 10:1. This means that the shadows are very dark. Because of this, a direct source is unsuitable for detailed work even if the illuminance of the highlighted areas is sufficiently high (see section 12.3). The most diffused form of lighting is achieved by directing the light source towards a wall or ceiling, rather than directly into the room. This produces a lighting effect with almost no shadow.

12.1.2 Light and Colour

As has been seen from section 12.1.1, a light source can consist of a narrow band of wavelengths, in which case it has a distinct colour. White light, on the other hand, is made up of approximately equal components of the visible spectrum. Many of the light sources we use are not pure white and may have a distinct colour cast. Light with a preponderance of long wavelengths has a yellow colour and is interpreted by the brain as a 'warm' colour because of previous association effects we have gained from experience. Light with a preponderance of short wavelengths has a blue colour and is seen as a 'cool' colour. One method of expressing the colour of a light source is to use the concept of *colour temperature.* This concept may be explained by considering the light produced by heating a metal filament (either directly in a flame or by passing an electrical current through it). As the temperature of the filament rises it gives off radiation. At low temperatures this is invisible infra-red radiation but at higher temperatures the light has a red colour. At even higher temperatures the filament is seen to glow with an orange, then yellow and finally white colour (if it could be heated to a high enough temperature without melting the filament would give off a bluish light). We can therefore describe the colour characteristics of a light source by matching it to the light from a filament of known temperature. This is known as the **colour temperature** of the light, and it is usually expressed in degrees Kelvin (K). (0°C ≡ 273°K, 100°C ≡ 373°K etc.). The colour temperature of some natural daylight sources is given in Table 12.1. The reader must avoid confusion between colour temperature and the psychological feeling of warmth or cold. As can be seen from Table 12.1 the 'warm' yellow-orange afternoon sunlight has a lower colour temperature than that of a blue sky, which is interpreted psychologically as a 'cold' light.

Table 12.1
Colour temperature of natural light sources

Afternoon sunlight	4000 K
Noon sunlight	5000 K
Cloudy sky	6500 K
Blue sky	10000 K

We have seen that when light strikes a surface it may be reflected, absorbed or transmitted. This is an over-simplification, however, as different wavelengths of radiation within the light source may react in different ways. For example, a red surface looks red because when it is illuminated with white light it absorbs all of the wavelengths other than those which the eye interprets as red. These wavelengths are reflected back from the surface. In a similar way a red filter over a light source absorbs all of the wavelengths other than red light; these are transmitted through the filter.

12.2 ARTIFICIAL LIGHT SOURCES

12.2.1 Gas filled incandescent lamps

If an electric current is passed through a metal filament, both heat and light are produced. A light source which produces the radiation in this way is known as an incandescent lamp (see Fig. 12.8). A coil of tungsten is sealed inside a glass

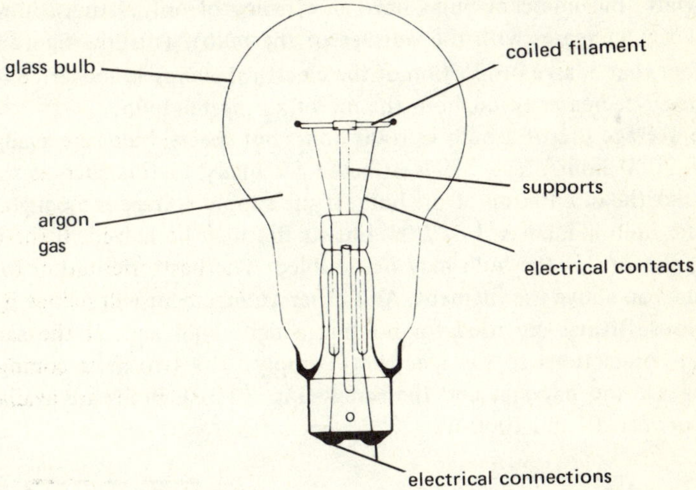

Fig. 12.8 – Construction of gas-filled incandescent bulb (Hall, 1976).

bulb with an atmosphere of an inert gas, such as argon. The inert gas both prevents oxidation, which would occur where there is any oxygen present, and reduces the rate of evaporation of metal from the filament. By coiling the filament, a higher temperature can be achieved; this increases the light output. An even greater light output can be achieved by having a double coiled filament as shown in Fig. 12.9.

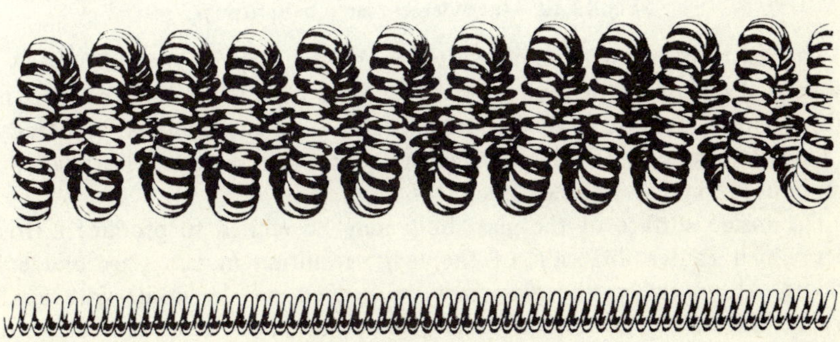

Fig. 12.9 – A double coil filament.

Gas filled bulbs are available in a wide range of shapes, sizes and light outputs. The light output is related to the *wattage*, which is the number of watts of electricity consumed under normal conditions. The relationship between electricity input to light output is given by the *efficacy* of the bulb. It is normally measured by the number of lumens of light output produced by 1 watt of electricity input. The theoretical maximum efficacy for a lamp would be 685 lumens/watt. Incandescent bulbs have an efficacy of only 7 to 18 lumens/watt (the efficacy increases with the wattage of the bulb). This low figure is caused by the fact that a large proportion of the electrical energy is converted into heat. The efficacy decreases throughout the life of a tungsten bulb.

The average life of a bulb is 1000 hours but special bulbs are available with a life of 2000 hours. The life is affected by many factors such as the supply voltage and the orientation of the bulb. If the supply voltage is 5% high, the output of the bulb is increased by 20% but its life may be halved. If the voltage is 5% low, the life of the bulb may be doubled. The best orientation for a bulb, is with the cap above the filament. Any other arrangement will reduce its life.

Various fittings are used for holding a light bulb and, at the same time, providing connections to the electricity supply. The two most common types of fitting are the bayonet and the screw (Fig. 12.10). Bulbs are available with ratings between 10 and 1000 W.

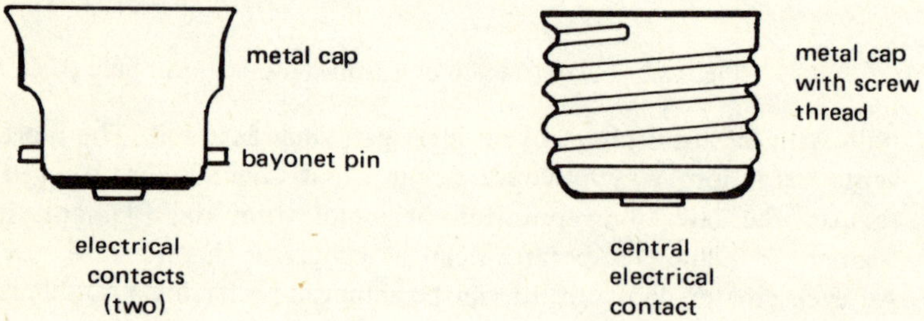

(a) bayonet fitting. (b) screw fitting.

Fig. 12.10 – Incandescent lamp bulb fittings.

The light output of a bulb is controlled by the shape of the bulb and by the nature of the glass. If the bulb is made of clear glass it produces a direct type of light which, because it is very bright, can dazzle anyone looking directly at the bulb. It also produces harsh shadows. For these reasons, clear glass bulbs are only used to create special effects.

The inside surface of the glass bulb may be etched to produce a frosted effect which causes diffusion of the light, resulting in less glare and softer shadows. Alternatively, the glass may be coated on the inside with a white powder. This has a greater diffusing effect than is the case with the frosted bulb, but results in a 10% loss in the light output. Bulbs with coloured coatings (amber, blue, pink, yellow and red) are available for creating special effects.

To produce a more concentrated beam of direct light, bulbs are available with part of the glass coated in a reflective metal layer. There are two types, the crown silvered bulb, which is used in a reflector, and the silvered reflector lamp (see Fig. 12.11).

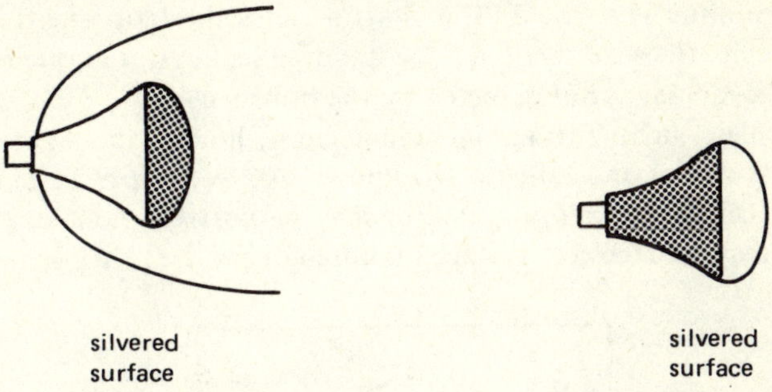

(a) Crown silvered lamp. (b) Reflector lamp.

Fig. 12.11 – Silvered reflector lamps.

A similar type of lamp is the parabolic aluminized reflector (PAR) which can be used for clear or coloured spotlights. Another type of incandescent bulb is the Tungsten halogen type which provides a constant output throughout its life. This is longer than with conventional bulbs. The presence of a halogen (iodine) in the vapour in the bulb controls the rate of evaporation of tungsten from the filament. In addition to pear and spherical shapes, incandescent lamps are available in many other shapes, a few of which are illustrated in Fig. 12.12.

Plain candle Twisted candle Pygmy Tubular Double cap tubular

Fig. 12.12 – Special shapes for incandescent bulbs.

The light output of an incandescent lamp can be controlled using a dimmer switch (Consumers Association 1976). These control the flow of electricity to the bulb and are used in place of a conventional on/off switch (section 5.4).

Control of light level is by a rotary or sliding control and it may, in addition, have an on/off switch. Dimmers contain an electronic device known as a *Triac* which effectively switches on and off the electricity during each half cycle of the alternating current. If the electricity is switched off for a large proportion of the cycle, the lamp only receives a small amount of electricity and therefore the filament temperature is reduced. The control varies the ratio of off to on time during each cycle. Because there are one hundred half cycles during one second, this continual switching is not detected by the human eye.

The complete light fitting, including lamp, holder and light modifying fittings such as shades and reflectors is known as a *luminaire.* In general, luminaires may be ceiling or wall mounted or may be portable floor or table lights. Ceiling lights may be recessed, surface mounted or pendant fittings (Fig. 12.13).

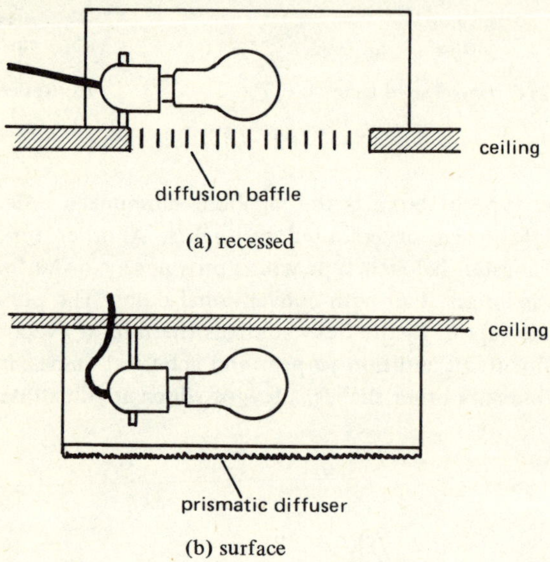

Fig. 12.13 – Recessed and surface fittings.

With the recessed and surface lamps, the diffuser may be either translucent or prismatic glass or plastic, or it may be made of concentric baffles. In the case of a pendant fitting the shade performs an important light control function. Of particular importance in creating a lighting effect is the balance of light in the three directions – upwards, downwards and sideways (Fig. 12.14). Similarly wall and table lights have a diffuser or shade. Alternatively, if used as a display light or a reading light they may have a reflector.

One method of obtaining flexibility in a lighting system is to use *track lighting*. This consists of a strip of extruded plastic which contains the live,

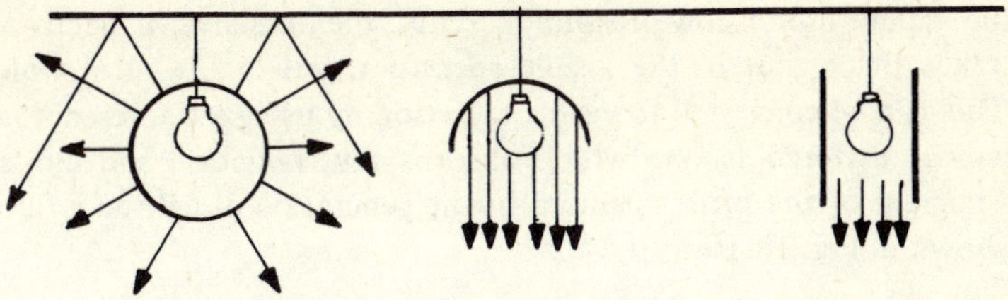

Fig. 12.14 – Light control in pendant ceiling fittings.

neutral and earth connections along its length. It is designed to accept a light fitting at any point along its length. These fittings may be spotlights or pendants, and each track can take a large number of fittings (Fig. 12.15). In situations where several fittings are required which can be separately switched on and off, multi-circuit track lighting is available. Up to four separately switched circuits can be used.

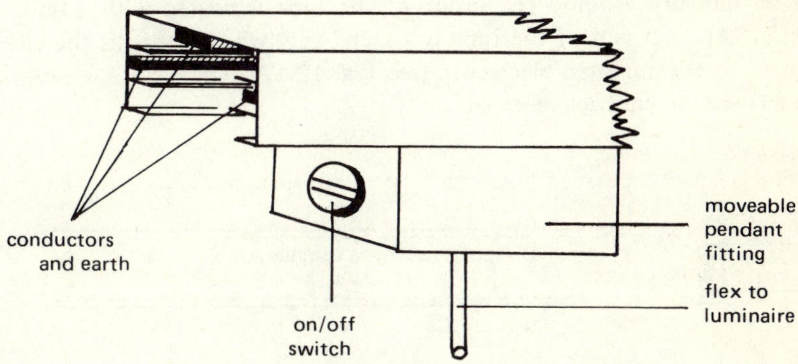

Fig. 12.15 – A single circuit track lighting system.

12.2.2 Fluorescent lamps

A fluorescent lamp uses a combination of two processes, *electrical discharge* and *fluorescence*. If electrons are passed through some gases and vapours, these electrons may be absorbed by atoms of the gas with which they collide. When this occurs, an atom is excited to a higher energy level. The atom releases this energy, as it returns to its normal energy state, as electromagnetic radiation. If this radiation is in the visible region of the spectrum, the *discharge* of light can be used for illumination. Discharge lamps produce high levels of light output, with a high efficacy. Sodium and high-pressure mercury discharge lamps, which are used for streetlighting and flood lighting, have an efficacy of between 60 and 140 lumens/watt, that is, up to ten times that of gas-filled incandescent lamps.

If the vapour used is low-pressure mercury, the discharge of electromagnetic energy takes place, not in the visible spectrum, but in the ultra violet (u.v.) region. This can be converted to visible radiation by using a fluorescent material. A *fluorescent* material is one which absorbs u.v. radiation and emits visible light. A diagram of the processes used in the generation of light in a fluorescent lamp is shown in Fig. 12.16.

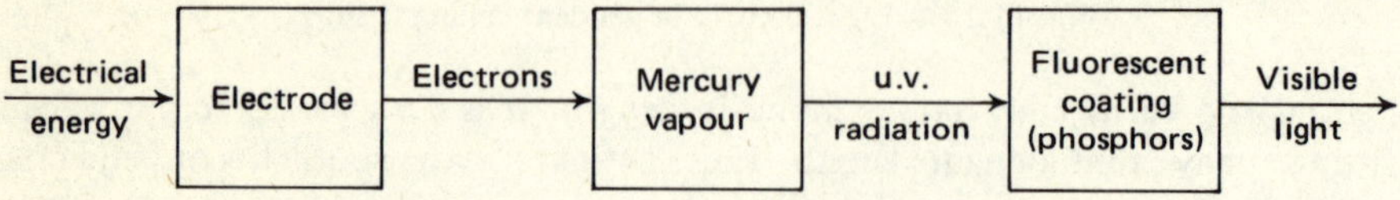

Fig. 12.16 – Energy conversions in a fluorescent lamp.

A fluorescent lamp consists of a long glass cylinder which contains low-pressure mercury vapour. The inside of the tube is coated with a fluorescent material. At each end of the tube is a metal cap, which contains the electrical contacts to the tungsten electrodes (see Fig. 12.17). The electrodes are coated with an electron emitting material.

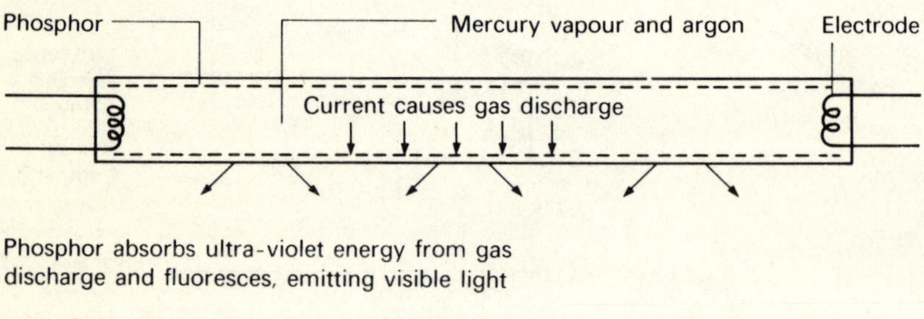

Fig. 12.17 – Construction of a fluorescent tube (Hall, 1976)

A fluorescent tube requires special starter circuits, because the action of an electric discharge from a cold electrode would lead to rapid evaporation of the coating. This would reduce the life of the lamp. To prevent this the electrode is heated by passing electricity in series through both electrodes (Fig. 12.18a). When the electrode is heated, one electrode is connected to the live wire and the second to the neutral wire (Fig. 12.18b). A starter is required to switch from the starting position to the running position after a few seconds delay. A number of different types of starter are used. In addition to this a ballast (choke) is required to control the flow of current between the electrodes. A capacitor is required to keep the current and voltage in phase, because the ballast causes them to be out of phase.

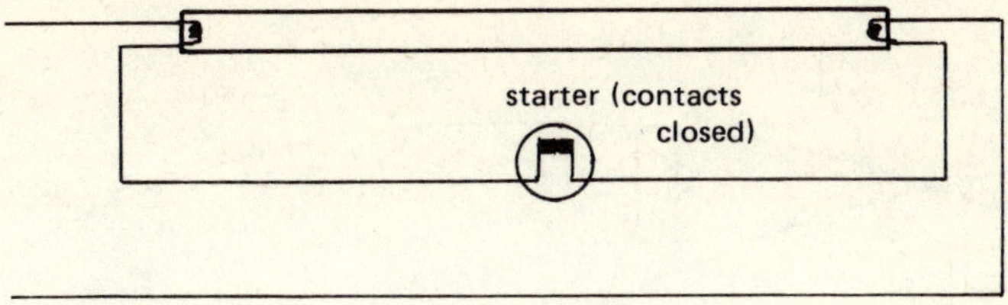

(a) starting position.

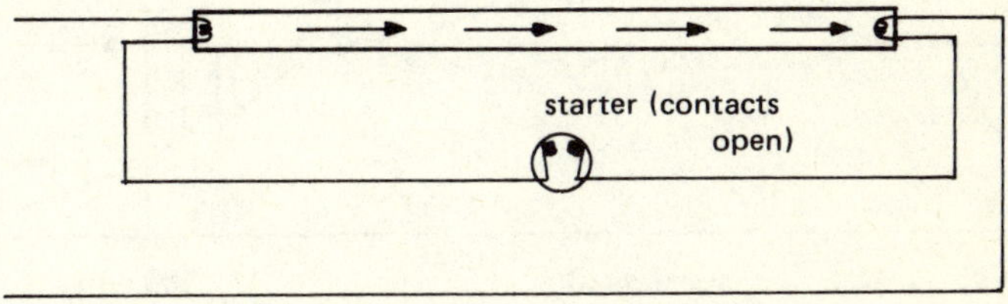

(b) running position.

Fig. 12.18 – The operation of a starter in a fluorescent tube.

Fluorescent lamps are available in a number of different types, with a range of colours. The characteristics of a selection of lamps are given in Table 12.2. Although a colour temperature is given the output is not in the form of a continuous spectrum, as shown in Fig. 12.19, which gives a comparison of daylight, tungsten light and 'plus white' fluorescent light. Normal fluorescent lights may cause distortion of some colours (particularly reds and purples) because of this continuous spectrum. Special fluorescent tubes are available for use where colour accuracy is important, such as for colour matching and food displays. These tubes however have a lower efficacy than the normal fluorescent tubes, that is, 50 lumens/watt.

Table 12.2
Characteristics of typical fluorescent tubes

Type	Colour temp (K)	Efficacy (lumens/watt)	Characteristics
White	3400	80	Poor colour rendering
Plus White	3600	76	Improved colour rendering
Warm White	3000	79	Warm rendering
Daylight	4300	76	Cool rendering

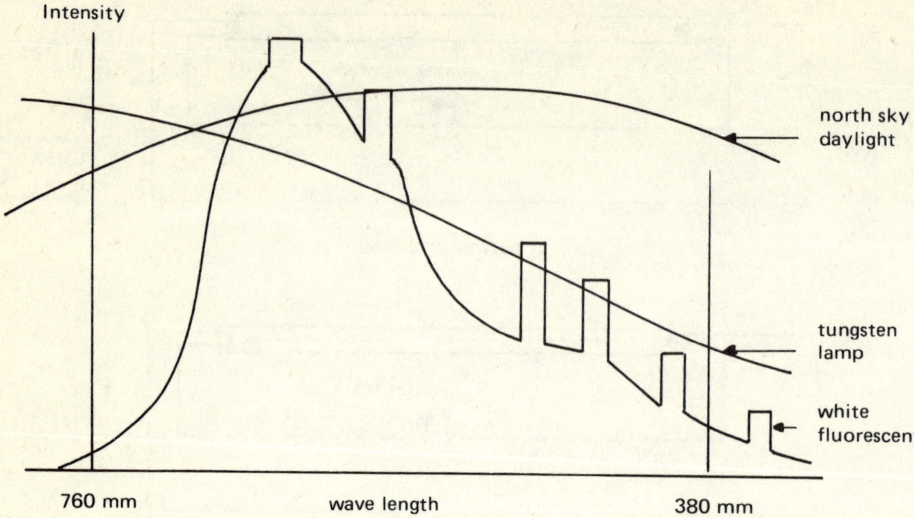

Fig. 12.19 – Spectra of daylight, tungsten lamp and fluorescent lamp.

The life of a fluorescent tube is between 5000 and 10000 hours compared with the 1000 to 2000 hours for an incandescent lamp. It is affected by the number of times the tube is switched on and off. Tubes are available with wattages between 15 and 125 W. A number of special types of tube are available such as reflector tubes (these have a built-in reflector and are used for display purposes), coloured tubes, U-shaped and circular tubes. Normal dimmer switches cannot be used with fluorescent lamps.

In the case of a fluorescent tube, the complete luminaire consists of the tube, its electrical contacts, the ballast and starter circuits and light controlling fittings such as reflectors and diffusers. The tubes are available in lengths from 0.5 m to 2 m and luminaires may hold one or a number of tubes.

12.3 DESIGN OF HOUSEHOLD LIGHTING SYSTEMS

The design of a lighting system for a house is complex since it must be capable of providing the correct level and type of illumination for a wide range of activities. It must also be capable of providing back-up lighting to compensate for low levels of daylight on dull days and in the early evening. It is therefore necessary for the lighting to have a degree of flexibility.

Specifically the lighting must provide:

(a) sufficient light for the task in hand to be carried out comfortably;
(b) correct balance of direct and diffused lighting;
(c) freedom from glare;

(d) correct colour of light to provide 'natural' colour for furnishings, food, skin tones etc.;
(e) the creation of atmospheres to suit activities of an area, for example, soft, restful lighting for relaxation.

Desirable lighting levels for various activities are given in Table 12.3. For detailed work the light should be direct, but shadows must be filled in by background lighting to give a contrast of no more than 3:1. The relationship between illumination levels and lamp wattage depends upon:

(a) the efficacy of the light source;
(b) the effect of light control devices – shades, diffusers etc.;
(c) the distance between source and surface (inverse square law);
(d) the reflectivity of the surface;
(e) the cleanliness of the lamp and its fittings.

Table 12.3
Desirable illumination levels

Background lighting, relaxation	150 lux
Non-detailed tasks	200–400 lux
Detailed tasks – reading, writing	400–600 lux
Very detailed tasks – drawing, sewing	700–1000 lux

The illuminance from a light source can be calculated, although the calculation is complex (Durrant 1977). A simple method of calculating background illuminance from a diffused source is to calculate the average illuminance over the horizontal surface of a room. As an example, if a room of area 10 m^2 has 300 W of tungsten lighting (that is, three 100 W bulbs) the lumen output may be calculated as follows.

$$\text{luminous flux (output)} = \text{watts} \times \text{efficacy (lumens/watt)}$$
$$= 300 \times 12 \text{ lumens}$$
$$= 3600 \text{ lumens.}$$

The quantity of light directed down on to the horizontal surface is known as the *utilization factor* (UF) and is a function of the type of luminaire and its positioning. In this example it is assumed that the utilization factor is 0.4. This means that 40% of the light illuminates the horizontal surface. Charts are available giving utilization factors for luminaires and for values of ceiling and wall reflectance (Durrant 1977). Using this utilization factor, with the three 100 W bulbs, allows the calculation of the light reaching the horizontal surface, that is,

$$= 0.4 \times 3600 \text{ lumens}$$
$$= 1440 \text{ lumens.}$$

This is spread over an area of 10 m^2, therefore the illuminance is:

$$\frac{1440 \text{ lumens}}{10 \text{ m}^2}$$

$$= \textit{144 lux.}$$

The eye is adaptable to a wide range of light levels, but discomfort is caused if both high and low levels are present in a single field of view, or if the light level changes suddenly from low to high or vice versa. Within a single field of view the contrast should not be greater than 10:1 between the centre and peripheral areas of view. In the centre of the field of vision the contrast should not be greater than 3:1. Glare is a special problem and a common cause is where an occupant of a room can, from some point in the room see directly into the light source. Here there is a need for shades and/or diffusers. Glare can also be caused by light reflecting from a white surface, such as the paper on a writing desk.

The colour temperature of light sources can cause problems, particularly where there is a combination such as daylight and incandescent light or fluorescent light and incandescent light. Fluorescent tubes can be selected to blend with either daylight or incandescent light, but not both. Tungsten lamps and fluorescent tubes can change the colour of fabrics and cosmetics and this may be an important factor in the selection of lighting for dressing tables and mirrors. Lighting levels depend upon the type of luminaire, the size of room, the siting of the luminaire and reflectance of furnishings. These can be calculated as in the example given above.

Some factors involved in the design of lighting for specific areas of the home are discussed below.

Living room: this area must have flexible lighting to suit the wide range of activities taking place. The illumination should be provided by several small fittings with independent switching, rather than by a single central pendant. Background lighting of 50–150 lux should be supplemented by portable direct light sources for detailed tasks (reading lamps, table lamps and floor lamps). Lighting can be used to highlight pictures, ornaments and furnishings using display and spotlights.

Dining room: flexibility is required to create different atmospheres from everyday meal times to an intimate eating environment for entertaining. Multiple direct sources of light provide an attractive sparkle to cut-glass and silver. This should be supplemented by a low level of background lighting. Pendants are available which can easily be adjusted in height to modify the lighting effect. Dimmers are useful for balancing the levels of direct and background lighting.

Kitchen: a level of illumination of 200–500 lux should be provided for all working surfaces, including the sink and cooking range. This lighting should be arranged to prevent shadows, particularly those of a person standing in front of a work surface. Fluorescent tubes provide a diffused light source and this should be chosen to match daylight, since a combination of daylight and artificial light is often required in a kitchen. Fluorescent lights can be mounted beneath wall cupboards to provide illumination for work surfaces. If this is done, a baffle is required in front of the tube to prevent glare (Fig. 12.20). Alternatively, tubes may be mounted at ceiling level, directly over the work surface.

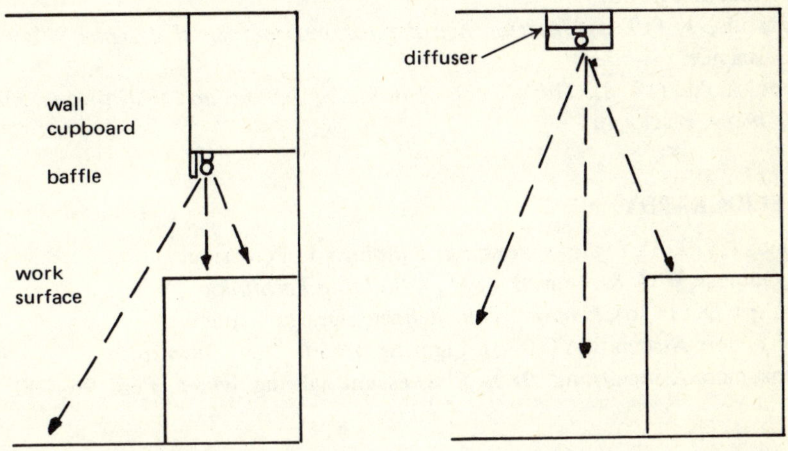

(a) fluorescent tube beneath wall cupboard. (b) flurescent tube on ceiling.

Fig. 12.20 – Lighting arrangements in kitchens.

Bedrooms: a background light level of 50–150 lux with specific lighting for dressing-table, full length mirror and bedside lighting are required. Lighting for mirrors is difficult since it must light the person looking into the mirror, but not the mirror itself. It must not cause glare. The most satisfactory arrangement is to have two fluorescent tubes with diffusers, one on each side of the mirror. Light for reading in bed can be provided by wall lights or bedside table lights. Bedrooms are often used for general activities (playroom, nursery, etc.) and the lighting should allow for these, as appropriate.

Bathrooms: the requirements are for the background lighting of 50 to 150 lux with specific lighting for the bathroom mirror. As mentioned previously, this is particularly difficult. Fluorescent tubes are available with built-in electric-shaver sockets for mounting below or above mirrors.

Halls and staircases: background lighting of 100–200 lux is required. Care is needed to avoid heavy shadow areas which can contribute to accidents. This uniform illumination is best provided by a number if small luminaires.

REFERENCES

Consumers Association (1976), Light dimmers, *Which,* **Feb.**, p. 26–29.

Durrant, D. W. (1977), *Interior lighting design,* 5th Ed., Electricity Council.

Grandjean, J. (1973), *Ergonomics of the home,* Taylor & Francis.

Milson, A. & Kirk, D. (1980), *The design and operation of catering equipment,* Ellis Horwood.

Nuckolls, J. (1976), *Interior lighting for environmental designs,* Wiley Interscience.

Rees, A. M. (1970), *The science of home economics and institutional management,* Blackwell.

BIBLIOGRAPHY

Bond, J. (1973), *Lighting design in buildings,* P. Peregrinus.

McGuiness, W. J. & Stein, B. (1977), *Building Technology,* Wiley.

Phillips, D. (1976), *Planning your lighting,* Design Council.

Consumers Association (1976), Lighting, *Which,* **Nov.**, 98–106.

Consumers Association (1980), Fluorescent lighting, *Which,* **Feb.**, 96–100.

Part III

KITCHEN EQUIPMENT

CHAPTER 13

Cookers and ranges

13.1 INTRODUCTION

When foods are cooked, they undergo changes induced by the action of heat. These changes may be chemical, physical, biochemical or microbiological. Some examples are given in Table 13.1. In addition to the above changes, cooking may also affect the nutritional quality of food. In some cases cooking makes the food easier to digest, but in other cases nutrients may be damaged by the action of heat or by leaching into the cooking fluid.

Table 13.1
Examples of changes occurring during cooking

Change	Example
Chemical	
Hydrolysis	Breakdown of pectin in fruit and vegetables
Denaturation of proteins	Setting of egg protein and gluten (in bread)
	Shrinking of meat fibres
	Destruction of enzymes and micro-organisms
Oxidation	Destruction of vitamin C
Non-enzymic browning	Browning of the crust of bread and cakes
	Development of colour and flavour in roast meats
Physical	
Phase change: dehydration	Baking biscuits
Phase change: melting	Loss of fat from cooked meats and sausages
	Thawing of frozen foods
Migration (diffusion)	Transfer of volatile oils from bay leaf on surface of pate into the centre

All of these changes are effected by both time and temperature. The rate of chemical reaction is accelerated by an increase in temperature. When foods with a high moisture content are cooked, they will not rise much above 100°C, the boiling point of water (except in the case of pressure cooking). However if the moisture content is low, or if moisture is lost by dehydration, then the temperature of the food can rise above 100°C. Because of this, reactions such as *non-enzymic browning,* which occur at high temperatures, take place at the surface of the food, where moisture has been lost by dehydration. However, the volatile flavours produced as part of the non-enzymic browning reaction can then migrate to the centre of the food. In this way the flavour of freshly-baked bread or roast meat travels from the surface of the food, where the flavour is developed, right through the food. However these volatile materials are just as easily lost and because of this bread soon looses its fresh 'crusty' flavour.

Dehydration is affected by both temperature and by the Relative Humidity (RH) of the cooking environment (section 11.3.1). If the moisture content of the environment is less than that of the surface of the food, moisture will be lost from the food (Milson & Kirk 1980). In some cases, dehydration plays an important part in the cooking process, such as in the baking of biscuits and pastry. Here the RH of the cooking environment must be kept low to encourage dehydration. However, moisture loss in roasting can lead to excessive weight loss and shrinkage and here the cooking environment should have a high RH. The atmosphere surrounding the meat can be given a high RH by covering the food with foil or with a plastic cooking film. If food is cooked in a moist atmosphere, the dehydration rate is slowed down and the surface temperature does not rise to such a high temperature. Because of this there will be less non-enzymic browning at the surface of foods cooked in a moist atmosphere compared with the same food cooked in a dry atmosphere.

Phase changes are also related to temperature. If ice is heated, its temperature rises until it reaches 0°C, the melting point of ice (Fig. 13.1). The temperature of the ice/water mix will remain at 0°C until all of the ice is melted, after which the temperature will again rise. When the water reaches 100°C, it boils and water is converted to steam. The temperature of the water does not rise above 100°C (at atmospheric pressure). The relationship between heat and temperature is more complex in the case of food materials, as shown in Fig. 13.1.

Biochemical and microbiological changes are also affected by temperature. As the temperature rises a biochemical (enzymic) or a microbiological process will take place more rapidly, up to a critical point, known as the optimum temperature. Beyond this point the process will slow down because the protein of the enzyme or micro-organisms is denatured by the heat and the enzyme is inactivated or the micro-organism destroyed (see Fig. 13.2). Each micro-organism and enzyme has its own characteristic optimum temperature.

To provide the heat required to produce these changes in the food, there

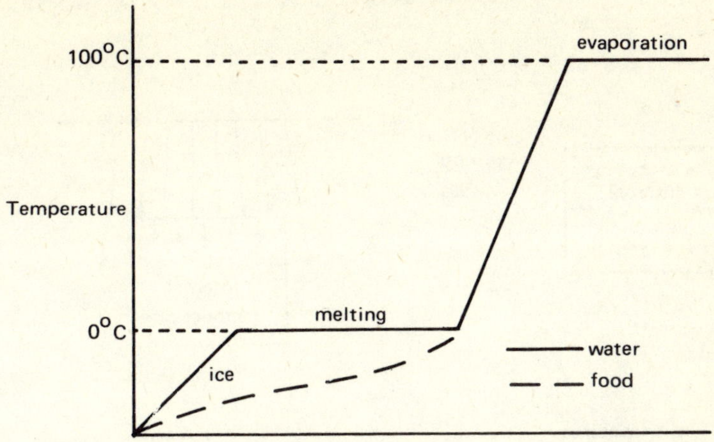

Fig. 13.1 – The relationship between phase change and temperature.

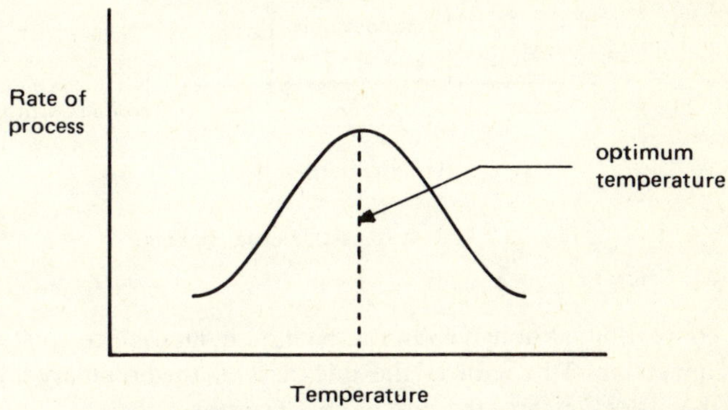

Fig. 13.2 – The effect of heat on biochemical and microbiological processes.

must be a transfer of this heat from the heat source (electrical heating element, flame etc.) to the food. Heat transfer (except when using microwave heating) to food takes place in two stages:

(1) absorption of heat by the surface of the food
(2) conduction of heat from the surface to the interior of the food.

With the exception of microwave heating (section 13.2.2.8) all of the heat used in the cooking process must enter through the surface of the food. This can be by convection, radiation or conduction (Fig. 13.3).

Convection (section 7.2.2): the fluid may be water (boiling), oil (deep fat frying), air (oven) or combustion products (gas oven). The combustion products from coal, smokeless fuels, wood and liquid fuels cannot be used directly because they would produce taints in the food. Heat transfer is usually assisted

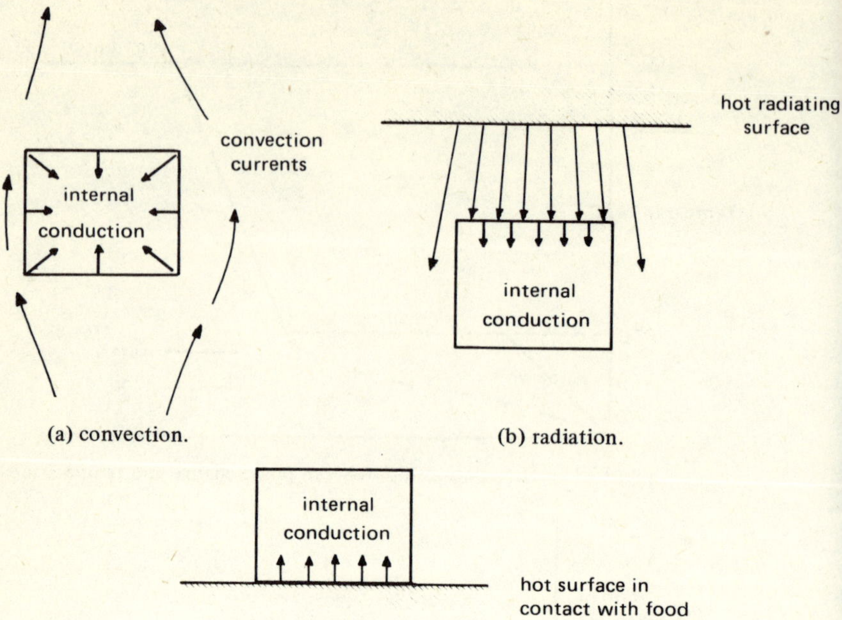

(a) convection. (b) radiation.

(c) conduction.

Fig. 13.3 – Heat transfer during cooking.

by natural convection, although a fan can be used in an oven to create conditions of forced convection. This reduces the thickness of the boundary layer (section 8.3.1) and therefore increases the rate of heat transfer.

Radiation (sections 7.2.2 and 12.1.1): this may be from a hot surface emitting radiant heat or from the air or gas inside an oven. The energy is absorbed by most foodstuffs within a fraction of a millimetre of the surface, and it is converted into heat. This absorption results from atomic vibration produced in certain types of molecules, particularly those containing O–H and C–H bonds.

Conduction (sections 7.2.2 and 8.3.1): foods may be heated by conduction during shallow frying and griddling. In an oven, heat will be transferred by conduction through the base of the food container.

Once the heat has been absorbed by the food, it is transferred from the surface to the interior by conduction. Since foods have a poor thermal conductivity (section 7.2.2), this process is relatively slow, particularly where the conduction must take place over a long distance, as in a large roast of meat. An important aspect of cooking is to achieve the correct balance between the rate of heat transfer at the surface and the rate of conduction to the centre. At low rates of heat transfer to the surface, heat can be conducted to the centre at the same rate at which it arrives at the surface, but since this is low, the process takes a long time. If the rate of heat transfer to the surface is increased, a balance can be achieved where the centre of the food is just cooked at the same point in time when the surface of the food has achieved the correct degree of dehydration and browning. However, if the rate of heat transfer to the surface is increased beyond this point, by the time the centre of the food is cooked the surface will be overcooked or even burned. For this reason, methods of cooking which utilise intense surface heating are normally carried out with thin slices or small pieces of food, as in grilling, frying and toasting.

In some cooking processes, such as baking, the balance of heat transfer to the surface and internal conduction of heat is critical. If the surface of bread or cake dries and sets too quickly before the interior has finished rising, the pressure from the expanding dough or sponge will crack this dried surface layer. For a fuller discussion of the physical processes involved in heat and mass transfer during cooking, see Milson & Kirk (1980).

13.2 COOKING APPLIANCES

13.2.1 The layout of cooking appliances

Modern cooking appliances have developed from the traditional range, itself a development from cooking over an open fire. (There is a difference in the use of the word 'range' between the UK and USA. In the UK a 'range' is a combination of cooking units, surrounding a single heat source, as shown in Fig. 13.4, which is usually constructed from cast iron and which, in addition to cooking, often provides hot water for domestic use or central heating. In the USA 'range' is the generic term for a cooking appliance, which in the UK is referred to as a 'cooker'). These cooking appliances may be classified into one of three types.

(1) *Oven:* an enclosed metal-walled box, with shelves for the food and providing heat by a combination of convection, conduction and radiation. The proportion of convected and radiated heat varies from one type of oven to another (Dunning 1953).
(2) *Surface heating unit/boiling plate:* a heated horizontal surface used to heat pans and other utensils through their base by conduction.
(3) *Grill/boiler:* a source of intense radiant heat which traditionally would have been from glowing coals, but which is now produced by a heating element placed directly over the food.

As the use of electricity and gas for cooking were developed, it was unnecessary to use a single source of heat and the above three cooking methods could be separated to give the type of cooker (range) shown in Fig. 13.5, which

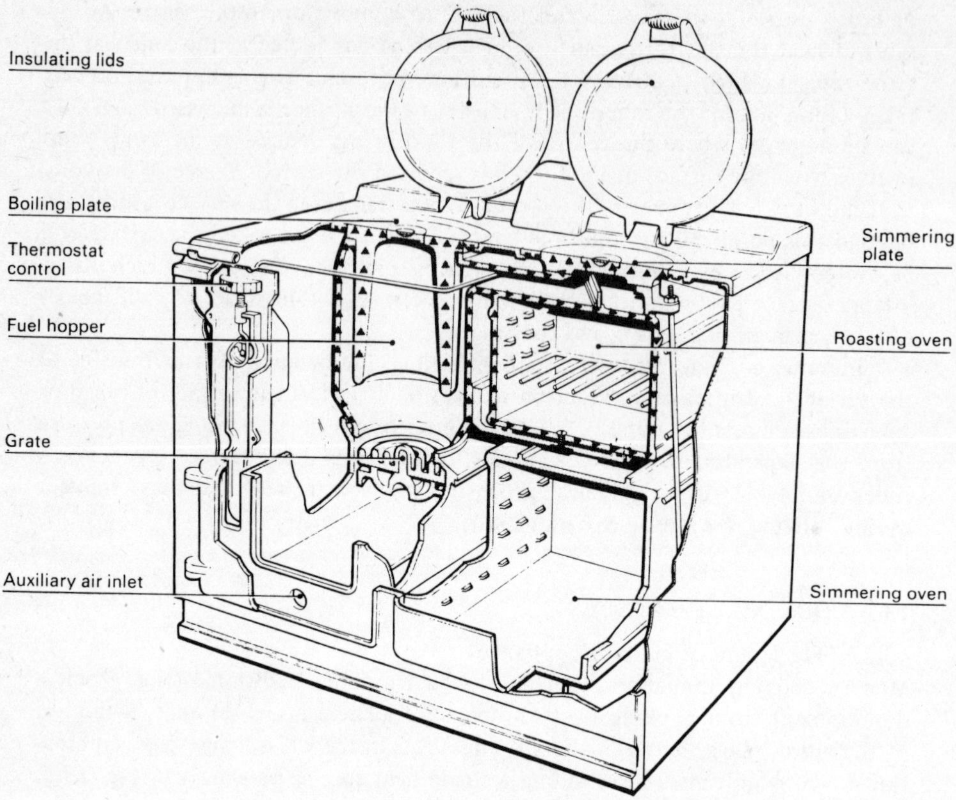

Fig. 13. 4 – The layout of a traditional solid fuel range.

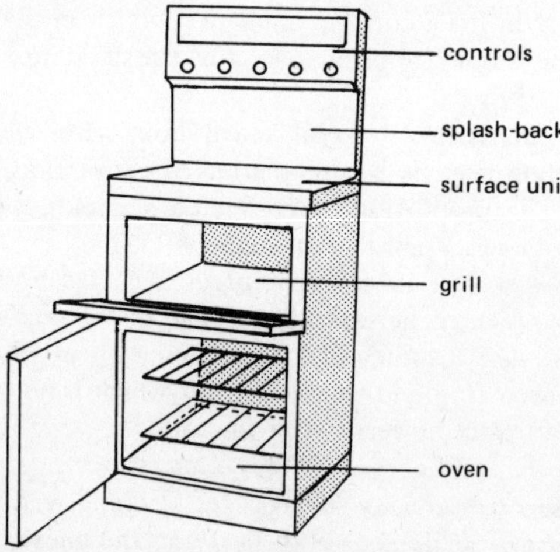

Fig. 13.5 – Construction of a basic cooker.

illustrates a free-standing vertical cooker. Many alternatives to this basic structure are now available:

(1) the grill/boiler and oven can be combined in a single unit
(2) the grill can be placed at eye level
(3) two ovens, one large, one small, may be combined in one cooker
(4) the cooker can be built into kitchen units (drop in or slide in)
(5) the oven, surface units and grill can be provided as separate units (Fig. 13.6), since they do not rely on a single heat source
(6) work top cookers and ovens are available for small houses and flats.

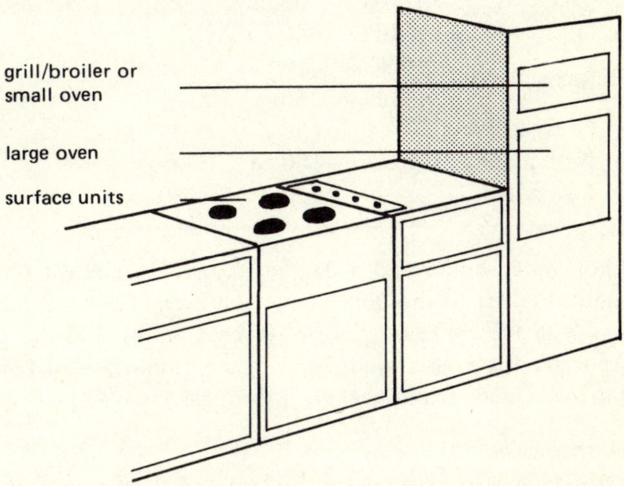

Fig. 13.6 – Split level built in cooker units.

The choice of fuel for use in cookers varies from country to country, as shown in Table 13.2.

13.2.2 Ovens

13.2.2.1 *Solid-fuel ovens*

The traditional solid-fuel range is constructed of cast iron, with heat transfer from the fire to the oven and surface units both by convection from combustion products along channels and by conduction along thick cast iron plates (Fig. 13.4). Normally the range is used to provide domestic hot water in addition to heat for cooking. A high standard of insulation is required both to improve fuel efficiency and from a safety view point. Ranges of this type are now available for solid fuel (including wood), oil and gas. There are normally two ovens, one operating at a high temperature for roasting and baking and the other at a lower temperature for slow cooking and warming. Separate flue passages are used for

Table 13.2
Fuels used in European cookers
(Hodgkinson *et al.* 1980)

		Gas %	Electricity %
Germany		70	30
Austria		58	42
Belgium	NG	43	23
	LPG	34	
France	NG	33	27
	LPG	40	
Spain	NG	9	5
	LPG	86	
Great Britain		56	44
Netherlands		100	–
Switzerland		20	80

the oven and hot water boiler and a damper is used to control the flow of combustion products either to the boiler or to the oven. The total heat output is between 15 and 25 kW. A large quantity of heat is stored in the thick iron parts of the oven and these also produce a higher proportion of radiant heat compared with an oven made from sheet steel, as in gas and electric ovens.

13.2.2.2 *The electric oven*

A modern electric oven is constructed out of pressed sheet steel fastened to a metal framework. It has two walls, the inner wall forming the oven compartment is finished with vitreous enamel, stainless steel, PTFE (non-stick plastic) or a special easy-clean finish (see below). The oven walls have liners which, in most cases, may be removed for easy cleaning and the side walls have built in shelf supports. A vent is provided in the oven compartment which allows the release of vapours from the oven space and induces convection currents. In many ovens an electric lamp is provided. This lamp, and its electrical fittings and wiring must be able to withstand high temperatures. The oven compartment is surrounded by a layer of thermal insulation, such as glass fibre covered in aluminium foil.

The outer wall of the oven is made either of stainless steel or steel coated with vitreous enamel. White enamel helps to contain heat within the oven.

A door to the oven compartment can be made, as is the rest of the oven, out of two layers of metal separated by a thick layer of insulation or it may have a double-glazed panel to allow the food to be inspected without opening the door. The door may have side or bottom hinges and the fitting between the door and the cabinet is provided with a gasket, to reduce heat loss. If the door is not glazed the oven may be fitted with a glass inner door.

The heating elements are in the base or at the sides of the oven compartment (section 5.7.1), and provide 1 to 2.5 kW of heat. The location and arrangement of the elements varies from manufacturer to manufacturer. The temperature in the oven compartment is controlled by a thermostat – usually of the vapour pressure type (section 9.4.1) with the sensing element mounted inside the oven compartment, near to the top of the oven. In some ovens a special oven thermometer is available for measuring or controlling the temperature at the centre of a joint of meat. A metal probe which contains a heat sensitive resistance can be inserted into the meat and the temperature at the tip of the probe is conveyed to a dial on the control panel of the oven. This can also operate a switch to turn off the oven and/or sound an audible buzzer when a pre-set centre temperature is achieved.

13.2.2.3 *Gas Ovens*

Gas ovens are of similar construction to electric ovens (section 13.2.2.2). A gas oven may be heated directly with the gas burners inside the oven compartment, as shown in Fig. 13.7a. This produces a temperature gradient inside the oven

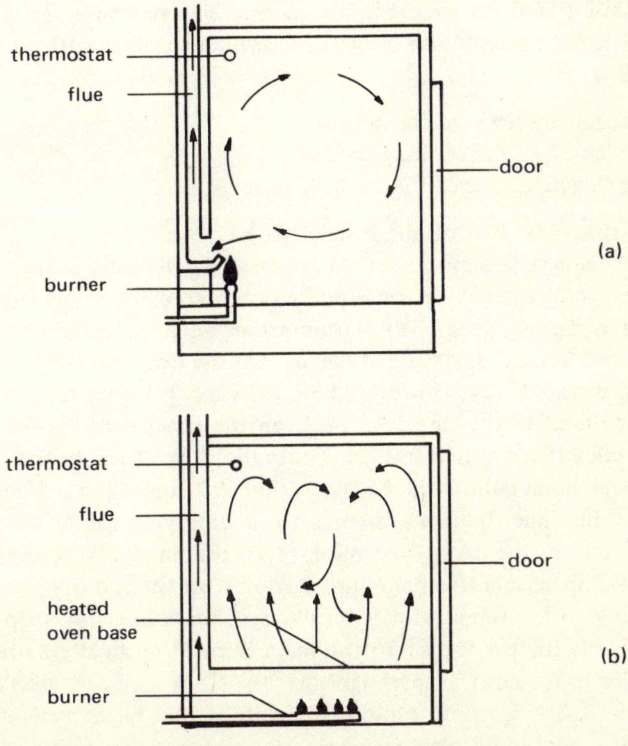

Fig. 13.7 – Directly and indirectly heated ovens.

with higher temperatures at the top of the oven and lower temperatures at the bottom. This temperature difference is about 40°C. This method is used for most ovens manufactured in the UK. Alternatively the oven may be indirectly heated, as shown in Fig. 13.7b. The burners heat the base plate of the oven and this produces a uniform temperature distribution from the top to the bottom of the oven. Indirectly heated ovens vary from a minimal temperature gradient to 20°C higher temperature at the bottom (Hodgkinson *et al.* 1980). Ovens from most European countries and the USA are heated in this way. Gas ovens require a flue to allow removal of combustion products (Taylor 1973) and a thermostat to regulate the temperature of the oven.

Modern gas ovens have spark ignition, rather than a permanent pilot (section 2.8) and in addition they may have a flame-failure device (section 2.9). The burner rating of gas ovens varies from 1.5 to 3 kW.

13.2.2.4 *Controls for Gas and Electric Ovens*

Controls for an oven consist of a combined on/off thermostat control, a manual/automatic selector and an automatic timer. In addition to these a gas oven will have ignition controls.

Automatic timers are available for gas and electric ovens. They consist of a clock (electric or, on some gas ovens, clockwork) together with an arrangement for controlling:

(a) the starting time for the cooker
(b) the length of the cooking process
(c) the finishing time of the cooking process.

In practice only two of these need to be set since the third will then be fixed. The timeswitch may be set via rotary central knobs connected by gears to the clock and switches or, on some modern cookers, using solid state electronics and a digital clock. This latter arrangement is more reliable since it involves no mechanical parts and it means that the oven can be 'programmed' in a more sophisticated way. However the cost is much higher than for a conventional clock timer. In the case of a gas oven the timer controls the flow of gas to a timer pilot flame which itself controls the flow of gas to the main burner by the use of bi-metallic strip FFD (section 2.9, Fig. 2.15). When the timer reaches the 'on' time, it opens the gas valve supplying gas to the timer pilot, which is ignited by the main oven pilot. Once the bimetallic strip is heated the main gas valve opens and the main burner ignites. At the end of the cooking time, the gas supply to the timer pilot is terminated, the bi-metallic strip cools down and switches off the gas supply to the main burner. As an alternative to this, in an oven which uses electrical spark ignition, the clock may activate a gas solenoid valve and the spark ignition. Failure to ignite would be detected by a flame-failure device. On some electric ovens the conventional controls have been replaced by touch-sensitive switches. The touch of a finger on one of these switches is detected by solid-state electronics which are also used to connect the electrical current to the appropriate heating element.

13.2.2.5 *Easy-clean ovens*
Three methods are used to simplify the cleaning of ovens and these are:

(1) non-stick coating
(2) catalytic coating
(3) pyrolytic cleaning.

A *non-stick coating* is made of the plastic polytetra-fluoro-ethylene (PTFE) which has non-stick properties and can withstand temperatures up to 290°C without melting. Because it is non-stick, soil can be removed easily from the oven by wiping with a cloth. There are, however, problems with the durability of PTFE, since it can easily be scratched and chipped.

In *continuous cleaning ovens,* the oven lining is coated with a special type of vitreous enamel which contains a catalyst (Denny *et al.* 1974). At temperatures above 200°C (that is, whenever the oven is used for cooking at a high temperature) the catalyst causes oxidation of food particles from the surface of the linings, so that they are continuously removed. However some staining of the linings does occur.

The *pyrolytic* cleaning method uses much higher temperatures (up to 500°C) to oxidise the food particles from the oven. To achieve these high temperatures a special thermostat is required and a higher standard of insulation of the oven must be used. An additional safety factor (an interlock) is required on the door to prevent it being opened while the oven is at the high temperature used for cleaning. The food particles are burnt off, leaving fine ash which can be removed by wiping with a damp cloth. The fumes produced during the pyrolysis are de-odourised by a filter unit mounted in the roof of the oven.

13.2.2.6 *Fan-assisted ovens*
If a fan is mounted in the back of the oven (Fig. 13.8) it:

(1) improves temperature uniformity in the oven;
(2) reduces cooking time, particularly with frozen foods (Milson & Kirk 1980).

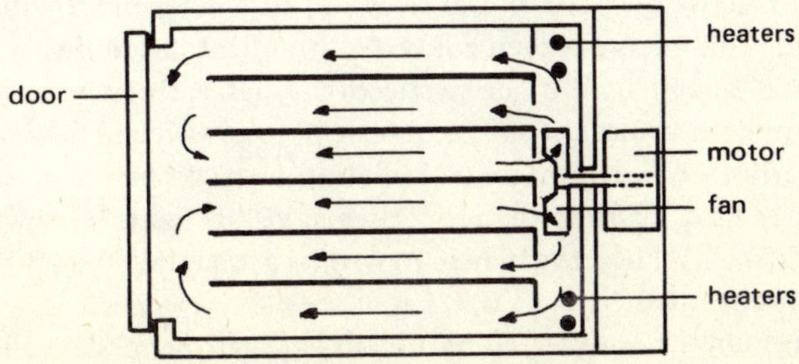

Fig. 13.8 – A fan assisted oven.

Fans have been used in catering ovens for many years but these are very noisy because they use high-velocity air-currents. For domestic ovens, a much fan speed is used. In a fan oven the whole of the oven can be used to cook large batches of the same food and it is therefore suited to the needs of the freezer owner. Fan ovens are available for electricity and indirect gas ovens. Cooking times are reduced using these ovens because the forced convection currents reduce the thickness of the boundary layer surrounding the food (section 8.3.1). This increases the rate of absorption of heat at the surface of the food. It is often necessary to use a lower thermostat setting with a fan oven.

13.2.2.7 *Oven-grill combinations*

Grill (broiler) electrical elements and oven elements can be combined in the same oven compartment. Grill elements (sectio 13.2.4) are sited at the top of the oven compartment and these may be operated independently of the oven elements or, in some designs of oven, the two may be used together.

Where the two sets of heaters can be used in combination, a selector switch is used to control the supply of heat to the top and bottom heaters.

A typical oven-broiler used in the USA will have the possibility of the following combinations:

(1) bake alone (controlled by thermostat)
(2) broil alone (controlled by infinite switch or high/low switch)
(3) bake and broil simultaneously (broil elements on for 20% of the time, with both controlled by a thermostat)
(4) preheat (both elements with broil on all of the time).

In practice there are many variations in the arrangements of controls provided in oven-grill combinations. In many ovens of this type the two sources of heat cannot be used in combination and, in this case, separate oven and grill controls are provided, with a two-way selector switch. During grilling the oven compartment is operated with the oven door open.

13.2.2.8 *Microwave Ovens*

Unlike other forms of heat used in cooking (section 13.1) microwave energy enters the surface of the food not as heat but as a very high frequency electromagnetic field. This energy is then converted into heat inside the food.

Microwave energy used in domestic ovens has a frequency of 2450 MHz (1 MHz ≡ 1 million cycles/second), equivalent to a wavelength of 122 mm. The general properties of electromagnetic radiation were discussed in section 12.1.1 and microwaves have many of the properties of visible light. Microwave radiation travels in a straight line, and when it strikes a material it may be reflected, transmitted or absorbed.

Microwave energy is reflected by metals, so that the metal walls of the oven can be used to contain the microwaves inside the oven. Because of the physical size of microwaves, compared with lightwaves, a perforated metal screen may be used for the oven door. This allows direct visibility of the contents of the oven, without the possibility of any emission of microwaves from the oven. A glass screen is used in combination with the metal screen to prevent small metallic objects being poked through these small holes.

Microwaves are absorbed by polar molecules, that is molecules which do not have an even distribution of electrical charge. A common polar molecule, found in most food materials, is water. As shown in Fig. 13.9, the electrons are not evenly distributed throughout the molecule, but are drawn towards the oxygen atom. This results in the oxygen having a slight negative charge and the hydrogen atoms a slight positive charge.

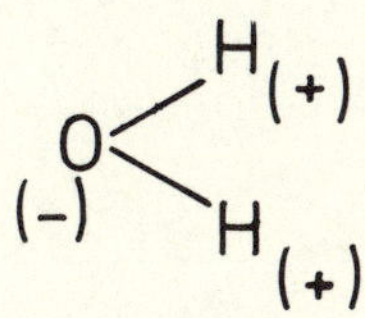

Fig. 13. 9 – The water molecule.

Microwave energy when absorbed by a food increases the energy in individual atoms of the food. In turn this higher energy level also varies the bondings or binding of the atoms to one another to form molecules (ionic strength). A molecule consists of two or more atoms bonded together. When the atoms carry unequal electronic charges the molecule is said to be **polar** (i.e. carries a dipole moment).

Higher energy levels in a molecule produce stretching or bending of the bonds between atoms and this causes vibration in a changing electromagnetic field. This vibrational energy is converted to heat. However in the case of a *free* polar molecule such as water (Fig. 13.9) in an electromagnetic field absorption of energy produces rotation of the molecule. Thus the positively charged hydrogen atoms are aligned to face the negative direction of the field and the negatively charged oxygen atoms are aligned to face the positive direction of the field. When the direction of the negative and positive poles of the magnetic field are constantly reversed then the free polar molecules reverse direction at the same rate at which the field reverses direction, that is, the molecule rotates (Fig. 13.10). In the case of domestic microwave ovens the direction of the field changes 2450 million times a second. This microwave energy absorbed by the rotating *free* polar molecules is converted to heat.

Foods which contain free liquid water absorb energy at a greater rate and thus become heated more quickly than solid or viscous foods. There are many restraining influences to vibrational and rotational activities within foods. Fat is a poor absorber of microwave energy in comparison to water so that energy is

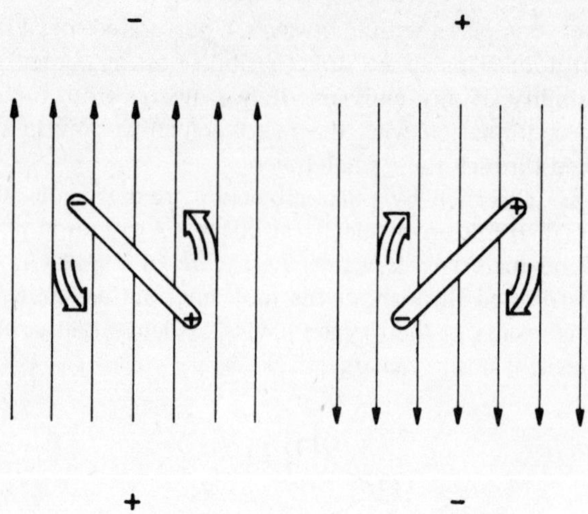

Fig. 13.10 – Movement of a polar molecule in a changing electrical field.

absorbed at different rates by various components of food and the heating is not uniform within the food. The conduction of heat from one component to another over a period of time tends to produce the required uniform temperatures within the food.

The mechanism of microwave heating is discussed in more detail by Milson & Kirk (1980).

Normally no surface browning occurs with microwave heating. This is because (1) the food is heated in a cool oven and therefore the surface of the food rapidly looses heat by convection and radiation and (2) the intense heating below the surface of the food produces a moist cooking environment. These two effects mean that the surface of the food does not rise significantly above 100°C and browning reactions occur only to a limited extent.

Where browning is required in order to produce acceptably-cooked foods, a browning dish can be used. This is a special ceramic plate or dish which will absorb energy from the microwave field. The ceramic dish is pre-heated in the microwave oven and the ceramic attains high temperatures so that when food is added browning takes place by radiation and conduction from the ceramic dish.

Alternatively microwave ovens are available with additional radiant heating elements to produce the surface browning of food where this is a desirable quality.

These ovens may be free standing work-top models or built-in units, but they may also be combined along with conventional gas or electrically heated ovens in a complete floor standing range.

Electrically heated conventional or fan assisted ovens may also have a microwave energy input so that food may be subjected to external heating from a conventional thermal source and internal heating from the microwave source.

Molecules may be restrained from movement by physical forces. This occurs with 'bound' water and with water molecules in the frozen state. In frozen foods there will always be a portion of the water which is unfrozen, and therefore free to rotate, so that when frozen foods are placed in a microwave field, some heating occurs. This heating takes place where the microwave field is at its strongest, which may be near corners or where the food is at its thinnest, resulting in localised thawing. Any thawed food will then absorb more energy from the microwave field than the frozen regions and the temperature in these areas will rapidly rise to boiling, leaving other parts of the food frozen. To simplify the thawing and re-heating of frozen foods it is necessary to allow conductive heat transfer to even out these non-uniform temperatures. This can be done manually by allowing a rest period of 10 to 15 minutes half-way through the heating period. Alternatively many ovens now provide a defrost facility which pulses the microwave power so that it is only on for 40–50% of the time the food is in the oven.

Any material which neither reflects nor absorbs microwave energy must be transparent and this type of material is ideal for use as a cooking container. In practice few materials are totally transparent to microwaves but the proportion of energy absorbed by the container is small and most heating of the container occurs by conduction from the hot food. A wide range of glass, ceramic and plastic materials are available (Colato 1978). Special packages are being developed by the frozen food industry which contain metal foil to shield parts of the food (which would otherwise overheat) from the microwave field.

Microwave ovens may be countertop models, built in units or, as already discussed, combined with conventional ranges. They are all of similar construction, as shown in Fig. 13.11.

The microwave energy is produced by a *magnetron,* which converts high voltage electricity (4 kV) into electromagnetic radiation of the required frequency. This radiation is transferred from the magnetron into the *oven cavity* in a *waveguide,* which is a cylindrical or rectangular metal tube. One of the problems with early designs of microwave ovens was the presence of localised 'hot' and 'cold' regions within the oven, that is areas with a greater or lesser heating effect than the rest of the oven (Ohlsson 1981). A number of methods have been devised to overcome this, but it can still cause problems. One of the most common methods, used in most ovens, is the use of a *mode stirrer.* This is a slowly rotating metal-bladed fan which changes the pattern of reflections within the oven. An alternative, though less common method, is to use a *rotating turntable* on which the food is cooked. Here the principle is that cooking uniformity is

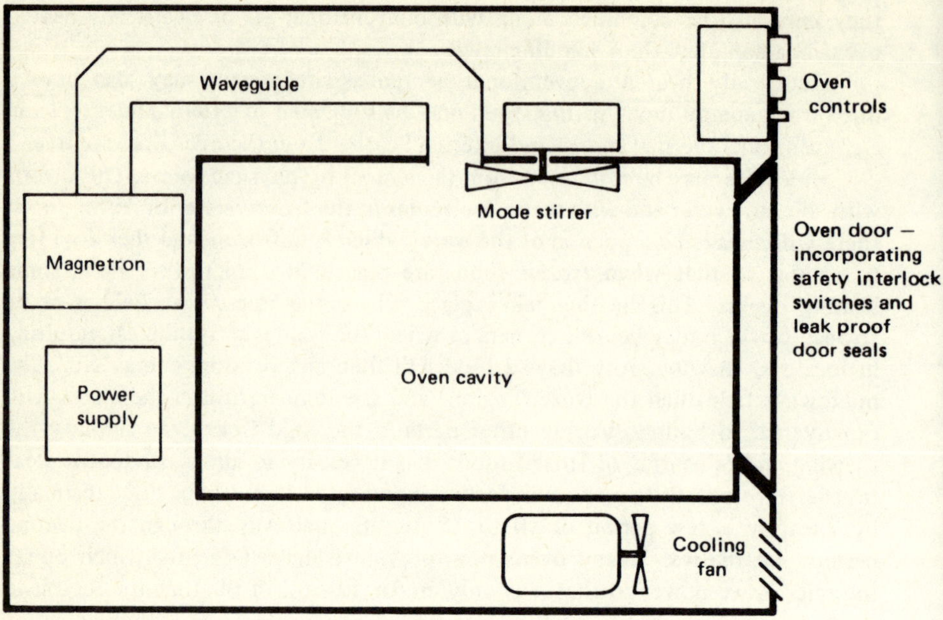

Fig. 13.11 – General construction of a microwave oven.

achieved because the food moves through both hot and cold regions. Turntables are built into some ovens, but they are also available as an accessory for most ovens.

An important part of the microwave oven is the door and particularly the design of the door seal which is the most likely source of leakage of microwave energy from the oven. The design of door seals has improved from early designs of ovens. In addition to the components described above microwave ovens usually have an oven light and an extractor fan for removing moisture and heat from the oven cavity.

Microwave ovens vary considerably in the sophistication of their controls and this is one of the major influences on the price of an oven (Rennekamp 1978). All ovens have an off/on/cook control and a timer, which may be clockwork or electrical. On many ovens it is now possible to vary the power level. This may, at its simplest, be a defrost/full-power switch or it may be a continuously variable control which allows from about 20% power, at its lowest setting, up to full power. These controls (including the simple defrost control) usually operate in the same way as an energy regulator (Figs. 13.15 and 13.16) switching the power on and off, with the control adjusting the ratio of 'on' to 'off' time.

The microwave oven timer at its simplest is a mechanical or electrical clock which is set to the required cooking time. At the end of this period power is switched off and an audible warning is provided. Some ovens now have an automatic timer which operates in the same way as a conventional oven timer (section 13.2.2.4) and they may incorporate a digital clock. Touch sensitive switches may be used in place of the rotary control.

As an alternative to the timer, the end of the cooking process may be detected automatically using either a temperature probe placed in the centre of the food or by measuring the humidity of the oven.

Microwave ovens may also be provided with programmers which, at their most sophisticated will provide instructions to the operator about quantities of food and provide reminders about when to add each component while, at the same time, controlling the sequence of time and power level. Various methods of programming have been used including a memory bank of popular and user-programmed recipes within the oven. An alternative is the use of recipe cards with written instructions and a magnetic strip which has coded instructions for the oven.

One of the continuing areas of concern is that of microwave safety standards which stipulate maximum leakage levels from ovens (Voss 1981). New ovens should not leak more than 1 mW/cm^2 at any point 50 mm from the oven and when the oven is operated with a standard water load. This level should not rise to more than 5 mW/cm^2 in use. Users of ovens can do much to contribute in the area of safety by preventing the build up of food in the gap between the oven cavity and door. Also, since the inverse square law applies equally well to microwave leakage as it does to light (section 12.1.1) users should be discouraged from putting their eye close to the door grille to observe the cooking process. Leakage levels should be checked as part of regular servicing. There is much research effort worldwide into the possible hazardous effects of microwave radiation and evidence from this will be incorporated into International Safety Standards.

13.2.3 Surface Units

Surface units represent a development of the solid cast-iron plate placed directly over a fire and used to provide heat to pans placed on them. A variation of this is the use of a griddle plate for the direct cooking of food.

Solid cast-iron surface units are still used on heat storage solid-fuel ranges (Fig. 13.4). They are covered with insulated hinged lids to reduce the heat loss when not in use. European gas and electric surface units are normally covered with a metal work top when not in use.

13.2.3.1 *Electric Surface Units*

Electric surface units use resistance heaters (section 5.7.1) to generate heat which is then transferred by conduction to the base of a pan. The resistance heater may be incorporated in a cast-iron solid plate or in a radiant ring. A solid surface unit has the heating elements set into a cast-iron plate which weighs

2–3 kg. (Fig. 13.12). These plates are usually circular (in some cases rectangular) and it is common for a surface unit to have a combination of plates of different diameter. For example, there may be two plates of 145 mm diameter (1 to 1.5 kW) and two plates of 180 mm diameter (1.5 to 2.5 kW). High-speed solid plates are used as a compromise between the good heat storage properties of solid plates and the rapid heating properties of radiant rings. High speed solid plates have a special control which, for a period of time after the plate is first switched on, supplies the plate with its full current.

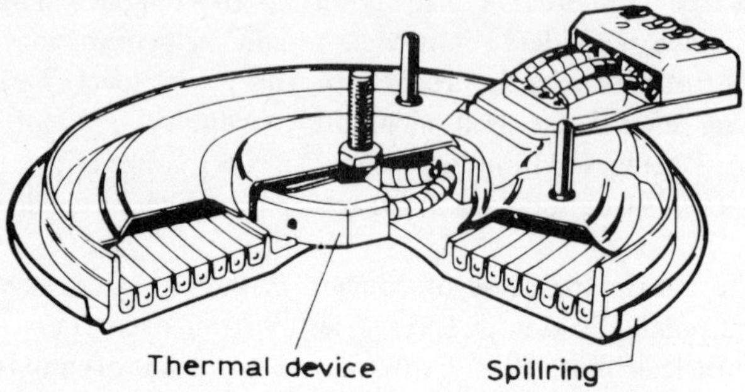

Fig. 13.12 – Solid surface unit (underneath view).

A radiant ring is similar to a sheathed wire element (section 5.7.1) which has been flattened and formed into a spiral. The flattened surface provides better contact for thermal conduction with the base of the pan. Elements are used with diameters from 150 to 200 mm and with an electrical loading from 1 to 2.5 kW. Dual element rings are also used. These have an independent inner coil 12 cm in diameter and an outer coil which is used only in combination with the inner coil (Fig. 13.13). The control for this ring allows for either the combined use of both rings or for the separate use of the inner rings, which is used to increase the efficiency when heating small-diameter pans.

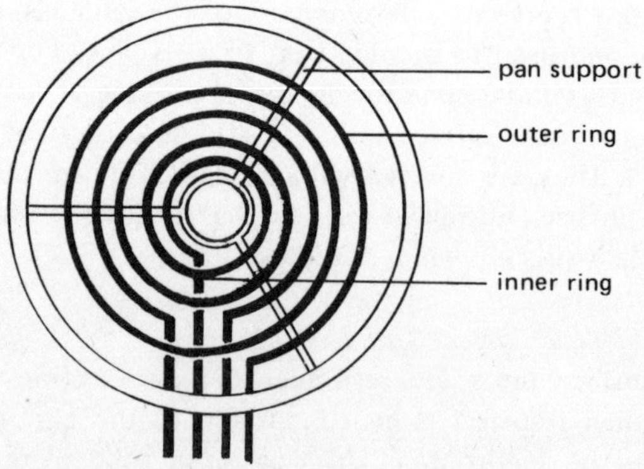

Fig. 13.13 – Dual-element radiant rings.

An alternative to solid and radiant rings is the ceramic surface unit. This is a sheet of heat resistance ceramic glass with electrical resistance elements (held in a fireclay bowl) fastened underneath and with a decorative pattern to indicate the position of these elements (Fig. 13.14). The ceramic plate is easier to clean than conventional surface units but it can be cracked if heavy pans are dropped on it and its surface can be scratched by abrasive cleaning materials.

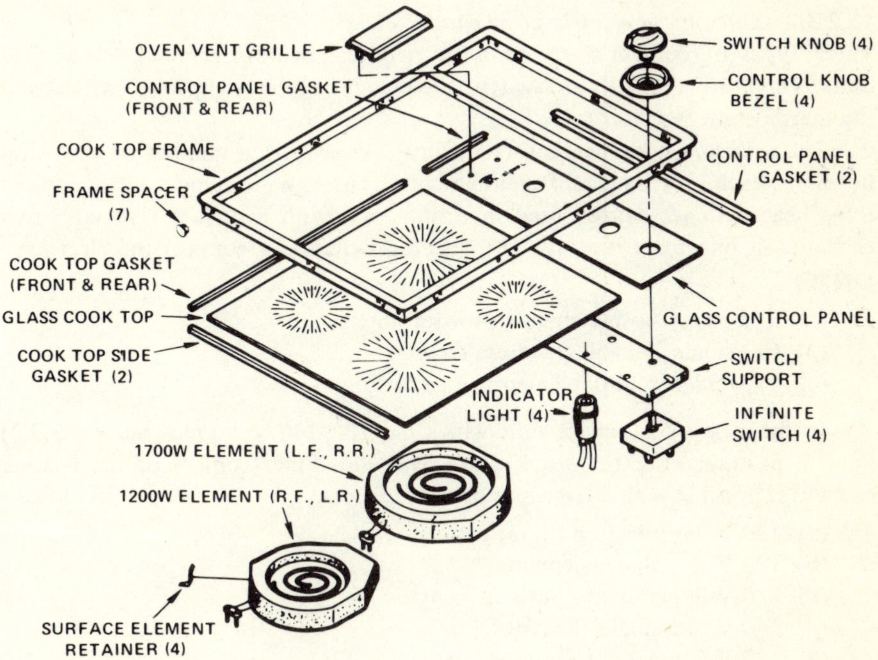

Fig. 13.14 – Ceramic surface units.

Magnetic-induction hobs have a similar appearance to ceramic surface units, but the principles governing their operation are different. Underneath the ceramic plate is an induction coil connected to an electronic circuit which produces an alternating current of about 250 kHz. This produces a high-frequency magnetic field, which passes unchanged through the ceramic plate. If a metal pan is placed on the ceramic plate the magnetic field is converted into heat. The heating effect results from the magnetic field inducing a flow of current in the metal of the pan, and this current leads to a heating effect as is the case when electricity flows through any conductor. The electrical conductivity of the pan is very important because, as with all resistance heating, power is given by:

$$W = I^2R$$

If the material has a very high conductivity, such as copper or aluminium, the heating effect is low and the maximum efficiency is achieved with ferromagnetic materials.

Magnetic induction hobs are relatively expensive, because of the electrical components required, and commercial viability will depend on the development of a large market to reduce the cost of components (as occurred with microwave ovens).

13.2.3.2 *Controls for electric surface units*

Three types of control are used with electrical heating elements on surface units. These are the selector switch, the energy regulator (infinite switch) and the thermostatic pan control.

The *selector switch* has a rotary control knob with a number of click-stop positions, each providing a different amount of heat. With a single 240 V supply, three heat switches can be used in conjunction with a surface unit with two elements in one plate or ring. The selector switch can connect the electricity supply:

(a) in series to both elements (low setting)
(b) to a single element (medium setting)
(c) in parallel to both elements (high setting).

In the case of a surface unit which has 120/240 V supply (section 5.3.2) 5 or 7 position selector switches are available. Finer control of the heating element is possible with a large number of positions:

(a) 120 V in series to both elements
(b) 120 V to a single element
(c) 120 V in parallel to both elements
(d) 240 V to a single element
(e) 240 V in parallel to both elements.

An *energy regulator,* or infinite switch, does not have a limited number of positions but is continuously variable as is a gas valve. It can be used with solid plate, radiant ring and ceramic surface units. The regulator is effectively a small timer which switches the heating element on and off as shown in Fig. 13.15. An energy regulator is more expensive than a selector switch and is a more frequent source of failure.

An energy regulator consists of a bi-metallic strip switch (section 2.9.1) which has an electrical heater wrapped around it. The supply of electricity to this heater and to the surface unit heating element goes through this switch which, when cold, is in the 'on' position (Fig. 13.16). When the flow of electricity through the bi-metal heater causes it to heat up, the strip bends and breaks the electrical contact, switching off both the bi-metal heater and the heating element. Once the strip cools down it straightens back into the on position. The control knob operates a cam which applies pressure to one part of the

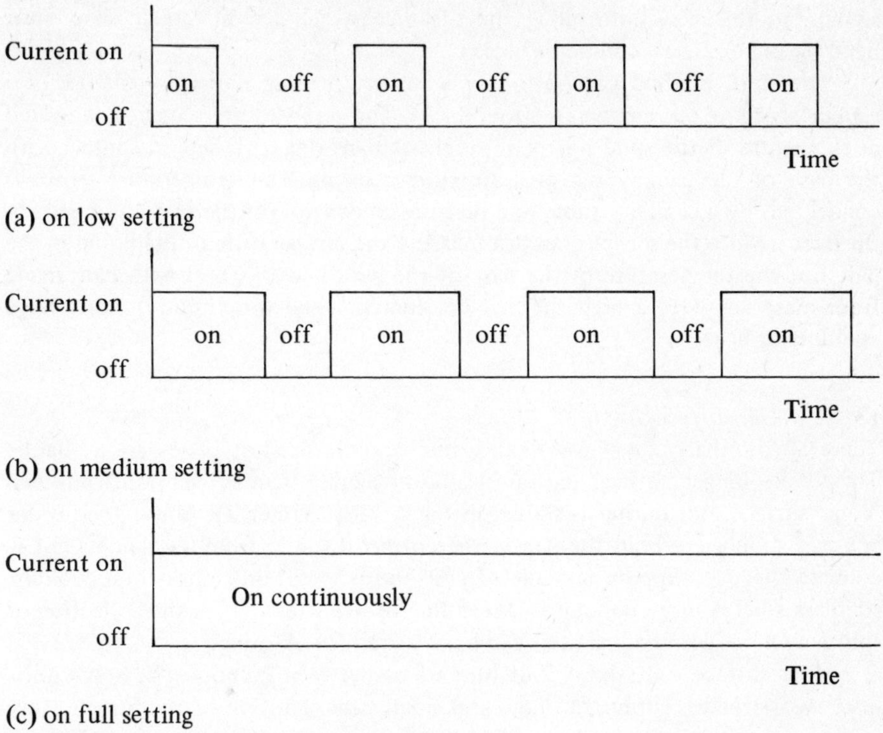

Fig. 13.15 – Method of control of an energy regulator.

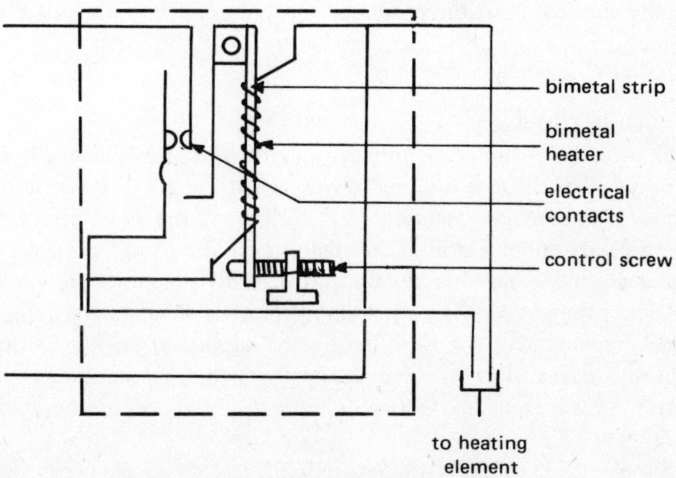

Fig. 13.16 – Energy regulator (infinite switch).

switch, in this way controlling the distance which the bimetallic strip must bend before electrical contact is broken.

The third method of controlling a surface heating unit is by the use of a *vapour-pressure thermostat* (section 9.4.1). The temperature sensor is mounted at the centre of the solid hot plate or the radiant ring. It is held in contact with the base of the pan by the pressure from a spring. This temperature sensor is connected, by a capillary tube, to a thermostat control (of the same type used in the oven). Since the sensor does not measure the temperature of the liquid in the pan, but the temperature of the base of the pan, it works best with pans made from materials with a high thermal conductivity and which do not store large amounts of heat.

13.2.3.4 *Gas surface units*

These are normally open gas flames, but ceramic gas hot plates are available. These take longer to heat and are less controllable than conventional burners. A gas surface unit burner is shown in Fig. 2.13 (Chapter 2). Above the burner is a pan support to hold the pan at the correct distance from the flame. On UK cookers the pan support is made of cast-iron whereas on Continental cookers, stainless steel is more common. Under the burner is a well for the collection of spillage. On modern cookers this can be removed for cleaning.

Most surface units have four burners of between 2 and 4 kW. Some units have two sizes of burner for large and small pans. Ignition of the burner is by permanent pilot or (on all new units) by spark ignition (section 2.8.2). Vapour pressure thermostats may be used on one or more of the burners. Burners on German hotplates incorporate a flame-failure device.

Unlike the electric surface units, heat transfer to the pan is by convection from the hot combustion products. Gas ceramic hot plates are also under development.

13.2.4 Grills (Broilers)

A grill provides a source of intense radiant heat, for which a high temperature is required. This heat is directed down on to the food. Because of the effect of the inverse square law (section 12.1.1) one of the most effective controls for cooking foods under a grill is the distance of the food from the radiant surface. To achieve this a number of alternative shelf positions are provided in some grills. The other means of controlling grilling is through the temperature of the radiant element. This is controlled by the supply of energy to the element and can be by means of a gas valve, a selector switch or an energy regulator. Since radiation upwards is ineffective, a reflector may be required to redirect this heat downwards.

An electric grill has tubular radiant resistance elements shaped to cover the area of the grill. In some cases two elements are provided each covering one half of the grill area and with a selector switch. When only small amounts of food are being grilled, greater economy is achieved through the use of a single element (Fig. 13.17).

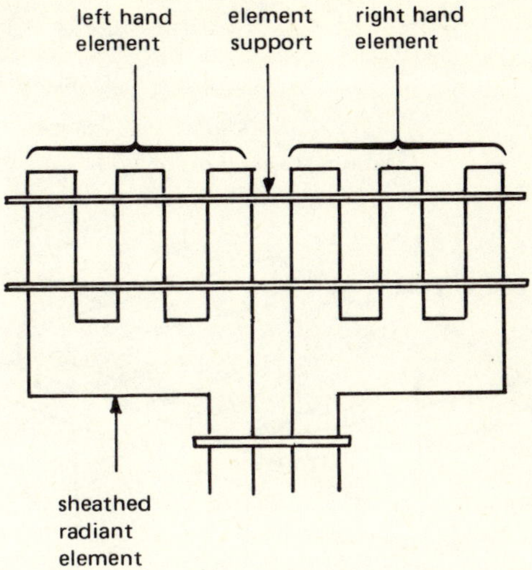

Fig. 13.17 – Dual element electrical grill.

In order to generate radiant heat from a gas flame, a ceramic or metallic material is heated by the flame to a temperature at which it generates IR and visible radiation (section 12.1.1).

A conventional gas grill consists of a gas burner producing a horizontal flame which is directed onto a light steel fret. These frets direct the radiation downwards onto the food. At low gas settings some unevenness of grilling may be found since the fret closest to the flame will be hotter than that at some distance away. To overcome this, a surface combustion burner may be used. In this type of burner the gas/air mixture is introduced into the grill through a metal gauze. Combustion takes place on the surface of this gauze. When the flow rate of gas is turned down all the surface falls in temperature. However at all gas settings a higher temperature is found at the centre of the burner than at the sides. To prevent this an H-shaped metal plate is fastened to the centre of the gauze (Fig. 13.18).

One feature now found on many grills, oven/grill combinations and ovens is the rotisserie. This consists of one or more stainless steel rods or spits which are rotated slowly by an electric motor. The speed of the electric motor is reduced by a set of gears.

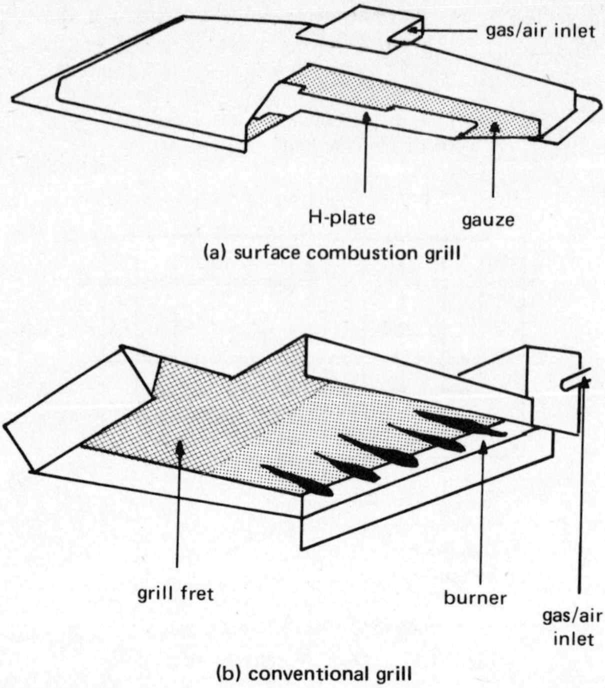

Fig. 13.18 – Gas grill burners.

13.3 THE USE OF ENERGY IN COOKING

Cooking represents the third highest use of energy in the home and therefore it contributes significantly to the total fuel bill (Table 1.3). Manufacturers have done much to improve the efficiency of cooking appliances through the elimination of permanent pilots, the use of higher standards of insulation and the provision of small ovens, dual element rings etc. However a large amount of energy can be wasted through bad practice in the use of cooking appliances.

In the case of surface units, energy is wasted if a small pan is heated on a large element or burner. With a gas burner, flames should not be allowed to burn up the sides of a pan. A saving of up to 50% can be made if the inner ring of a dual element is used when heating a 13 cm saucepan. Particularly in the case of electric rings, pans should have a flat base to give maximum contact with the element. If a liquid is boiling vigourously, this uses almost double the energy of a simmering liquid and similarly a pan without a lid wastes energy. Energy can be saved by anticipating the end of the cooking period and turning off the heat source before this occurs.

With a grill, energy can be saved by not pre-heating the grill. It is more efficient to have the grill at a low setting close to the food, than to have the grill at a high setting with the food at some distance away.

An oven requires a large amount of energy to pre-heat and to maintain it at the required temperature (with the exception of microwave ovens) and therefore the best use should be made of this energy by operating the oven with as much food in it as possible. The efficiency of an oven increases with the load of food in the oven. Similarly, if, a small oven is available, this should always be used in preference to the large oven if only small amounts of food are being cooked. The large oven should be reserved for those occasions when large amounts of food are being cooked or for bulky items such as turkeys. The traditional practice of having a 'baking day' is very efficient from an energy point of view, since the pre-heat energy is spread over many batches of food. In many cases it is not necessary to pre-heat the oven before putting food in the oven and this can save energy.

Frequent opening of the oven door can lead to the loss of much energy (and a hot kitchen). Faults in the oven, such as an inaccurate thermostat, or a worn door gasket can also lead to a loss of energy.

In general, energy can be saved by the use of independent cooking appliances, such as toasters, kettles, contact grills (see Chapter 14), but these are usually more expensive to buy than the general purpose cooker (Watt Committee 1978). Also, cooking a complete meal in the oven can take up to 30% less energy than by the use of a combination of hot plate and oven (Singer & Hunt 1976). In general microwave ovens use less energy than a conventional oven, because the oven does not need to be pre-heated.

REFERENCES

Colato, A. E. (1978), *Microwave Energy Applications Newsletter,* **11** (6), p. 3--6.

Denny, R. J., Kay, F. & Reed, R. A. (1974), *Communication No. 950,* Institution of Gas Engineers.

Dunning, E. W. B. (1953), *Gas Council Research Communication CC 12.*

Hodgkinson, A. J., Roberts, J. and Ross, N. C. (1980), *A comparison of the features and national standards of European cooking appliances, Communication 1120,* Institution of Gas Engineers.

Milson, A. & Kirk, D. (1980), *Principles of the design and operation of catering equipment,* Ellis Horwood.

Ohlsson, T. (1981), Methods of measuring temperature distribution in microwave ovens, *Microwave World,* **2** (2), p. 14–17.

Rennekamp, R. (1978), *Microwave Energy Applications Newsletter,* **11** (6), 7–13.

Singer, D. D. & Hunt, R. O. (1976), *MAFF.*

Taylor, B. (1973), *Le Fevre's Domestic Utilisation of Gas,* Ernest Benn.

Voss, W. A. G. (1981), Microwave safety, *Microwave World,* **3** (2), 10–11.

Watt Committee (1978), *The rational use of energy,* report No. 3. HMSO.

BIBLIOGRAPHY

Bradbury, K. (1973), *Solid fuel in the home,* 2nd Ed., Women's Solid Fuel Council.

Consumers Association (1979), Microwave ovens, *Which,* **November,** p. 599–604.

Consumers Association (1981), Gas and electric cookers, *Which,* **September,** p. 490–497.

Electrical Association for Women (1971), *The EAW Electrical Handbook,* 9th Ed., English University Press.

Houghton, J. (1973), What's Cooking, *Institution of Gas Engineers Communication 898.*

I.E.C. (1971), *Methods for measuring the performance of electric ranges and ovens,* International Electrotechnical Commission Publication No. 350.

Peat, L. J. Pickett, M. S., Arnold, M. E. & Wolf, L. H. (1978), *Household equipment,* 7th Ed., Wiley.

Plant, E. M. (1971), *Gas and its domestic uses,* 2nd Ed., Women's Gas Federation.

Scharff, R. (Ed.) (1976), *Refrigeration, air conditioning, range and oven servicing,* McGraw-Hill.

VanZante, H. (1973), *The Microwave Oven,* Houghton and Mifflin.

Webb, J. M. (1977), *Microwave: the cooking revolution,* Forbes.

CHAPTER 14

Small electrical cooking appliances

14.1 INTRODUCTION

In Chapter 13 we considered the traditional methods of cooking which developed from the use of a single source of heat for all cooking purposes. With the introduction of electricity as a source of heat for cooking, it became possible to incorporate heating elements into small portable appliances which could be used independently of the cooker (range). Some of these appliances, such as the kettle and the toaster, have been used for many years. However, many small cooking appliances are now available which, together, can perform the majority of the duties of the traditional cooker.

In most cases a purpose-designed appliance can offer the following advantages over the use of a cooker:

(a) speed
(b) lower energy consumption (Watt Committee 1978)
(c) better performance than general purpose design
(d) greater convenience of use.

These electrical appliances can either be free standing or built-in. As an example of the latter, some small appliances such as deep-fat fryers, griddles and warming plates can be combined with built-in surface cooking units. Alternatively, these electrical appliances can be free standing portable units fitted with a flexible cable and plug. Small ovens and rotisseries are available but, since the design of these are very similar to conventional ovens, they are not described in this chapter.

14.2 WATER-HEATING KITCHEN APPLIANCES

14.2.1 Kettles (electrically heated)

A kettle consists of a stainless steel, chrome plated or plastic water container, with a handle and a spout for pouring. It has an electrical immersion heater (section 5.7.1). The kettle is filled with water either through its spout or through a removable lid. Typically kettles hold a maximum of 1½ to 2 litres. They can be operated with less than the maximum volume of water provided that the heating element is covered with water. The rating of the heating element is between 2½ and 3 kW (Fig. 14.1).

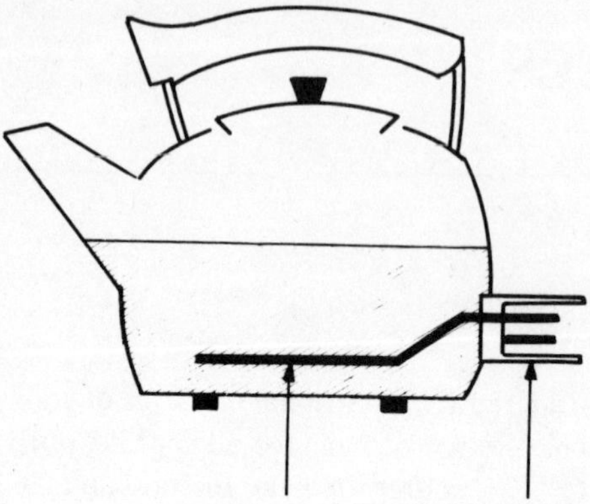

Fig. 14.1 – The electric kettle.

Many kettles are now provided with an automatic switch to cut off the supply of electricity to the heating element once the kettle boils. Alternatively, the kettle may keep the water simmering once it has boiled. The mechanism of a typical automatic kettle is shown in Fig. 14.2. When the kettle is in the off position, the peg (A) on the on/off switch holds apart the electrical contacts B and C. Switching on the kettle by pushing in the on/off switch, closes contacts B and C, causing electricity to flow to the heating element. When the water boils, steam is produced which passes through a hole (D) and heats up the bi-metallic strip (E) (section 2.9.1). As this heats up, it bends pushing out the on/off switch. This opens the contacts B and C.

Kettles are equipped with a mechanism to switch off the kettle if it is operated without water or allowed to boil dry. This may be an ejector which pushes out the plug from the socket, or it can be a bi-metal strip. In the latter case, this may be incorporated into an automatic kettle as shown in Fig. 14.2. A safety cut-out bi-metallic strip (F) is connected at one end to the heating element with the other end resting on the on/off switch when this is in the 'on' position. If the heating element is operated without water, its temperature rises significantly above the normal operating temperature. Heat is conducted along the bi-metal strip which then bends, pushing out the on/off switch. The switch cannot be operated again until the element cools down.

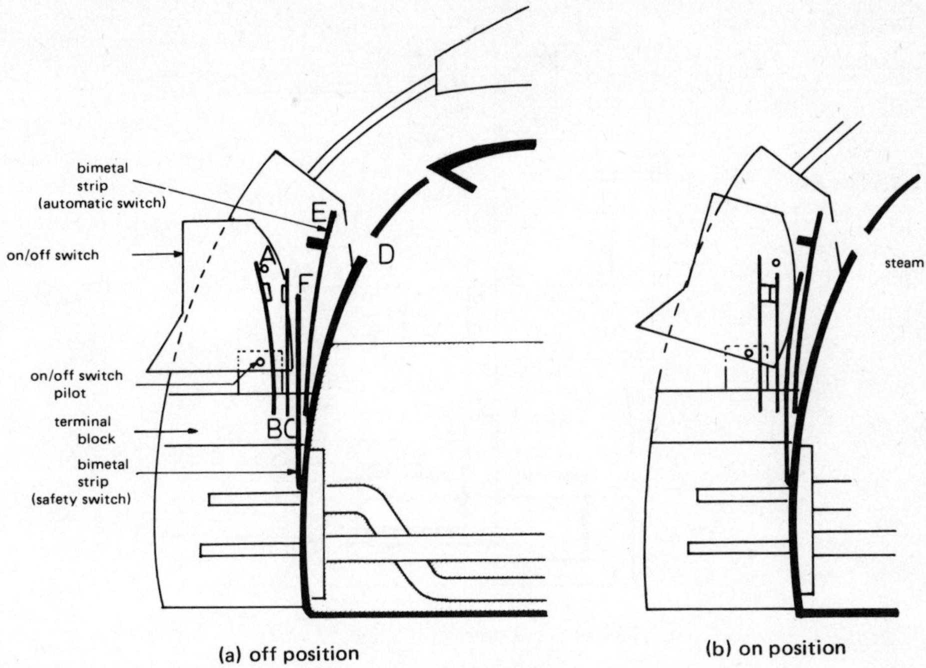

Fig. 14.2 – The mechanism of an automatic kettle.

14.2.2 Coffee Percolators

Coffee making is an example of mass transfer by leaching (Milson & Kirk 1980) – the extraction of soluble components from a solid. The rate of extraction from coffee depends upon the size of the grind of the coffee beans and on the temperature of the water used for extraction. As the temperature of the water increases a faster rate of extraction is achieved. Volatile coffee flavours are rapidly lost however if boiling water is used for extraction and chemical breakdown of the coffee extracts occurs, this spoils the flavour. It is usually recommended that the extraction is carried out at between 90 and 98°C. This gives a fast rate of extraction and minimises the loss of volatile flavours and the development of bitter tasting chemicals. For this reason methods of coffee making involve the use of temperatures below the boiling point of water.

In a percolator (Fig. 14.3), the coffee is held in a perforated metal basket which is mounted on the end of a long stem terminating in a heating chamber. The percolator is filled with cold water and the water in the heating chamber is rapidly heated to boiling by the primary heating element (immersion heater). Small pockets of steam push the hot water up the stem and onto the ground coffee beans. As this hot water trickles through the coffee it dissolves soluble

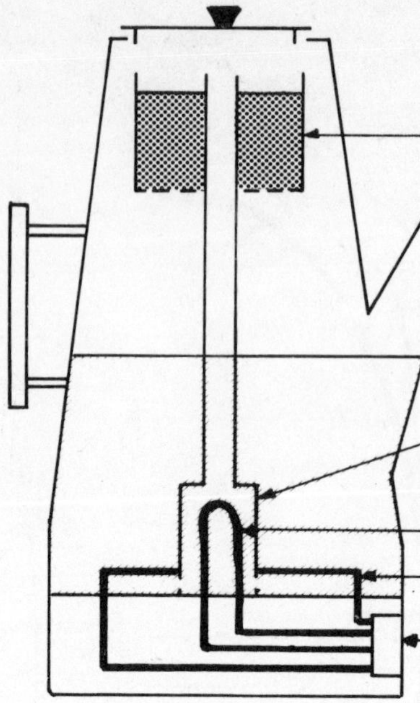

Fig. 14.3 – A coffee percolator.

extracts and returns them to the bulk of water in the percolator. In this way the strength of the coffee gradually increases. The end of the process is usually controlled by a thermostat situated in the base of the percolator. When the desired temperature has been achieved a second heating element (the warming element) is connected in series with the primary heating element. This has the effect of reducing the temperature in the heating chamber, stopping percolation, but the two heating elements then keep the coffee warm. On some percolators this transition from percolating to keeping the coffee warm is indicated by a pilot light being turned on as an indication that the coffee is ready.

On some percolators it is possible to control the strength of the coffee produced by the use of a variable thermostat operated by a lever on the base of the percolator, a low final percolation temperature results in a low concentration of coffee extracts and no thermal degradation. A high percolation temperature increases the strength of the coffee but will also result in a greater loss of volatile flavour compounds and the development of bitter tasting chemicals as a result of thermal degradation.

Coffee percolators may be made from stainless steel, chrome plated steel or ceramic and have capacities from ½ to 1 litre. The rating of the heating element is from 400 to 800 W.

14.2.3 Filter Coffee Makers

In a filter coffee maker, the ground coffee is held on a filter paper and water which is just off boiling, is poured slowly through the grains of coffee. It collects in a serving container. Manual filter coffee makers are available which are simply a holder for the filter paper through which hot water from a kettle is poured.

A typical automatic filter coffee maker is shown in Fig. 14.4. It consists of a reservoir for cold water, a heater for the water and a holder for the filter paper held over a service jug. The jug sits on a heater which keeps the coffee warm.

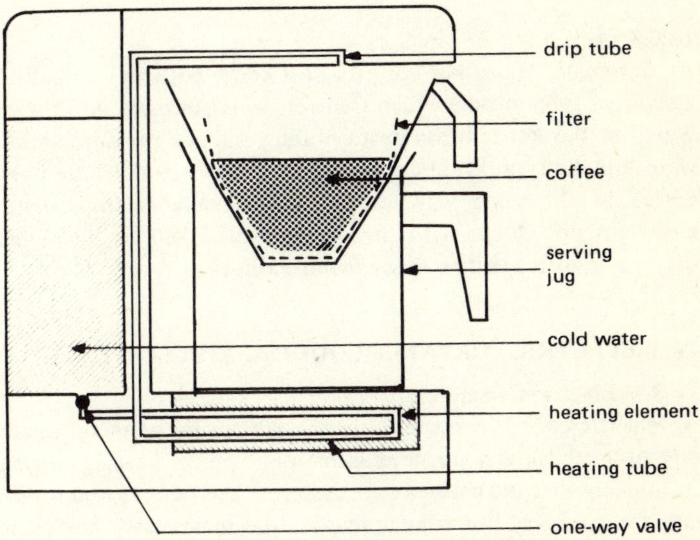

Fig. 14.4 – A filter coffee maker.

The cold water reservoir may be sited above the filter paper, so that water can flow by gravity from the reservoir into the filter. If the reservoir is on the same level as the service jug a mechanism is required to induce a flow from the reservoir and out of the drip tube. The outlet for water from the reservoir has a simple one-way valve which allows water to flow out of the reservoir into the tube but will not allow water to flow in the reverse direction. Cold water flows from the reservoir into the heating tube. (In some models (Fig. 14.4) a single heater is used for both the heating tube and for the warming plate. In other models two separate heaters are used.) Because the heating tube holds only a small volume of water it is rapidly heated to boiling and steam is produced. Since the one-way valve prevents water flowing back into the reservoir it is pushed along and out of the drip tube. This process continues until the reservoir is empty.

The heating element has a bi-metal strip thermostat to control the temperature of coffee in the serving jug. There is much less danger of overheating the coffee in a filter coffee maker than in a percolator, since water passes through the coffee only once and the coffee, once it has been extracted is not heated to boiling.

On some filter coffee makers the heating tube can be easily dismantled to remove scale (section 6.5). The capacity of filter coffee makers varies from ¾ to 1 litre. It is possible to make less coffee than the maximum quantity, but in most cases the appliance works best with the maximum volume of water. The heating elements are rated from 600 to 800 W.

14.2.4 Automatic Tea-makers

An automatic tea-maker consists of a kettle with an air-tight lid with, instead of a spout, a syphon tube which transfers water directly into the spout of a tea pot. Water in the kettle heats up to boiling and the pressure from the steam forces water up the syphon tube into the tea pot. It can also be used to make instant coffee in the same way. Automatic tea-makers are usually combined with electric alarm clocks with the mechanism linked so that when the alarm goes off, tea or coffee will be ready for drinking.

14.3 ELECTRIC PORTABLE COOKING APPLIANCES

14.3.1 Electric Casseroles.

An electric casserole consists of an earthenware or metal casserole with a capacity of 1 to 3 litres, surrounded by a low-power heating element (60 to 200 W). In some models the earthenware casserole can be removed for service at the table and for washing. For similar reasons the mains lead may be removed on some models.

A typical electric casserole is shown in Fig. 14.5. In this model the heating elements are in two sections, controlled by a switch. With the switch in the 'low' position only the lower section of the heating elements are used but both of the elements are used when the switch is in the 'high' position.

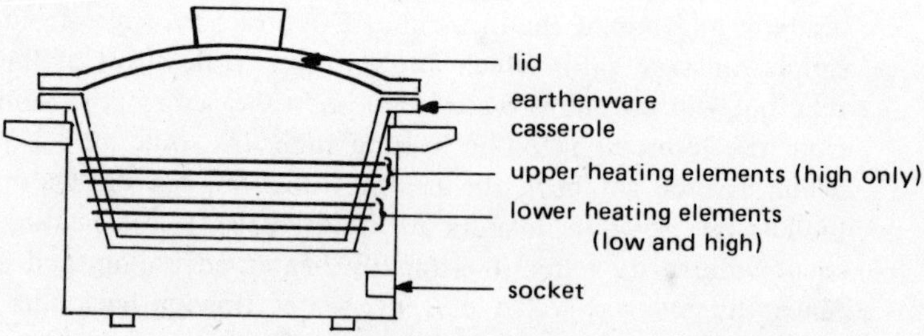

Fig. 14.5 – An electric casserole.

Electric casseroles are used for foods which require cooking slowly (long cooking times). Used in this way they consume less energy than would a casserole cooked in a conventional oven. Tests have shown that an electric casserole uses 15-25% of the energy consumed by an oven if a casserole is being cooked on its own (Watt Committee 1978).

14.3.2 Electric Deep-Fat Fryers

An electric deep-fat fryer consists of an aluminium-alloy or steel pan with heating elements sealed in the base. The pan is surrounded by insulation covered by an outer metal skin. The temperature of the pan is controlled by a thermostat which can be set to the desired temperature within the range of 90°C to 190°C A safety thermostat protects against the failure of the adjustable thermostat. This ensures that the vapour over the oil cannot reach ignition temperature, in this way reducing the risk of a fat fire. In addition to frying, the pan may also be used for stews, soups and casseroles.

The electrical ratings of deep fat fryers vary from 1 to 1.7 kW. Some models are provided with a replaceable charcoal filter in the lid which helps to reduce contamination of the air in the kitchen with frying odours. Deep fat fryers are also available as part of built-in hob units.

14.3.3 Electric multi-cookers

Electric multi-cookers are also referred to as electric frying pans and electric skillets. They consist of an aluminium alloy pan with a heating element moulded in the base (Fig. 14.6). The electrical supply to the heating element is brought into the pan through a plug, which also contains the thermostat sensing element. The thermostat and its control knob are contained in this plug. Since the electrical components of the pan are sealed, it can be washed in water without damage, provided that the supply plug has been removed. Some models have a

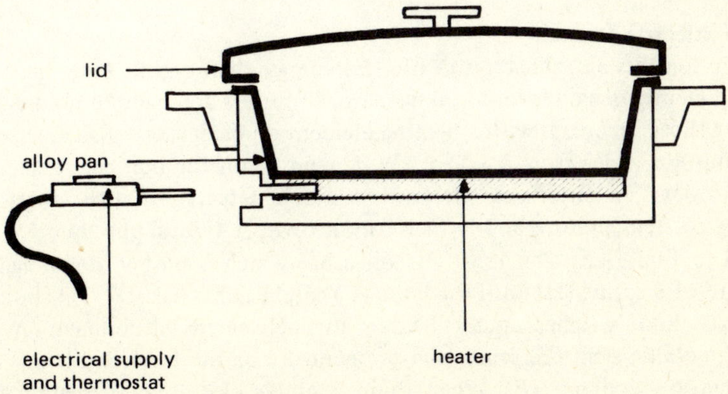

Fig. 14.6 – Electric multi-cooker.

broiling element built into the lid of the appliance. Multi-cookers can be used to perform a wide range of cooking duties including baking, roasting, frying and grilling.

14.3.4 Electric Contact Grills

An electric contact grill consists of two hinged PTFE covered cooking plates with thermostatically-controlled heating elements (Fig. 14.7). On some models both plates can be used in the horizontal position allowing double the area for griddling. Baking, casseroling etc. can also be carried out by using a baking dish between the two heating elements. Various sizes are available from 100 mm × 200 mm up to 200 mm × 300 mm and electric ratings from 900 W to 2 kW. Some models have interchangeable plates for making toasted sandwiches and waffles.

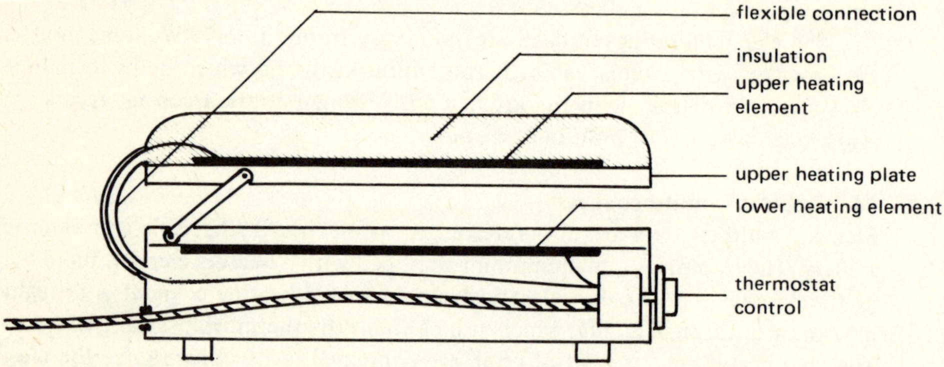

Fig. 14.7 – Electric contact grill.

14.3.5 Electric Toasters

Electric toasters are quicker and use less energy than a grill. They have vertical heating elements wound onto an insulator. Toasters are made to hold either two or four slices of bread, with a heating element on both sides of each slice. Power consumption varies from 1¼ to 2 kW depending on the number of slices. Most toasters have an automatic mechanism which detects when the toast reaches the correct temperature and is therefore brown. A typical pop-up mechanism is shown in Fig. 14.8. The toast sits on a bar which is pushed down against the pressure of a spring (H) until the latch (C) catches on a bar (D). This holds down the toast and switches on the heaters through electrical contacts (A and B). The bi-metallic strip (E) senses the temperature of the toast and bends towards the electrical contacts (F). When these contacts close they activate a solenoid (G) (section 5.7.2), which pulls back the bar (D), releasing the latch (C). This causes the toast to be ejected under the force of the spring (H). The sensing element (E) is sited in only one slot and this slot must be used if the mechanism is to work correctly. For this reason one of the toast slots will be marked for the use of a single slice.

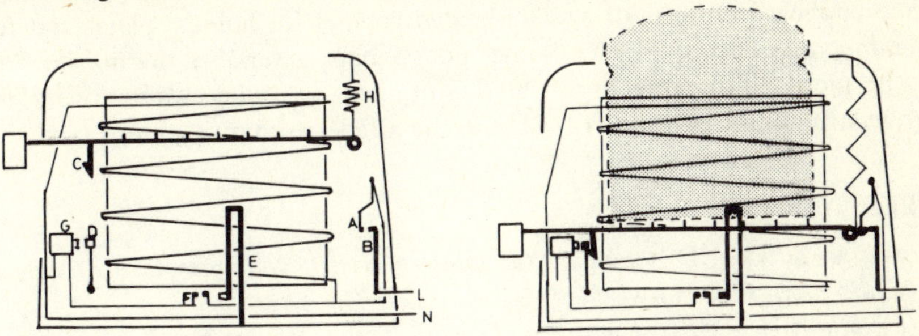

(a) toaster off (b) toaster on

Fig. 14.8 – The mechanism of an automatic toaster.

The degree of browning can be adjusted through a lever which adjusts the positions of the contacts (F). On a low setting, the bi-metal strip only has to move a short distance before it closes the contacts (F). At a high setting the contacts are moved away from the bi-metallic strip so that a higher temperature is required to switch off the heaters and eject the toast.

Toasters are also incorporated into miniature ovens. These use quartz tubes (section 5.7.1) to provide heat both to the toaster and to the oven. The toast may be held vertically through a slot in the top of the oven, or it may fit horizontally as in a conventional grill.

14.4 APPLIANCES FOR KEEPING FOOD HOT

14.4.1 Heat Trays

Heat trays are glass or stainless-steel trays with electrical heaters mounted underneath. Simple heat trays maintain the same fixed temperature (100–120°C) over the whole of the surface. More expensive models have two areas, one which has a low fixed temperature (100–120°C) and the other with a higher and adjustable temperature (up to 140°C). These trays normally maintain temperature only whilst connected to the electrical supply. Some types act, however, as heat storage trays. These are pre-heated by connecting them to the electrical supply for 15–20 minutes, after which they maintain their temperature for up to one hour. A ceramic top to the tray has heat-retaining material below it. A thermostat is used to control the initial heating. Some models have inset spaces for serving dishes with lids rather than a flat surface. The serving dishes, which are made to fit the size of the recesses in the tray, are made of stainless steel and have lids with an insulated handle. The heat trays have insulated wooden or plastic handles to facilitate carrying to the table. Alternatively, they may be mounted on a service trolley.

14.4.2 Hot Cupboards

A hot cupboard consists of a small heated cabinet for holding plates and food at temperatures up to 150°C. Some models have a variable thermostat. They may be mounted underneath a heat tray or a serving trolley. Alternatively, they may be a free-standing cabinet for use in the kitchen or dining room.

REFERENCES

Milson, A. & Kirk, D. (1980), *Principles of design and operation of catering equipment,* Ellis Horwood.

Watt Committee (1978), *The rational use of energy,* Report No. 3.

BIBLIOGRAPHY

Consumers Association (1978), Toasters, *Which,* **Nov.,** p. 599-601.

Consumers Association (1978), Coffee Makers, *Which,* **Dec.,** p. 704-705.

Consumers Association (1980), Toasters, *Which,* **Dec.,** p. 701-704.

Consumers Association (1980), Electric Kettles, *Which,* **March,** p. 115-117.

Cox, L. (1976), *Home Economics,* **July,** p. 2-4.

Electrical Association for Women (1971), *The EAW Handbook,* 9th Ed., EUP.

Gaddis, B. (1976), *How to repair home kitchen appliances,* Tab Books.

Page, E. A. (1978), The microbiological safety of slow cooking in electric casseroles, *Journal of Consumer Studies and Home Economics,* **2** (2), p. 91-98.

Peet, L. J., Pickett, M. S., Arnold, M. E. & Wolf, H. L. (1978), *Household Equipment,* 7th Ed., Wiley.

CHAPTER 15

Refrigeration equipment

15.1 INTRODUCTION

It has been known for many hundreds of years that a reduction in temperature will slow down the processes of decay and spoilage in many foods. The traditional location of food storage areas within a house is in the cellar or on the coolest side of the house. In areas of the world where ice is freely available it has been used to cool and therefore extend the life of perishable commodities. However, it was not until the introduction of mechanical refrigeration that chilling and freezing became viable techniques for the preservation and storage of food.

In order to understand the preservative effects of reduced temperatures, it is necessary to consider the ways in which foods spoil and then to consider how these processes are affected by temperature (Rolfe 1967). There are three main reasons for the spoilage of food, although in many cases these may operate in combination:

(1) changes caused by micro-organisms
(2) changes caused by enzymes
(3) chemical and physical changes.

15.1.1 Micro-organisms

As described in Chapter 13, each species of micro-organism has an optimum temperature for growth and its rate of growth decreases as the temperature deviates from this optimum (see Fig. 13.2). A primary effect of the reduction in temperature is to alter the balance of micro-organisms present on or in a food (its micro-flora) and, in general, low temperatures will favour the growth of *psychrophilic* bacteria. There are bacteria which are capable of growth at 5°C or below. Some of these are capable of growth even down to temperatures as low as −7°C (moulds −10°C). Below this temperature growth, in the sense of reproduction, ceases although some metabolic activity may continue down to much lower temperatures. The reduction in growth rate with temperature is also partly caused by the reduction in the amount of free water available below the freezing point of a food. As the water in food freezes it does so in the form of pure water (particularly if the freezing rate is slow) leading to an increase in the concentration of electrolytes and other water-soluble material in the remaining free (unfrozen) water. These increased concentrations of electrolytes can inhibit micro-biological growth.

When food is frozen, some of the micro-organisms present are destroyed whilst others survive the freezing process. One of the factors which affects the rate of mortality of micro-organisms is the rate of freezing. Unfortunately conditions which maximise the rate of distribution of micro-organisms in food also cause maximum damage to the food itself (see section 15.2.2).

Micro-organisms spoil food because they produce off-odours or flavours in the food as by-products of their normal metabolism. This may involve the conversion of carbohydrates to acids, the breakdown of proteins to hydrogen sulphide and other sulphur containing chemicals, and the hydrolysis of fats to fatty acids. All of these processes can change the sensory properties of food and render it unacceptable. Also the presence of micro-organisms can lead to visible mould growth on the surface of food or slime on the surface of meat.

Table 15.1 shows minimum growth and toxin-forming temperatures for some food poisoning bacteria (Rieman & Bryan 1979). From these data it can be seen that freezing prevents growth and toxin formation (although the organism and toxin may survive the freezing process). With the exception of Type E *Cl. botulinum*, a temperature of 5°C or below will normally allow the safe storage of foods. Type E is relatively rare and is associated with certain sea-foods, particularly if packaged anaerobically.

Table 15.1

Minimum temperatures for growth and toxin development for some food poisoning bacteria (Jarvis 1972)

Organism	Minimum temperature for growth °C	Minimum temperature for toxin production °C
Staphylococcus aureus	7	10
Clostridium botulinum (Types A & B)	10	10
Clostridium botulinum (Type E)	3.3	3.3
(Type F)	4	4
Salmonella species	7	–
Clostridium perfringens	6.5	–

15.1.2 Enzymes

As with micro-organisms, enzymes have an optimum temperature at which they are most active (see Fig. 13.2). Enzymes, if not destroyed, can remain active at normal frozen-food storage temperatures and, although the rate of reaction can be very slow, foods can spoil. In some cases the freezing process ruptures cell walls causing mixing of enzyme and substrate and accelerating the rate of spoilage. Table 15.2 gives some typical enzymes and the types of spoilage associated with them. Where enzymes are likely to be a problem they can be inactivated by heat; this process is known as blanching and is carried out by immersing the food in boiling water or steam for a short period of time.

Table 15.2

Enzymes associated with food spoilage

Enzyme	Spoilage
Lipoxidase	oxidative rancidity of fats
Phenolase	browning of fruits and vegetables
Chlorophyllases	loss of green colour in vegetables
Proteinases	breakdown of the texture of meats
Lipase	hydrolytic rancidity of fats
Peroxidase	hay like flavour in vegetables

15.1.3 Chemical and physical spoilage

There are many chemical reactions which can contribute to the spoilage of foods. These include oxidation reactions which cause rancidity of oils and fats and loss of colour in vegetables; and hydrolysis, which causes the breakdown of fats, pectins and starches. The rate of a chemical reaction is affected by temperature and, for a typical chemical reaction, a reduction of 10°C will reduce the rate of a reaction by one half or two thirds. Therefore if spoilage of a food were due solely to chemical reactions, a reduction in temperature from 20°C to 10°C would double its storage life.

Physical changes associated with storage include:

(1) the loss or gain of water
(2) loss of volatile flavour components
(3) absorption of volatile flavours from other foods.

The rate of moisture loss/gain from a food depends upon the difference between the *equilibrium relative humidity* (ERH) of the food and the relative humidity of the air surrounding the food (section 11.3). If the ERH of the food is higher than the RH of the air, the food will dry out as it loses water to the atmosphere. On the other hand, the food will absorb moisture. The RH of air in refrigerators and freezers is low because moisture from the air freezes on the evaporator. Where moisture loss or gain is a problem foods can be protected by the use of moisture-proof packaging materials or containers. Similarly loss or gain of flavours by foods during storage can be controlled by wrapping them in a material which acts as a barrier to the migration of volatile chemicals.

15.2 STORAGE CONDITIONS FOR FOODS

15.2.1 Chilled foods

There is no precise definition of the word 'chilled' in relation to the storage of food. The minimum temperature for chilled storage is the freezing point of the food (usually −1 to −3°C). Foods which are likely to contain food-poisoning bacteria should be stored below 5°C (for example, dairy products, meat and fish).

When considering the storage of foods at chilled temperatures it is possible to differentiate between fresh fruits and vegetables, which are still biologically active, and foods in which natural bio-chemical activity has ceased, even though enzymes and chemical reactions may be taking place. For fresh fruits and vegetables the storage conditions (temperature and humidity) can be used to control the natural ripening processes. All fruits and vegetables have different optimal conditions for storage (Duckworth 1966) but in practice conditions in a domestic refrigerator must be controlled to suit a range of foods. In general a low temperature results in a low rate of respiration, which maintains the condition of the food. For example the rate of respiration of lettuce is 11 kJ/kg at 0°C, 15 kJ/kg at 5°C and 46 kJ/kg at 16°C. In certain cases however, too low a temperature can disturb the balance of bio-chemical processes. For example bananas should not be stored below 10°C and potatoes not below about 5°C.

Similarly the control of humidity must be a compromise. Too low a humidity will result in the wilting of leaf vegetables and shrinkage of root vegetables and fruit, but if the humidity is too high, mould growth may result.

15.2.2 Frozen foods

The quality of frozen food depends upon many factors, the most important of which are:

(1) cultivar or type of product (variety, maturity etc.)
(2) pre-freezing treatments (cooking, blanching etc.)
(3) rate of freezing
(4) storage conditions (humidity, temperature, packaging)
(5) temperature fluctuations in storage and during transport
(6) rate of thawing.

Many books have been written on the home freezing of foods and these provide recommendations of varieties of fruits and vegetables which are suitable for freezing. These books also give recommendations of pre-freezing treatments and blanching times.

The rate at which a food freezes has a pronounced effect on its quality (Heldman 1975, Milson & Kirk 1980). As food freezes, two processes are taking place. The first of these is the development of minute new ice crystals (known as nucleation) and the second is the growth of existing ice crystals. A rapid rate of freezing leads to a maximum rate of nucleation, resulting in the presence of a large number of very small crystals throughout the food. A slow rate of freezing favours the growth of existing ice crystals, resulting in a small number of relatively large ice crystals in the food. In general, the presence of large ice crystals results in more tissue damage than does the presence of small ice crystals.

As a further complication, slow freezing causes damage to cellular food materials because of differences in the osmotic pressure between fluids in the cell. When cellular foods are frozen slowly, ice crystals first develop in the fluid between cells. Since it is only the water which freezes, the concentration of salts in the remaining fluid is increased. This means that there is a greater osmotic pressure outside the cell than inside. Water passes by osmosis to balance these pressure differences causing shrinkage of the cell. When the food is thawed this fluid does not pass back into the cell but remains in the tissue resulting in spongy, water-laden fruits and vegetables. In the case of meats this fluid exudes from the tissue leading to the familiar pink, water drip-loss from meats.

In domestic freezers it is impossible to generate conditions for fast freezing of foods so that it is impossible to freeze susceptible foods (such as strawberries) to the same level of quality as the commercially frozen product.

Foods are usually packaged before they are frozen. The packaging acts as a moisture barrier to prevent loss of moisture from the food to the relatively dry air inside the freezer. It may also act as a barrier to oxygen (reducing the rate of oxidative rancidity) and to volatile chemicals (reducing flavour loss and preventing absorption of taints from other foods).

A major influence on the storage life of frozen foods is the temperature of storage (Van Arsdel *et al.* 1969). Because of the relationship between temperature and the rates of chemical reactions (section 15.1.3) a low storage temperature will increase the storage life. The relationship between storage temperature and storage life for a green vegetable is shown in Fig. 15.1. This diagram applies to storage times at constant storage temperatures. If the storage temperature is allowed to fluctuate then the storage times will decrease, e.g. if the temperatuee of the food is raised from −40°C to −20°C and cooled again, the effective shelf life is the same as if it had been held at a constant temperature of −20°C.

Domestic freezers usually operate at between −15 and −20°C. Frozen-food compartments in refrigerators operate at various temperatures. These are often marked with one, two or three stars which can be linked to storage times recommended by frozen food manufacturers (see Table 15.3) provided storage temperatures remain constant.

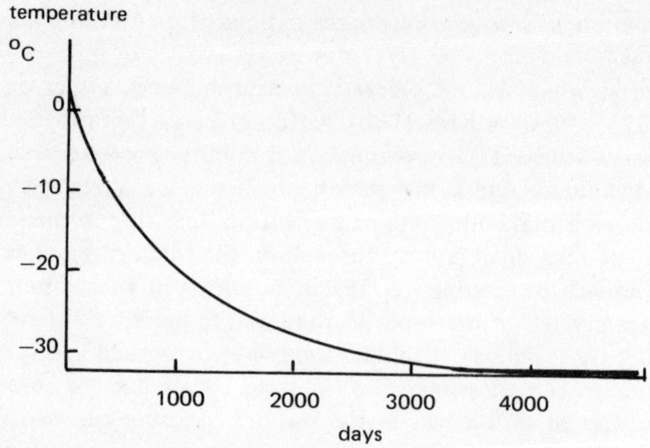

Fig. 15.1 – The storage life of a green vegetable.

Table 15.3
Storage temperatures of refrigerator frozen-food compartments

Temperature	Star rating	Typical storage life
− 6°C	*	1 week
−12°C	**	1 month
−18°C	***	3 months

15.3 REFRIGERATION EQUIPMENT

15.3.1 Principles of refrigeration

Normally heat flows from a region of high temperature to a region at a lower temperature, whether by conduction, convection or radiation (see section 7.2.2). In a refrigerator, this natural process is reversed and in order to do this it is necessary to supply energy (see Fig. 15.2).

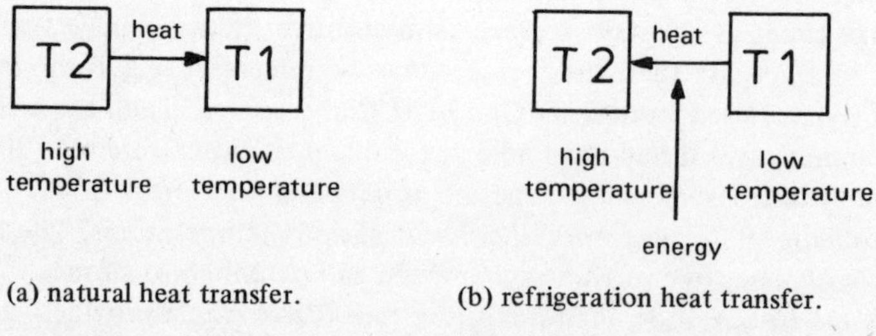

Fig. 15.2 – Comparison of natural heat transfer and refrigeration.

In a refrigerator a refrigerant (a volatile liquid) is used to transfer the heat from the low-temperature region to the high-temperature region. Heat from the low temperature region is used to evaporate the refrigerant and the resultant vapour is transported along a pipe to the high temperature region. Here the liquid is condensed back to a liquid, releasing heat as it does so. To force the liquid to evaporate at a low temperature, a low pressure is required, whereas to condense the liquid at a high temperature, a high pressure is needed (Fig. 15.3). There are two types of refrigerator: *the mechanical refrigerator* which uses a pump (driven by an electrical motor) to provide the necessary energy; and *the absorption refrigerator* in which energy is supplied in the form of heat.

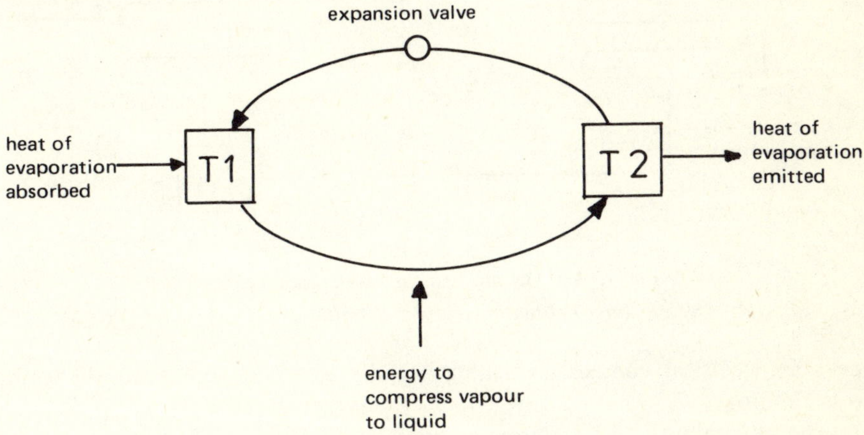

Fig. 15.3 – The refrigeration cycle.

15.3.2 The mechanical refrigerator

A schematic diagram of a mechanical refrigerator is shown in Fig. 15.4. It is made up of two halves one of which operates at a high pressure (generated by an electric pump or compressor) and the other which operates at a low pressure. On domestic refrigerators the expansion valve consists of a length of capillary tube which restricts the flow of high pressure liquid into the low pressure (suction) side of the compressor. It acts like a spray nozzle on an aerosol can, converting the liquid refrigerant into a very fine spray which quickly evaporates and falls in temperature. Because of this low temperature heat is extracted from the vicinity of the evaporator, reducing the temperature. The vapour is then sucked into the compressor and discharged at a higher pressure. Because of the high pressure it condenses, rising in temperature as it does so. This heat is given out to the air surrounding the condenser. Control of the refrigerator is maintained by a thermostat on the evaporator. When a pre-set low temperature is achieved, the thermostat (which is usually of the vapour pressure type as shown in Fig. 9.14),

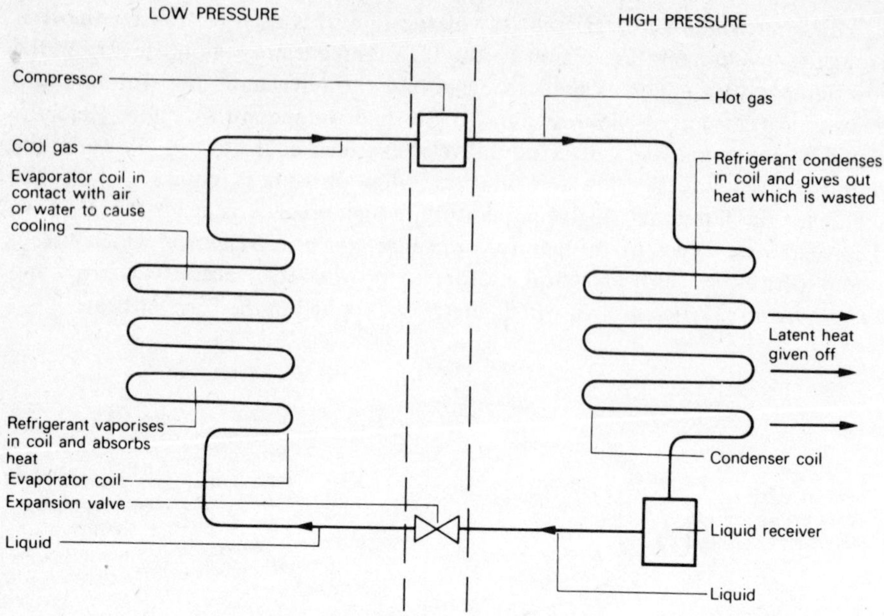

Fig. 15.4 – The mechanical refrigeration cycle.
(F. Hall 1980, *Heating, Ventilating and Air Conditioning,* Construction Press).

opens an electrical contact, switching off the electrical supply to the compressor motor.

The refrigerant used determines the temperature of the evaporator. In domestic refrigerators and freezers the normal refrigerant is Freon 12 (dichlorodifluoro methane). Some freezers use Freon 22 (mono-chlorodifluoro-methane) which boils at a lower temperature (–40°C) and therefore results in a lower evaporator temperature.

15.3.3 The absorption refrigerator

A diagram of an absorption refrigerator is shown in Fig. 15.5. Unlike the mechanical refrigerator it has no moving parts and is silent in operation. In the boiler (stage 1), a strong ammonia solution is heated causing ammonia vapour to boil off and rise to the condenser. At the same time the residual weak ammonia solution is returned to the absorber.

In the condenser (stage 2) the ammonia vapour gives up its latent heat of vapourisation to the air and condenses to a liquid. From here it runs into the evaporator.

The evaporator (stage 3) contains hydrogen gas which reduces the partial pressure of the ammonia, causing it to evaporate. As it does so, it falls in temperature and attracts heat from the air surrounding the evaporator. The ammonia vapour and hydrogen then pass to the absorber.

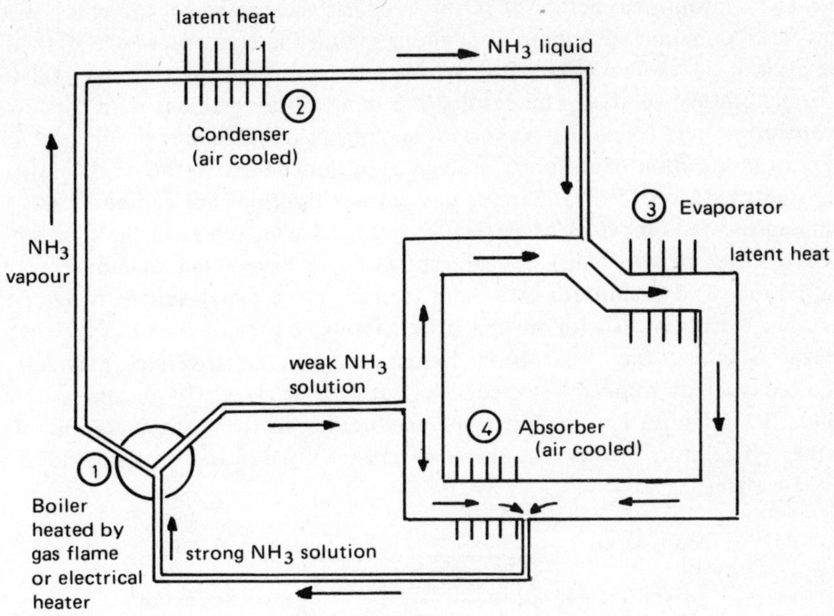

Fig. 15.5 – The absorption refrigerator.

At the absorber (stage 4) the ammonia is dissolved in the weak ammonia solution as it returns from the boiler. As the ammonia is dissolved it gives out heat which is removed by an air cooler, and the strong ammonia solution passes back to the boiler ready to repeat the cycle. The hydrogen circulates back to the evaporator.

15.3.4 The refrigerator

A refrigerator consists of sheet steel panels constructed to form an inner and outer cabinet separated by insulation. The metal layers are finished with a white or coloured enamel. The insulation may be fibreglass or foamed plastic which can be injected into the panels where it expands to fill all gaps (using the same principles as cavity wall insulation – section 8.3.3). The plastic used can be polystyrene or polyurethane. They have double the insulation value of conventional insulation.

The door is also made of a double metal wall separated by insulation. A plastic strip is used to seal the gap between the cabinet and door. It often contains a magnetic strip which holds the door closed. The inner and outer metal walls do not join directly but have a plastic thermal barrier between them to prevent conduction of heat into the refrigerator.

The conventional design of refrigerator has the evaporator in the top of the unit with a compressor mounted in the base and the condenser coil fastened to the back of the cabinet (Fig. 15.6). With the evaporator at the top, air circulation is by natural convection. This results in a temperature gradient with the lowest temperatures just below the evaporator and higher temperatures in the base. This allows the selection of different storage conditions to match the needs of different foods (section 15.2.1). Shelves and storage locations are chosen to suit the temperature requirements of particular foods (that is, salads in the base where they will not be too cold). Larger cabinets may have a fan to induce forced convection. The evaporator is usually formed into a metal-walled freezer compartment which is used for storing frozen foods and making ice cubes. Where a freezer is in use the frozen-food compartment in a refrigerator may not be required so that some refrigerators do not have a frozen food compartment. Similarly, fridge/freezer combinations may not have a frozen food compartment in the refrigerator. In this case the evaporator will be in the roof, back or walls of the cabinet.

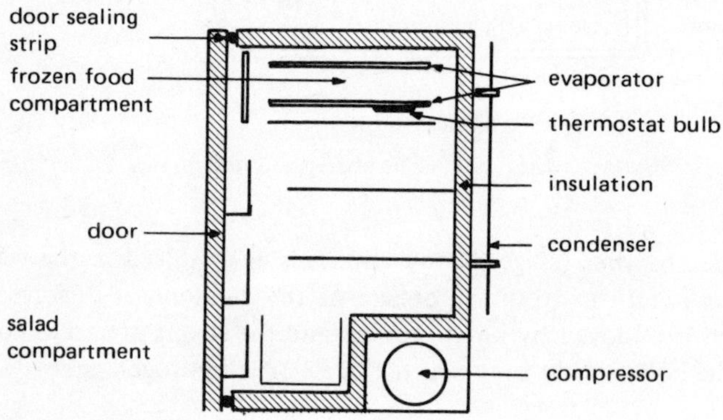

Fig. 15.6 – A typical refrigerator compartment.

Over a period of time moisture from the air in the cabinet and from any warm food placed in the refrigerator condenses on the evaporator and freezes. This reduces its efficiency. Because of this a refrigerator needs frequent defrosting. This may be manual, semi-automatic or automatic. In manual defrosting, the refrigerator is switched off to allow the layer of ice on the evaporator to thaw. In the case of semi-automatic defrosting, the refrigerator has a push-button which, once activated, starts the defrost cycle. Defrosting may simply involve switching off the compressor for a period of time or until the evaporator temperature rises above 0°C. An electrical heater may be used to accelerate this process. Alternatively the refrigeration cycle may be reversed so that the evaporator is heated by the refrigerant. Automatic defrosting is activated by a time switch, usually set to operate during the night when the door of the refrigerator is unlikely to be opened, using one of the two methods described for semi-automatic defrosting. In all methods of defrosting, the water, once it has been thawed must be removed. If defrosting is carried out at irregular intervals (manual and semi-automatic) the water is trapped in a tray held below the evaporator. With automatic defrosting which takes place once every 24 hours only small volumes of water are produced. This may be piped to a tray outside the cabinet from where it will evaporate, or on to a plate above the motor from where the water evaporates.

Single door refrigerators have a capacity of 150–350 litres usually with a frozen food compartment of between 13 and 25 litres. The refrigerator may be free-standing, under counter or built-in. Energy consumption for a refrigerator without a frozen-food compartment is 3 to 6 kWh/week and with a frozen food compartment 4 to 10 kWh/week. The frozen food compartment may be marked with one (−6°C), two (−12°C) or three (−18°C) stars.

Refrigerators may contain automatic or semi-automatic ice-makers. Automatic ice-makers must be supplied with a plumbed-in cold water supply.

15.3.5 Freezers

Freezers are constructed in a similar way to refrigerators but usually with a higher standard of insulation. There are three types of freezer: the chest freezer; the upright freezer; and the upright fridge/freezer combination.

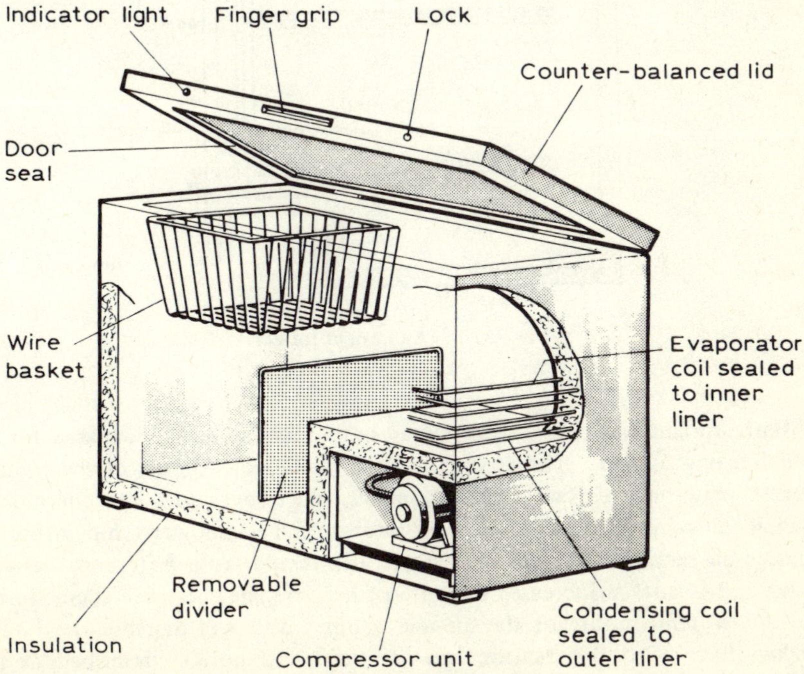

Fig. 15.7 – A chest freezer.

A chest freezer is shown in Fig. 15.7. The evaporator coil is fastened to the insulation side of the inner metal wall. The condenser coil may either be mounted as a separate coil at the back of the freezer or it may be fastened to the outer metal wall, as shown in Fig. 15.7. Chest freezers take up more floor area than upright freezers but are cheaper to buy and run and retain their temperature for longer during a power failure. However food is often difficult to locate. Organisation can be improved by the use of plastic-coated wire baskets.

Upright freezers have the evaporator coil running around the roof, back or sides. Additionally the coil may run under one or a number of the shelves as shown in Fig. 15.8. Upright freezers hold less food for a given volume than chest freezers but are easier to organise and take up less floor space.

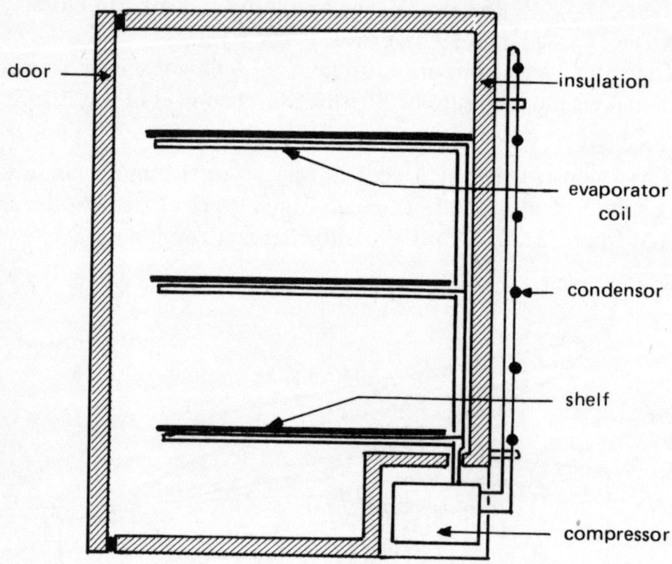

Fig. 15. 8 – An upright freezer.

Both upright and chest freezers may have a special area designated for the home freezing of food. This is usually sited at the coldest part of the cabinet. Freezers may have a 'fast freeze' control, which over-rides the thermostat, allowing the temperature of the evaporator to fall to a lower temperature. As discussed in section 8.3.1, an increase in the temperature difference between a fluid and a surface increases the rate of heat transfer between them. In this case, a low temperature of the air and cabinet wall will increase the rate of freezing. Even with this feature however, it could not be described as fast freezing (section 15.2.2). Freezers should not be overloaded with large amounts of unfrozen foods as this will cause a rise in the temperature of any frozen food being stored which will affect its quality.

As with refrigerators, freezers require regular defrosting. This may be manual or automatic. They may also have a visual and/or audible warning if the temperature of the cabinet rises too high. Warning devices may also be fitted as an accessory.

If there is a power failure, a full freezer with the door closed will hold the food at a satisfactory temperature for a long period of time. An upright freezer will take over 12 hours for the temperature to rise to −9°C (Consumers Association 1978). A chest freezer will perform much better than an upright freezer because of the higher standards of insulation.

15.3.6 Refrigerator/freezer combinations

The advantage of a fridge/freezer combination is that less floor area is needed than for two independent appliances. They consist of a two door double cabinet with a refrigerator section at the top and freezer below. A frozen food compartment is not provided in the refrigerator. The two sections may share the same refrigeration equipment with the evaporator split between the two, or they may have two separate mechanical refrigerators.

REFERENCES

Consumers Association (1978), *Which,* Upright Freezers, **August**, p. 431.

Duckworth, R. (1966), *Fruit and Vegetables,* Pergamon.

Heldman, D. R. (1975), *Food Process Engineering,* AVI.

Jarvis, B. (1972), in *Health and Food,* Eds. Birch, C., Green, L. & Plaskett, L., *Applied Science.*

Milson, A. & Kirk, D. (1980), *Principles and design of catering equipment,* Ellis Horwood.

Rolfe, E. J. (1967), in *Biological and Biochemical engineering,* Ed. Blakeborough, N., Academic Press.

Riemann, H. & Bryan, F. L. (1979), *Food-borne Infections and Intoxications,* 2nd Ed., Academic Press.

Van Arsdel, W. B., Copley, M. J. & Olsen, R. L. (1969), *Quality and stability of frozen foods,* Wiley.

BIBLIOGRAPHY

Consumers Association (1978), Chest Freezers, *Which,* **March**, p. 196.

Consumers Association (1979), Refrigerators, *Which,* **August**, p. 249.

Consumers Association (1981), Small Chest Freezers, *Which,* **May**, 268–273.

Peet, L. S., Pickett, H., Arnold, M. C. & Wolf, L. H. (1975), *Household Equipment,* 7th Ed., Wiley.

Scharff, R. (1976), *Refrigeration, air conditioning and range servicing,* McGraw-Hill.

CHAPTER 16

Motor driven kitchen appliances

16.1 INTRODUCTION

In addition to cooking appliances, described in Chapters 13 and 14, and refrigeration, described in Chapter 15, many other electrical appliances may be used in the kitchen. Most of these are labour-saving devices, in which an electrical motor (the principles of which are described in section 5.7.2) is used to replace a manual process. These appliances fall into three groups: food preparation machines, which are used for mixing, kneading, blending etc.; dishwashing machines, which reduce the chore of hand washing-up; and waste-disposal appliances.

16.2 FOOD PREPARATION MACHINES

16.2.1 Food Mixers

Food mixers consist of an electric motor connected by gears to one or two mixing elements which are usually interchangeable. The motor can be operated either over a range of fixed speeds or it may be continuously variable.

In a motor with fixed speeds, these are selected by connecting the stator coils (Fig. 5.25) in various ways (Fig. 16.1). The combined speed-selector/

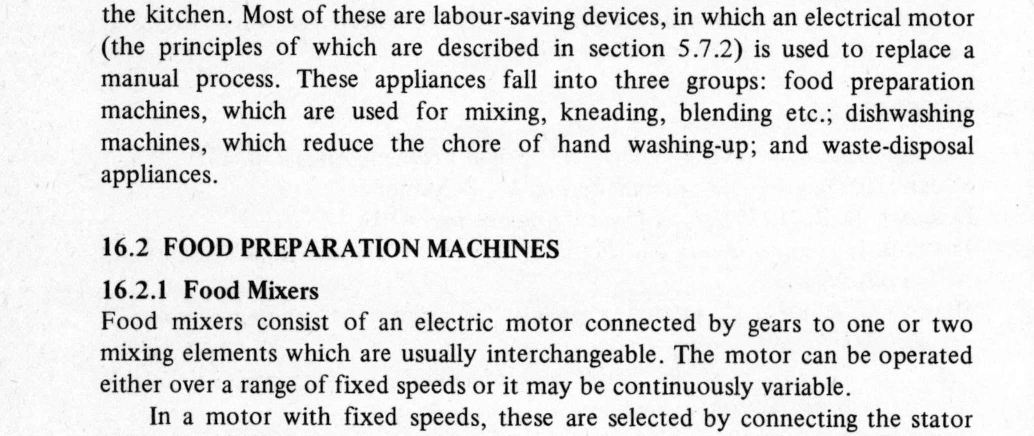

(a) low-speed single coil.

(b) medium-speed-coils in series

(c) high-speed-coils in parallel

Fig. 16.1 – Speed selector using stator coil windings.

on-off switch is located on the handle and has a number of click-stop positions. Its location on the handle is arranged to facilitate operation by the fingers.

Where continuously-variable mixer speeds are required these use either a governor or a triac. A governor uses feedback control of the motor speed (section 9.4) by measuring the centrifugal force produced by the rotation of the motor to open electrical contacts. The speed control adjusts the amount of centrifugal force required to open the contact (Fig. 16.2). A triac is an electronic device which controls the supply of electricity to the motor, as used in electric-light dimmers (section 12.2.1).

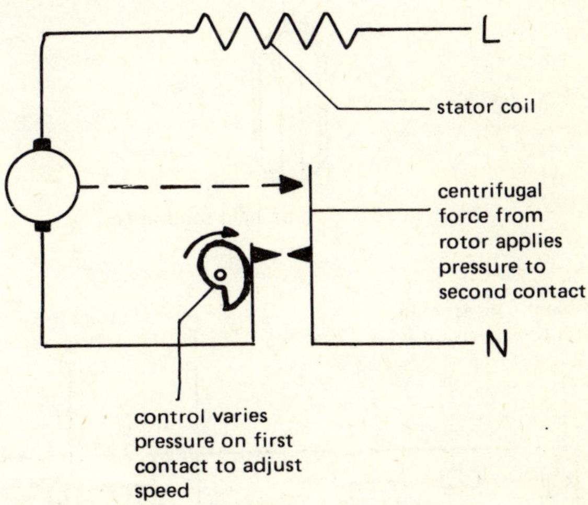

Fig. 16.2 – Centrifugal speed controller showing centrifugal force opening contacts.

A typical hand-held food mixer is shown in Fig. 16.3. Normally they have two mixing elements which rotate in opposite directions so that food is drawn between them. On some machines different attachments may be selected for whisking and mixing. The low power of the motor (100–150 W) means that only small volumes of food can be mixed and they cannot be used for heavy-duty mixing such as kneading of bread dough. The mixers have double insulated bodies (section 5.4.3). Normally these mixers have two, three and four fixed speeds.

Some hand-held mixers can be used with a stand and bowl so that they can be used as a table mixer. In addition, many of these hand-held mixers have liquidiser (section 16.2.2) and vegetable preparation (section 16.2.3) attachments.

Table mixers may have the mixing head, with a continuously-variable speed control, entering the mixing bowl from above (as shown in Fig. 16.4) or from below through the base of the mixing bowl. Interchangeable mixing attachments are used for whisking, mixing and kneading. Normally table mixers

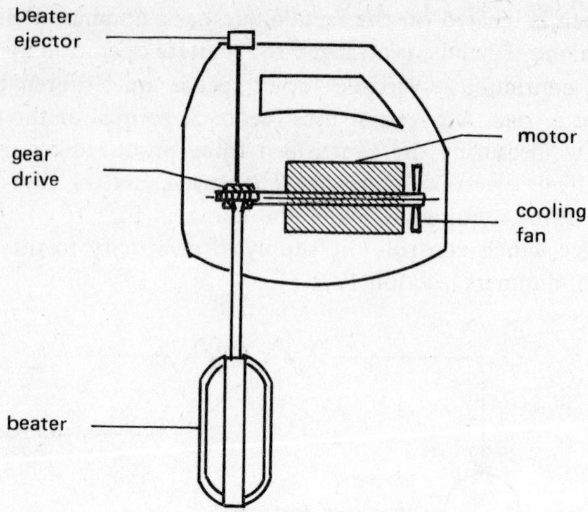

Fig. 16.3 – Hand-held food mixer.

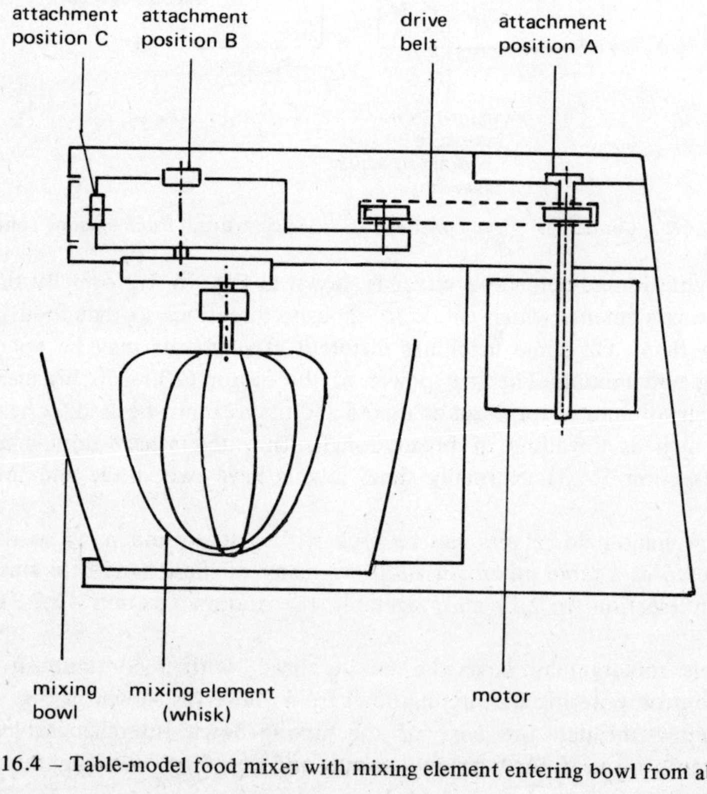

Fig. 16.4 – Table-model food mixer with mixing element entering bowl from above.

use a planetary action (Fig. 16.5). This may be achieved either by having the rotating mixing element mounted on a rotating mixing head (Fig. 16.5) or the mixing element may be off-set from the centre of the mixing bowl which itself rotates. Table mixers normally have more powerful motors (150–500 W) than hand-held mixers so that they can be used with larger volumes of food and with foods of much higher viscosities.

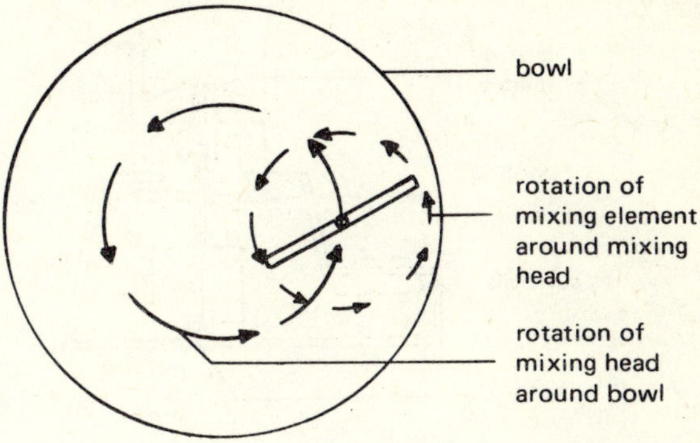

Fig. 16.5 – Planetary mixing action.

The same motor can also be used with a wide range of accessories such as liquidisers (section 16.2.2), vegetable preparation attachments (section 16.2.3), can-openers, mincing machines, potato peelers, etc. The mixing machine shown in Fig. 16.4 has a number of attachment points with different speed outputs. For example, the liquidiser requires a high speed and connects directly onto the end of the motor shaft at point A. Other attachments requiring slower speeds mount on the machine at points B or C.

16.2.2 Liquidisers (blenders)

Liquidisers (blenders) consist of a glass, transparent plastic, or stainless-steel goblet which has a handle and tight fitting lid and a capacity of ½ to 1 litre. The goblet has rotating knives mounted in its base (Fig. 16.6). It fastens to a motor base which may be part of a food mixer or may be an independent unit. The motor (200–300 W) is connected directly, without reduction gears, to the cutting blades to provide very high speeds (up to 3000 rev/min). They may provide a single speed, 3 to 7 selected speeds or continuously-variable speed control. To prevent damage to the motor they should not be used continuously for long periods of time (manufacturer's recommendations vary from 45 seconds to 3 minutes maximum). Liquidisers can be used for soups, purees, batters, grinding cereals and sugar and making breadcrumbs. They are often supplied with a separate coffee grinding attachment.

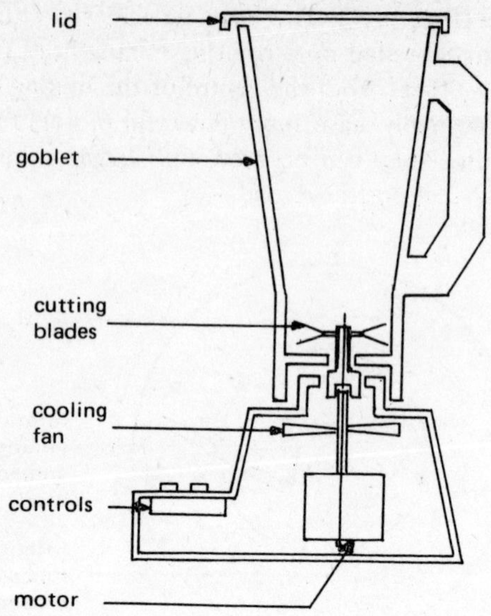

Fig. 16.6 – A liquidiser/blender.

16.2.3 Food Processors

Food processors are similar to a mixer with the mixing element mounted from below except that they use knife blades rather than a mixing element (Fig. 16.7). The mixing bowl, with a capacity of ¾ to 1 litre can be used for mincing, cutting, pureeing, kneading and beating. A powerful motor (400 to 650 W) is

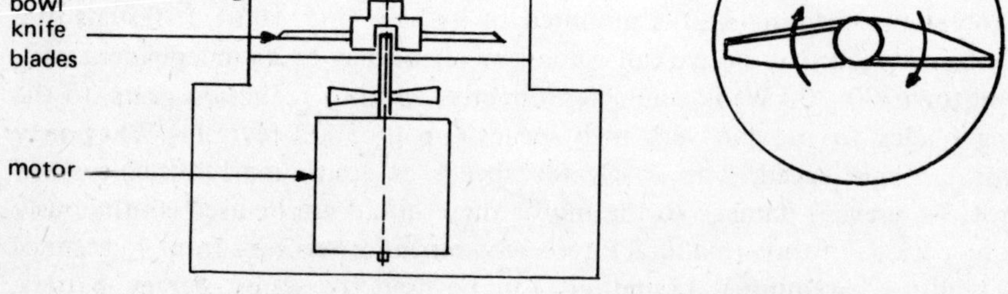

(a) side view. (b) top view.

Fig. 16.7 – Food processor-mixing bowl attachment.

used which may have a single speed or a choice of two speeds which are intermediate between those of a mixer and liquidiser (1200 rev/min). On some models a choice of metal and plastic blades is provided.

Food processors have an attachment for slicing and shredding (Fig. 16.8). This is very similar to the attachment used for table-model food mixers. Normally blades are provided for slicing and shredding. An ejector plate removes the food from the attachment once it has been cut. Some models do not have an ejector plate, the cut food being retained in the base of the mixing bowl.

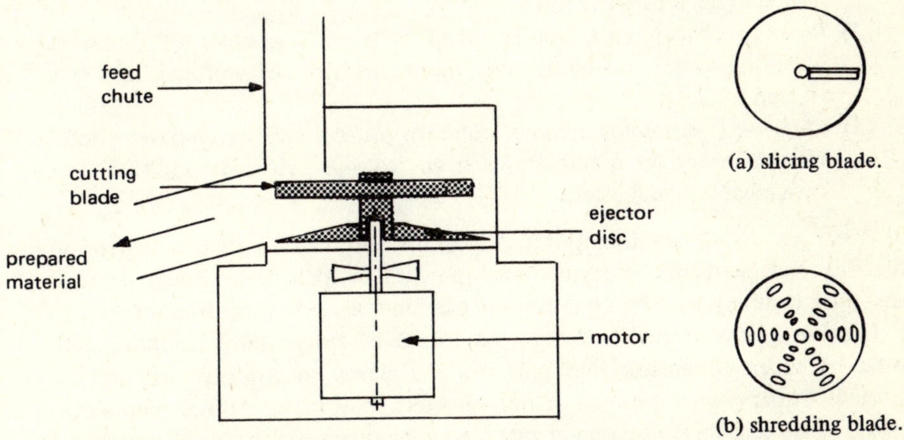

Fig. 16.8 – Food processor – slicing and shredding attachment.

16.2.4 Electric carving knives

Electric carving knives have a motor in the handle which may be powered by mains electricity or by rechargeable batteries. This motor drives a pair of reciprocating blades moving to give 30–50 cutting actions per second. The two blades move in opposite directions. Some models have interchangeable blades to suit the cutting of bread, meat and frozen foods.

16.3 DISHWASHING MACHINES

In a typical household 10 hours/week will be spent washing-up by hand unless a dishwashing machine is used, which reduces the time to 2 or 3 hours/week, the time taken to load and unload the machine. Additionally, because of the more powerful detergents, high final rinse temperature and air-drying, the process is more hygienic.

16.3.1 Principles of dishwashing

Food residues are removed from food by a combination of hot water (60 to 65°C), mechanical force from jets of water and the use of powerful detergents.

The wash is preceded by a cold or hot rinse and is followed by a number of cold rinses and final hot rinse (65°C) and air drying.

The detergent used is specifically developed in powder form for use in dishwashers. It uses a low-foaming, high-alkaline formulation. The major components are described below.

(1) *A surfactant* (non-ionic wetting agent) which reduces the surface tension of the water allowing easier wetting of dirty surfaces. It also acts as an emulsifying agent which helps to remove oily and greasy food residues (see section 18.2.1).
(2) *Builders* (phosphates and silicates) perform a number of functions including water softening and the suspension of soil particles (see section 18.2.2).
(3) *Chlorine compounds* remove stubborn protein and starch-based foods. They also act on tannin stains from tea and coffee. In addition they have a bactericidal effect.

Other chemicals are incorporated into the formulation such as corrosion inhibitors and perfumes. Enzyme-based powders, which break down protein-based food residues, may be used in some machines as a pre-wash treatment.

Because of the strength of detergents used in dishwashing machines, care should be taken to ensure that patterns and glazes on crockery will not be damaged. Cutlery with wooden or bone handles may suffer damage. Similarly, glassware may lose its transparency and sparkle by repeated washing in a machine. Crockery and glassware which is safe for dishwashing can be purchased. This usually carries a 'dishwasher-proof' label.

In addition to the detergent, most dishwashers incorporate a water softener and a final rinse aid. The water softener is an ion-exchange resin (as described in section 6.5) and needs periodic re-generation with salt. The rinse aid is a non-ionic wetting agent which is added to the final rinse. It prevents the water forming into droplets as it dries, ensuring that it forms an even film on the surface. Otherwise, drying may leave spots and streaks from minerals in the water on the surface of dishes.

16.3.2 The construction of a dishwasher

Dishwashers (Fig. 16.9) may be free-standing, built-in or bench-top models. They are usually front loading but some models load from the top. The outer cabinet is of sheet steel coated in enamel. It is separated by thermal and sound insulating materials from the inner cabinet. The door, which is constructed in the same manner as the cabinet, has a gasket which forms a water-tight seal. A micro-switch in the door is linked to the controls, switching off the machine if the door is opened. The door also has locations for the detergent and rinse-aid (and, if it is used, a pre-wash powder). These are activated automatically during the washing cycle (using a solenoid or bi-metal device) releasing the chemicals

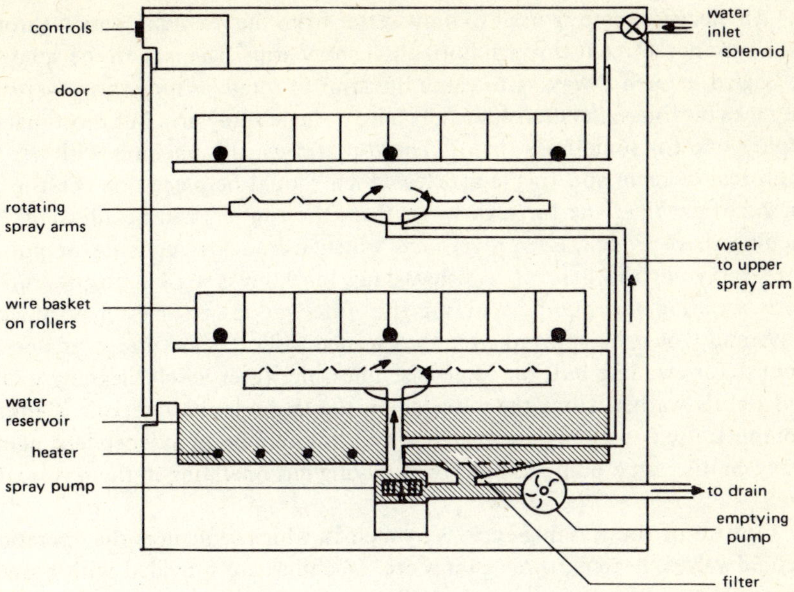

Fig. 16.9 – A dishwashing machine.

at the correct stage in the sequence. A reservoir in the base of the cabinet is used to hold salt which is used to re-generate the water softening medium. Many machines have an indicator to show when more rinse-aid and salt must be added. Plastic covered wire racks are provided for holding dishes. These slide into and out of the machine on telescopic arms with low-friction nylon rollers. The racks provide locations for plates, cups, saucers and bowls. Cutlery is held in small plastic-covered wire baskets. Dishwashers vary in size from four to twelve settings.

Dishwashers are provided with a plumbed-in water supply (in some cases tap connectors may be used). Machines may be cold fill only or hot and cold fill. Hot-fill machines may be more economic if a cheap source of hot water is available and the total time of the wash cycle is also reduced by up to one third. The inlet water supply is activated by solenoid valves and the reservoir has a water-level detector to close these valves once the correct amount of water has entered the machine. Most machines have a safety water valve should the main solenoid valve become jammed open. Without this there is a risk that the machine will flood. In the base of the reservoir is an immersion heating element of 2 to 3 kW, which is used to achieve and maintain the correct water temperature, under the control of a thermostat. The same heaters are, on some machines, used to dry the plates once the reservoir has been emptied. On other machines separate drying heaters (600 W to 1 kW) are provided. The reservoir holds 8 to 15 litres of water.

An electric pump is used to draw water from the reservoir, pass it through a filter and then force it through horizontal spray arms. The jets on the spray arms are angled in such a way as to cause the arms to rotate as they spray water onto the racks of dishes. Small dishwashers have a single spray arm but most machines employ two (or sometimes three). The top arm usually operates with less force to protect delicate and fragile articles which should be placed on the top rack. On some machines the force of spray form the top jets can be adjusted. Some machines have a 'delicate' programme which uses a low pressure or pulsating water spray. In the base of a dishwashing machine is a filter to prevent food debris entering the pump. Without this filter, food particles may block the sprays and would be sprayed over the cleaned dishes. This filter needs periodic removal for cleaning but, on some machines, the filter is self cleaning with the food debris washed from the filter down the drain as the reservoir is emptied. A pump is used to empty the machine and this may be an independent pump or it may be the same pump as used for spraying but operating in the reverse direction.

Control of the washing cycle is by a clock which sequences the operation of solenoid valves, heaters, spray pump etc. Machines are provided with a number of programmes. The normal or standard programme consists of a cold or hot prewash rinse, washing at 65°C, 2 or more cold rinses, a final hot rinse at 65°C followed by air drying. This programme takes 75 to 105 minutes and uses 2.3 to 3.2 kWh of energy and 50 to 70 litres of water (Consumers Association 1980). These times and energy requirements can be reduced using a hot-fill machine. In addition to the basic programme alternative programmes provide: intensive programmes for pots and pans; energy saving programmes using a wash temperature of 55°C; and a rinse and hold programme to allow rinsing of dishes when they are not going to be washed for some time. Machines may provide between four and twelve programmes.

16.4 WASTE-DISPOSAL MACHINES

A waste-disposer fits into a 90 mm sink opening and can be used to dispose of waste food material (excluding bones). The motor rotates the grinding blades at a high speed (1000 to 2000 rev/min). On some models the direction of the blades can be reversed. When not in operation a plug is used in the sink outlet. When in operation water is run through the machine and food is thrown by centrifugal force against the shredding rings. As it breaks down, the food is washed along the drain, suspended in the flow of water. The flow of water also prevents waste collecting in the sink trap. Many waste-disposal machines have a safety cut-out which operates if the blades should become jammed. Once activated the cut-out must be re-set manually.

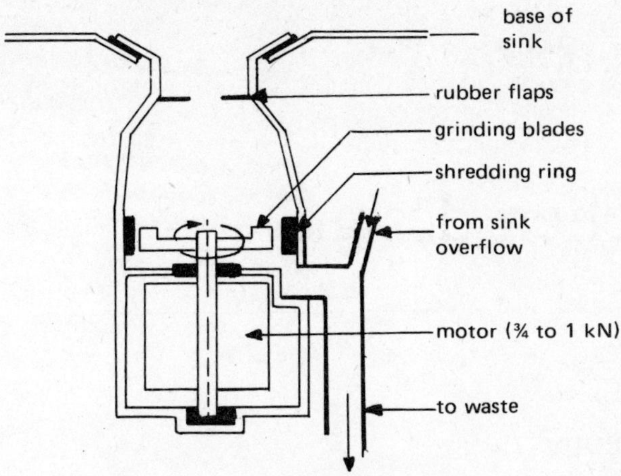

Fig. 16.10 – A waste-disposal machine.

REFERENCES

Consumers Association (1980), Dishwashers, *Which,* **Sept.**, p. 534–8.

BIBLIOGRAPHY

Consumers Association (1979), Food Mixers, *Which,* **August,** p. 435–41.

Consumers Association (1981), Detergents and Rinse Aids for Dishwashers, *Which,* **August,** p. 481–7.

I.E.C. (1973), *Methods for measuring the performance of electric dishwashers,* International Electrotechnical Commission Publication No. 436.

Morrison, B. (1971), *Food Mixers,* Forbes Publications.

Peet, L., Pickett, M., Arnold, M. & Wolf, I. (1975), *Household equipment,* Wiley.

CHAPTER 17

Kitchen design

17.1 INTRODUCTION

The kitchen represents one of the most used work centres within the house. Research in Sweden (reported by Grandjean 1973) showed that between the hours of 7 am and 7 pm a housewife spends over half her time (52%) in the kitchen and a quarter of her time either preparing, consuming or cleaning up after meals. Whilst these proportions of time are decreasing with the use of labour-saving devices and convenience foods the design and layout of a kitchen is still very important in terms of minimising the time spent and physical energy used in this area of the home. A poorly designed kitchen can lead to fatigue and tiredness, caused by unnecessary walking, reaching and bending. Also the design of the environment is very important, particularly ventilation (Chapter 11) and lighting (Chapter 12). It is the site of a large proportion of domestic accidents (Chapter 20) and therefore the design of a kitchen must minimise the risk of these. A wide range of activities takes place. It is used for the storage of food and utensils, for food preparation, cooking, food service and dishwashing. Many kitchens are also used for laundry. Additionally, a kitchen may have a dining area which varies from a simple breakfast bar to an integrated kitchen-dining room. The utilisation of a kitchen varies from one family to another depending on the number and ages of people in the family and their life style. Ideally its design should be based upon the needs of a particular family, including the selection of work-surface heights and shelf heights to suit the size of person using the kitchen (section 17.3). In practice, since more than one person uses a kitchen, compromises must be made. In some cases, such as the design of a kitchen for a handicapped person, a degree of personal design is essential.

17.2 THE SHAPE AND DESIGN OF A KITCHEN

With an existing house the basic shape and design has already been established by the architect who designed the house. However, the shape and size can usually be modified. A kitchen which is too small to provide satisfactory space for working may often be enlarged by an extension or by knocking down a wall. Local authority planning permission may be required for structural alterations. A kitchen which is too large also causes problems because of the large walking distances between work centres. This may be improved by the use of island work surfaces to reduce walking distances between the major work centres (section 17.5), or by the use of a partition wall. It is difficult to generalise about the optimum size of a kitchen because of its dependence on the life style of the family. Grandjean (1973) recommends a minimum of 8 m^2 for an enclosed kitchen for a family of four and 12 m^2 in the case of a kitchen–dining room. Similarly it is difficult to generalise about the storage space required since it depends upon the amount of crockery owned, whether the house is close to shops, whether the family carries out bulk buying, etc. For the purpose of this Chapter, storage is considered as a part of the design of the work centre (section 17.4).

Kitchens may be classified into three basic types as follows:

(a) The isolated or enclosed kitchen: this is self-contained and is connected to other parts of the house by doors. The link to the dining room may be through a door and/or service hatch.
(b) The combined kitchen–dining area: this is similar to the isolated kitchen but it incorporates a dining area within the self-contained area. The dining area may be a table or bar for informal meals, with a separate formal dining room or it may contain a dining table for all occasions.
(c) Open-plan living space incorporating a kitchen area: the kitchen forms a part of a larger living space. The link between the two may be an island work surface, a breakfast bar, a dining table, a moveable screen or a room divider.

The arrangement of work surfaces within all three types of kitchen may be in one of three forms (Fig. 17.1). A single-sided kitchen is suitable only for small kitchens because, in a large kitchen, it would involve long walking distances between work centres. A galley kitchen provides excellent relationships between work centres providing that the aisle is neither too wide nor too narrow and also providing that the aisle does not form a main passage or corridor from one part of the house to another. An L-shaped kitchen can also provide good relationships between work centres, but the corner (and particularly storage space in the corner) is difficult to use. The same is true of a U-shaped kitchen, except that here there are two corners.

17.3 ERGONOMIC PRINCIPLES OF KITCHEN DESIGN

17.3.1 Use of Vertical Space

Ergonomics is concerned with the relationships between man and his environment and, more particularly, with designing working environments to suit the

(a) one sided kitchen.

(b) galley kitchen.

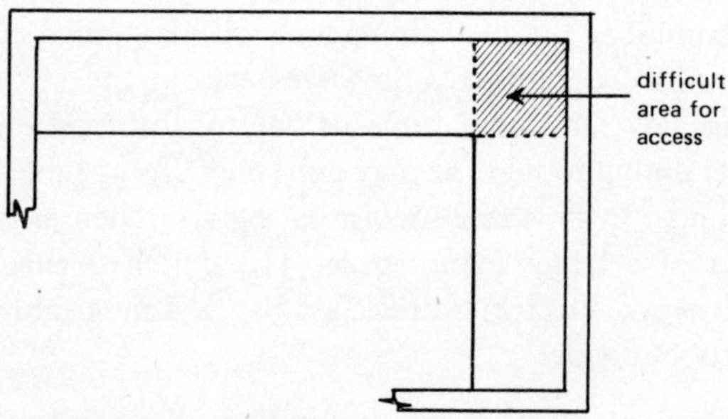

(c) L-shaped kitchen.

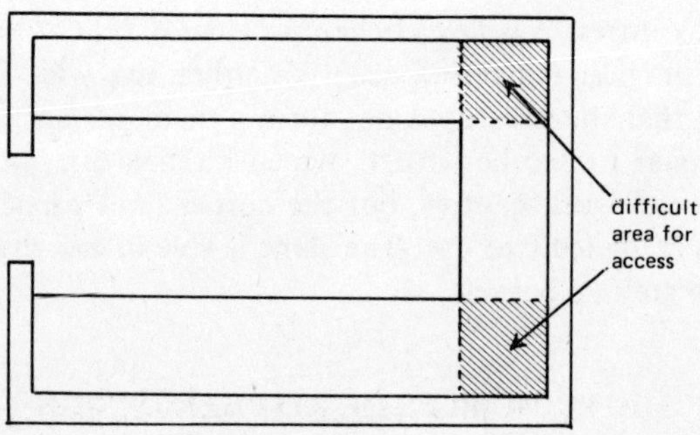

(d) U-shaped kitchen.

Fig. 17.1 – Basic arrangements of work surfaces in a kitchen.

characteristics of the human body. Ideally a kitchen should be designed to match the habits and physical characteristics of a single person who is the sole user of a kitchen. In practice design must be more generalised and bases on the average characteristics of a population based on statistically valid data of shapes and sizes of people. Flexibility can be built into a design by, for example, the use of variable-height shelving. Specialist design may be necessary in the case of a kitchen for a handicapped person or for a person of abnormal size. Standardised designs allow mass production of units which lowers their cost. Individually designed units would be prohibitively expensive for most people.

Within a kitchen the whole vertical height between floor level and ceiling height (2 to 3 m) is used. This space must be used to minimise the amount of movement required. A hierarchy (i.e. a system arranged in graded order) has been established in relation to the most economic body movements (Table 17.1), and an objective is to keep movement to the lowest level possible in this hierarchy.

Table 17.1

Hierarchy of economic movement.

Least economic movement	Walking
↑	Bending at waist, crouching and squatting
	Upper arm movement
	Lower arm movement
	Wrist movement
Most economic movement	Finger movement

The maximum upward reach for a person is 1.24 times their height (Grandjean 1973). Since the average height for women is about 1.6 m and for men 1.7 m, the maximum height for reaching, without the use of steps, is 1.98 m for women and 2.11 m for men. This gives the maximum height of a shelf as shown in Fig. 17.2. The height must be reduced for access to a deep shelf or if the shelf is situated over a cupboard. (These figures are based upon the limit of reach for a woman of average height; adjustable shelves are of great advantage to suit occupants of different heights.)

If wall cupboards are used rather than shelves, the same relationships apply. Additionally, clearance must be left between the top of the work surface and the bottom of the wall cupboard. For this reason the bottom of the wall cupboard should be at least 1350 mm high.

The height of a work surface should theoretically be based on the height of the elbow of the person using the work surface, whether seated or standing. For manual work, where pressure must be applied from the upper torso and arms, the vertical distance form the elbow to the work surface should be between 75 mm and 200 mm. For skilled work, the forearms should be on the same

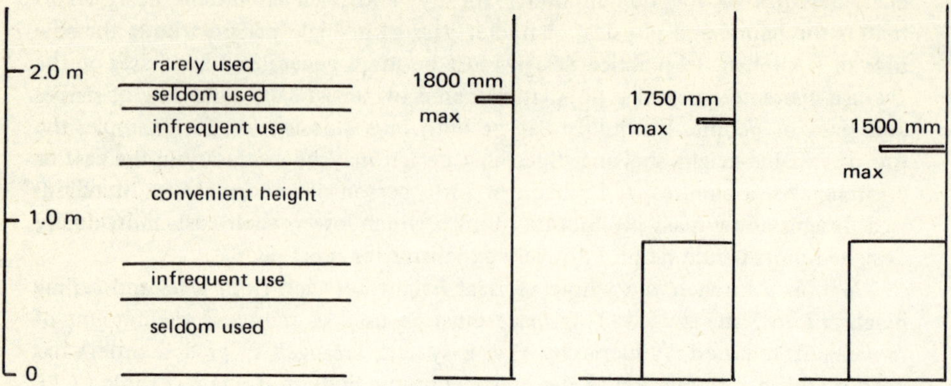

(a) use/height relationship. (b) narrow shelf. (c) narrow shelf over cupboard. (d) deep shelf over cupboard.

Fig. 17.2 – Maximum heights for storage space in a kitchen.

level as the work surface. These two factors, coupled with average heights for men and women give the data shown in Table 17.2 (Grandjean 1973). Many kitchen units are now standardised on a height of 900 mm, which is satisfactory for skilled jobs but is rather high for manual jobs, particularly for a short person. It may be useful to have at least one work centre with a height of 800 mm; this would be most logically the area which is used for kneading bread, rolling pastry and hand mixing (section 17.4). For a seated person, the height of chair or stool should be chosen to provide the correct relationship of elbow height to working surface height.

Table 17.2
Heights of Working Surfaces

	Male	Female
Manual activity	750-900 mm	700-850 mm
Skilled activity	900-950 mm	850-900 mm

17.3.2 Use of Horizontal Space

The correct planning of horizontal space in a kitchen is just as important as the use of vertical space. Room must be provided for the body and limbs of a standing or seated person at the work centre and there must be space for the easy and safe operation of equipment. Additionally, there must be space for the circulation and movement of people, particularly where the kitchen forms a main traffic route through the house.

The horizontal reach of an average male is 550 mm and for a female 500 mm (Grandjean 1973). Therefore, a work surface should be at least 500 or 600 mm deep. If it is any deeper than this it makes access to cupboards below and shelves above difficult. The width of a work centre depends on the activities of that work centre, but a minimum value is 450 mm. However, an area by the side of an oven may be as small as 250 mm if it is used as a resting place for pots, baking tins, etc. Table 17.3 gives minimum dimensions of work spaces in front of a work surface and of dimensions for passages and traffic routes.

Table 17.3
Working spaces and clearances in a kitchen

Width for working next to counter	800 mm
Freespace in front of opening appliance	1000 mm
Clearance for bending from waist	1200 mm
Clearance for squatting	1100 mm
Width of restricted passage	500 mm
Width of unrestricted passage	700 mm
Width of passage used for carrying large objects	900 mm

17.4 KITCHEN WORK CENTRES

Each work centre should be designed to be as self-contained as possible, each with its own storage locations and work surfaces. However, links between centres are important and are discussed in section 17.5.

Work centres are associated with a major activity within the kitchen and the location of materials and equipment used in that activity. The working height and area required (including both work surface, floor space and storage space) must be related to the needs of the activity and also how frequently it occurs.

The three main work centres within a kitchen are food preparation, sink and cooker. In a large kitchen it is usual to sub-divide these work centres, and include additional ones, as shown in Table 17.4.

17.4.1 Food Storage

The amount and type of storage space required depends upon factors such as closeness to shops and buying habits (for example, bulk buying, use of a freezer centre, etc.). The food storage area should be used to store 'ambient' foods (that is, stable at room temperature such as dry goods and tins), for larder or refrigerated storage (6 to 12°C for fruit and vegetables), refrigerated storage

Table 17.4
Division of kitchen activities into work centres

Main Work Centres	Sub-divisions of Main Centre
Sink	Dishwashing Laundry
Food preparation	Food Storage Food Preparation Food Mixing
Cooking	Cooking Food Service

(2 to 6°C for perishable foods such as meat and dairy products) and for frozen storage (Chapter 15). Frequently used foods may be more conveniently stored close to an appropriate work centre. For example, flour, sugar and dried fruit would be better near the mixing centre and salts, herbs and spices close to the food-preparation or cooking centre. Associated with the food-storage area should be a work surface for unpacking shopping and storage locations for wrapping materials, polythene bags, etc.

Ambient foods can be stored on open shelves, wall cupboards or storage cupboards under work surfaces. Shelves are useful for frequently used items but closed cupboards are better for most foods. Cupboards can have sliding or hinged doors. Sliding doors use less kitchen floor area but are not so convenient for access. Also the grooves of a sliding door fitting can collect dirt and spilled food and are difficult to clean. It is often difficult to use the full depth of a cupboard under a work surface because of difficulty of access to the back. Plastic covered wire baskets can improve the utilisation of space. These may pull out on tracks or rollers or may be fastened to the back of the door. The latter type are particularly useful for access into corner cupboards in U- and L-shaped kitchens (Fig. 17.1).

17.4.2 Food Preparation

In a small kitchen there may only be sufficient room for a single work surface for food preparation, food mixing (section 17.4.3) and food service (section 17.4.5). In a larger kitchen it is advantageous to divide these activities into separate work centres.

As discussed in section 17.3, the minimum length of work surface should be 450 mm but a better width would be 600 mm. This space should have convenient power points and storage for measuring cups and jugs, knives, scissors, etc. Many of these items can be located in draws beneath the work surface or on hooks behind the work surface for ease of access.

17.4.3 Food–Mixing Centre

A food-mixing centre requires 600 to 1000 mm of working surface at a convenient height for working (section 17.3). Location is required for food mixers, blenders and food processing machines. Storage space is required for accessories to these machines and also for mixing bowls, measuring jugs and baking tins. Foods which are usually associated with this area (such as flour) can also be located here. A power point is required in the food-mixing centre.

17.4.4 Cooking Centre

The layout of the cooking centre depends upon the type of cooker (range) chosen (section 13.2). There should be sufficient work space associated with this area for putting down pans, baking tins, etc., from the hob and oven. Additionally, space is required for independent cooking appliances (section 14.3). Because of their frequent usage, storage space with easy access is required for pots and pans. Salt and seasonings can be stored in a wall rack adjacent to this area. The cooker centre requires adequate services (electricity and/or gas and ventilation).

17.4.5 Food-Service Centre

In this area there is a requirement for a work surface for plating meals or placing food in service dishes. There should also be space on the working surface for appliances associated with this area such as kettles, coffee makers, hot cupboards. Power sockets will also be required for these appliances. Storage space is needed for crockery, cutlery and other table ware, table linen, place mats, etc.

17.4.6 Sink Centre

Sinks are used for food preparation, dishwashing, water for cleaning surfaces and floors and, in some cases, laundry. The sink is central to most of the activities which take place in a kitchen and its location is therefore very important (section 17.5). In a large kitchen the sink area can be divided and sinks associated with, for example, the food preparation centre, leaving the main sink to be used for dishwashing. A number of combinations of single or double sinks (with and without waste disposers) and draining boards are possible. Where a dishwashing machine is used, a draining board is not essential. The sink centre requires the location of a convenient drain (section 6.7), cold water (section 6.4.2) and hot water (Chapter 10).

Storage space should be provided near to the sink for buckets and washing up bowls and detergents. A rack should be provided for hand towels and drying-up cloths. The sink centre also requires a work space on either side and a location for a dishwashing machine, if this is to be included in the design. The sink centre also provides the most convenient location for a waste bin, which can be in a cupboard under the sink.

17.5 RELATIONSHIPS BETWEEN WORK CENTRES

Work centres should be arranged to minimise the amount of unnecessary movement caused by transferring food, utensils or waste from one centre to another. To facilitate this, the first principle of kitchen design is that work centres which are closely related should be sited close together. The second principle of design is to arrange the flow of material in the kitchen from raw food to cooked food ready for service and dirty dishes to clean dishes in such a way that major work flows do not cross. This will allow two or more people to work simultaneously in the kitchen without being in each other's way. In connection with this, it is important that corridors or traffic flows through the kitchen do not pass across work flow between major work centres. With a large number of work centres, many theoretical arrangements of these centres are possible. For example, there are 720 different combinations of six work centres.

Guidelines are available, based upon observations of a large number of kitchens, to help in the relative location of work centres. In the USA, the three key work centres are taken to be sink, cooker and refrigerator. Observation has shown that the highest frequency of movement is between the sink and the cooker. Peet *et al* (1975) recommend the following relationships:

(a) the sink, cooker and refrigerator should be arranged in a triangle whose perimeter is no greater than 8 m
(b) 1 to 2 m between sink and cooker
(c) 1 to 2 m between sink and refrigerator
(d) 1 to 3 m between refrigerator and cooker.

Senior (1975) makes use of the same work centres as above and recommends a minimum of 3.6 m and a maximum of 6.6 m as the perimeter of the triangle between these work centres. Grandjean (1973) recommends a left to right work sequence of sink, main working surface, cooker, small working surface. The Parker-Morris standards for UK public sector housing suggest a sequence of work surface, cooker, work surface, sink, work surface (Manser & Manser, 1976).

These recommendations are of use, but because they are generalisations and only give information for a limited number of work centres, they will not always give the best results for a specific kitchen and family. A design, based upon personal observation of the utilisation of a kitchen, will lead to a more satisfactory relationship between work centres. To do this, observations must be made of the movements between work centres for the major activities which take place in the kitchen. The number of work centres chosen is a compromise, since the larger the number of work centres used for the study, the more satisfactory the design is likely to be, but the more complex will be the study. The activities studied should represent those which occur most frequently. A record can be kept, as shown in Table 17.5, which shows the movement of a person from one work centre to another. By summing movements between work centres, relationships between work centres can be categorised as shown in Table 17.6. These relationships can then be entered into an activity-relationship chart of the type shown in Fig. 17.3.

Table 17.5
Record of movement between work centres

Centre	Activity
Food preparation	Select chopping board
Refrigerator	Select onions
Food preparation	Peel and chop onions
Refrigerator	Remove fat
Cooker	Place fat in pan
Refrigerator	Replace fat
etc.	etc.

Table 17.6
Categorisation of relationships between work centres

Category	Relationship
4	Very important
3	Important
2	Infrequent
1	Unimportant

Once the important relationships have been identified, they can then be translated into a kitchen plan. For an existing kitchen, a plan should be drawn on squared paper, showing doors, windows, electric sockets and other services. Templates of appliances can be made to scale out of card and moved around on the plan. This process should start by arranging these work centres with the most important relationships (categories 3 and 4). Lines can be drawn on the plan to indicate flows of people and materials between these centres. Main traffic flows through the kitchen should also be identified in this way. This will indicate any cross-over of work flow between centres and also between traffic routes and work centres. Constraints of services must also be considered before a final compromise plan can be arrived at.

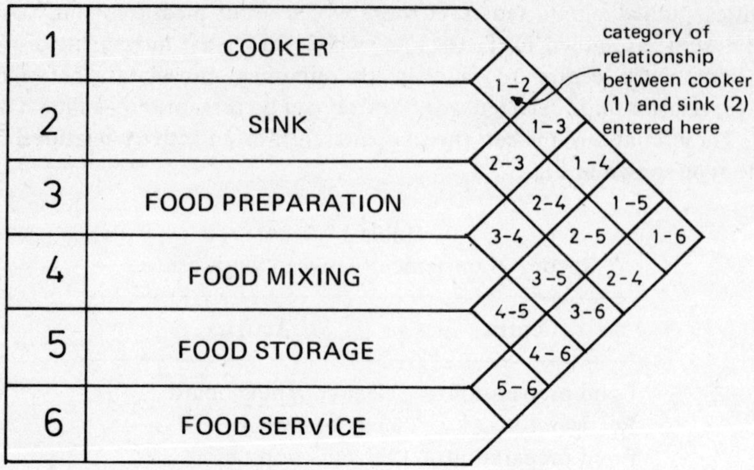

Fig. 17.3 – Activity-relationship chart.

REFERENCES

Grandjean, E. (1973), *Ergonomics of the Home*, Taylor and Francis.

Manser, M. & J. (1976), *Planning Your Kitchen*, Design Council.

Peet, L., Pickett, M., Arnold, M. & Wolf, I. (1975), *Household Equipment*, Wiley.

Senior, J. (1975), *Work Study in the Kitchen*, Pitman Publishing.

Part IV
CLEANSING EQUIPMENT

CHAPTER 18

Laundry equipment

18.1 INTRODUCTION

The objective of the laundry process is to remove dirt and soil from fabrics whilst, at the same time, minimising any damage to the fabric and ensuring that it keeps its 'feel' and 'look'. The activities associated with laundry represent 10 to 15% of the total household working time (Grandjean 1973).

Dirt is not a single material but consists of a number of substances, the most common of which are dust, adhering soil (which is often protein based), fatty material (exuded by the skin) and various types of staining. There are two main techniques for removing dirt from fabrics: washing in hot water and detergent or dry cleaning in an organic solvent. Normally dry cleaning is not carried out in the home, although dry cleaning solvents (contained in an aerosol spray) are available for use as an aid to washing and stain removal. The normal washing process consists of the following stages:

(a) washing – a combination of hot water, detergent and mechanised action
(b) rinsing – the removal of soil and detergent
(c) coarse drying – using rollers or spin drying
(d) finish drying – using air (that is, hanging clothes outside the house), a drying room, a drying cabinet or a tumble drier
(e) ironing.

There are many different types of textiles (both natural and synthetic) and care is required if the washing process is not to damage the fabric in any way (Collier 1980). Because of this there is a need for advice to be given to the householder about the best method for laundering a fabric. This advice should cover the temperature of washing, the amount of agitation, the method of water extraction and drying, recommendations for ironing and special recommendations for the use of bleach and dry cleaning solvents. There is now a code, the International Textile Care Labelling Code (ITCL), which is used as the basis for this information. This information may be included on the label attached to the fabric. In addition to the international labelling scheme there are a number of national schemes such as the Home Laundry Consultative Committee (HLCC) code used in the UK. An example of a care label is shown in Fig. 18.1. This includes information about the best programme for use in automatic washing

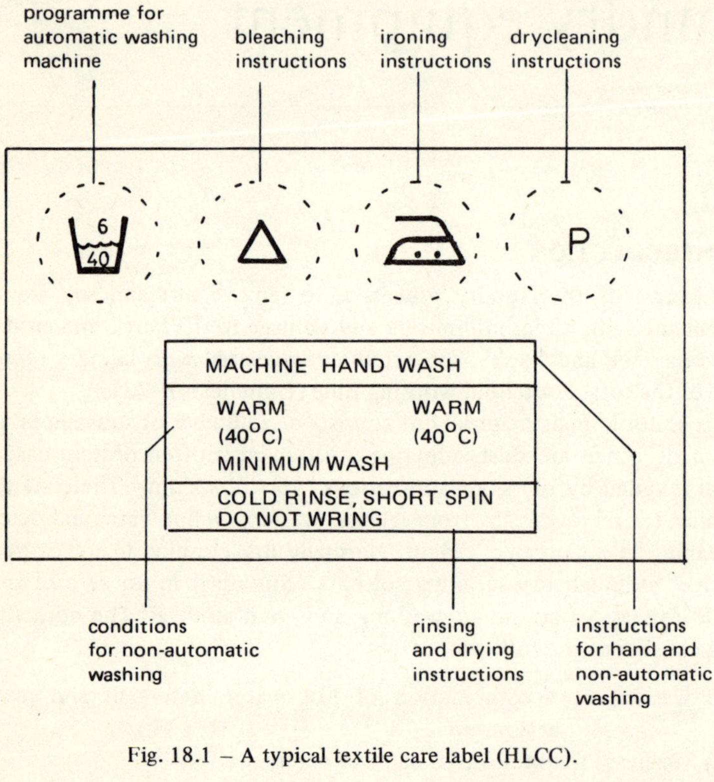

Fig. 18.1 – A typical textile care label (HLCC).

machines (section 18.3.1), non-automatic machines (section 18.3.2) and also hand washing and drying instructions. A method is used for indicating the maximum ironing temperature:

. 120°C (cool)
.. 160°C (warm)
... 210°C (hot).

18.2 DETERGENTS AND THE REMOVAL OF DIRT

The most common types of dirt are solid particulate material and fatty deposits. Their removal depends upon a combination of hot water, a washing powder and mechanical agitation. A washing powder contains many materials, the most important of which are described below.

18.2.1 Detergent

A detergent is a material which acts both as a wetting agent and as an emulsifying agent. It may be natural (based on soap) or synthetic but in both cases it consists of a molecule which has two parts, a hydrophilic (water-loving) head and a hydrophobic (water-hating) tail. When a detergent is dispersed in water it collects at the surface with the hydrophobic part of the molecule protruding out of the surface as shown in Fig. 18.2(a). The presence of this molecule at the surface reduces the surface tension of the liquid which also has the effect of allowing surfaces to become wet much more easily. This allows water to penetrate the fibres and loosen dirt particles. Within the bulk of the water, the detergent molecules form micelles (small globules of molecules with their hydrophilic heads facing out into the water) as shown in Fig. 18.2(b).

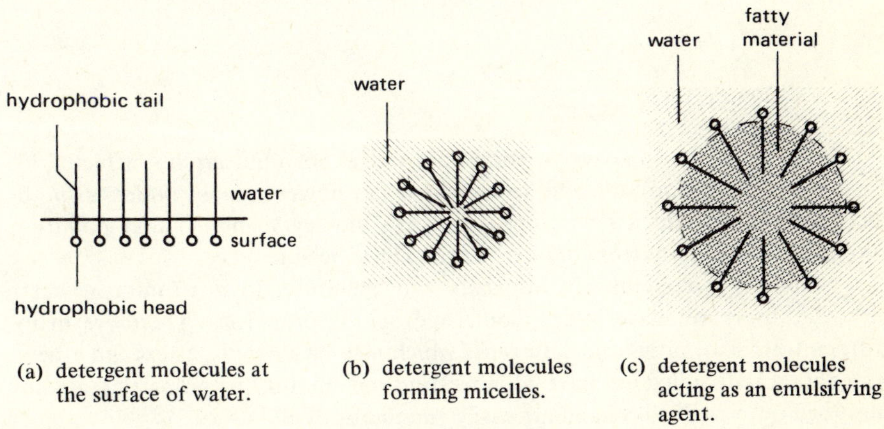

(a) detergent molecules at the surface of water. (b) detergent molecules forming micelles. (c) detergent molecules acting as an emulsifying agent.

Fig. 18.2 – Properties of detergent molecules.

If fatty material is present in a fibre, the detergent molecules collect at the surface of the fatty material, with the hydrophobic tail in the fat and the hydrophilic head in the water. The fatty material then forms into small droplets (Fig. 18.2(c)) with the detergent acting as an emulsifying agent.

Natural detergents are based on soaps, which are the salts of fatty acids.

They are manufactured from oils and fats. The hydrophobic part of the molecule is a saturated or unsaturated aliphatic hydrocarbon and the hydrophilic part is the sodium salt of a fatty acid. In water they ionise to give an anionic molecule with a negative charge. Typical soap molecules are:

$C_{15}H_{31}COONa$
sodium salt of palmitic acid

$C_{17}H_{35}COONa$
sodium salt of stearic acid

$C_{17}H_{33}COONa$
sodium salt of oleic acid (unsaturated).

The length of the carbon chain affects the properties of the soap. Long-chain soaps have poor solubility in water and are difficult to rinse out of fabrics. All soaps suffer from the disadvantage that they form scum with hard water (section 6.5).

Anionic detergents are similar to soap except that they are not carboxylic acid salts but sulphonates. A typical anionic detergent is alkyl benzene sulphonate:

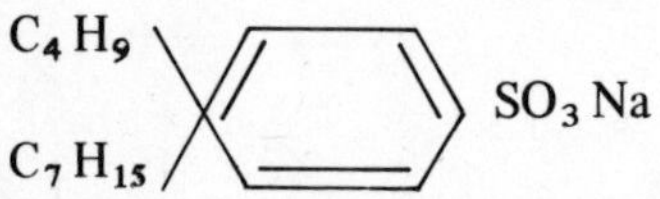

In solution they behave in the same way as soap but are not affected by hard water. They form a stable foam which can however cause pollution problems. Anionic detergents in current use are bio-degradable so that pollution problems (particularly from industrial users) have been reduced.

Non-ionic detergents are not salts and therefore do not ionise in water although they still have hydrophobic/hydrophilic properties. There are many different kinds of non-ionic detergents which may be alcohols, esters and ethers. They are low foaming but have good wetting and emulsifying properties, are not absorbed onto wool fibres and are easily rinsed out of all fabrics.

18.2.2 Builder

This is usually a phosphate. Its two major roles in a washing powder are to remove calcium ions and to aid the suspension of dirt particles. In hard water, phosphates act as softeners by forming complexes with calcium and magnesium ions. By doing this they maintain the detergency properties of the washing powder. Phosphates can cause pollution problems because they encourage the growth of algae in lakes and waterways. Low phosphate powders are available for use in regions where this is a problem.

18.2.3 Bleach

General purpose washing powders contain a bleach. This is usually sodium perborate which is only active at high temperatures (over 75°C). Clothes which will be detrimentally affected by bleach can be washed at a lower temperature (60°C) in the same washing powder as that which is used for whites (at 90°C). Light-duty washing powders without bleach are available for delicate clothes.

18.2.4 Blueing agents and optical whiteners

Blue dye may be added to a washing powder. This blue colour counteracts the yellow,pigments which remain in a fabric after washing. The combined effect of the blue and yellow pigments is to enhance the whiteness of a fabric. Optical whiteners are chemicals which absorb ultra-violet light and emit visible light at the blue end of the spectrum (fluorescence is discussed in more detail in section 12.2.2). The blue light enhances the whiteness of the fabric.

18.2.5 Other components

Foam control agents are used to stabilise and control foaming. In most powders a stable long lasting foam is desirable but in powders for front-loading automatic washing machines foaming is not necessary and can cause a poor washing action and also lead to leakages from the machine.

In addition to the builder, other *soil suspension agents* such as sodium carboxymethyl cellulose may be incorporated into a washing powder. These stabilise the suspended soil particles, preventing their redeposition on the fabric. *Corrosion inhibitors* are added to the powder to protect the washing machine. *Perfumes* are added to the powder to give an attractive smell to washed fabrics.

Special washing powders are available which contain *proteinase* (a proteolytic enzyme), (sections 13.1 and 15.1.2). If soiled fabrics are soaked in a powder containing enzymes, they will break down protein food such as blood, egg and milk. Many soils are bound by protein to the fabric and the enzyme will destroy the protein, loosening the soil. Since they are de-activated by heat they are destroyed during washing at temperatures over 65°C. Some people have skin which is sensitive to these enzymes.

18.2.6 Cold water detergents

There has been much research into the development of washing powders which will operate at low temperatures to reduce the energy consumption during washing. Detergents have been developed which are effective at 25 to 30°C (so that in many areas some heating is required, but considerably less than that for the normal types of washing powder). The effectiveness of most detergents is considerably reduced if the washing temperature falls below 25°C (Mathews 1978).

18.3 WASHING MACHINES

18.3.1 Automatic washing machines

An automatic washing machine is one which can be programmed to carry out the full washing cycle for a batch of clothes. The washing sequence, which is carried out in a single tub, includes the following stages.

(a) *Filling* with water to the correct level. Machines are available for either cold fill only or hot and cold fill. (The latter can be more economic if a cheap source of hot water is available.) It is normal to have automatic washing machines plumbed in. The filling sequence is controlled by electrical solenoid valves (Fig. 5.23) and the filling level is detected by a pressure switch. During the filling sequence detergent is flushed into the tub from a dispenser which is situated either in a drawer in front of the machine or from a compartment in the door. With both types it is common to provide three compartments for pre-wash, main wash and conditioner. The latter is added to the wash during the rinsing process.

(b) *Heating* the water to the correct temperature selected by the programmer (see below), by means of an electrical immersion heater (2–4 kW). In the case of a hot and cold-fill machine a balance of hot and cold water is used to achieve the desired temperature.

(c) *Mechanised agitation* during the wash cycle. The length and severity of mechanical agitation may be programmed. There are two types of automatic washing machine: top loading and front-loading. Top-loading machines usually have a vertical drum with a central agitator which may be made of aluminium, stainless steel or thermosetting plastic. The agitator turns ½ to 1 full turn in alternate directions at the rate of 60 oscillations/minute. An agitator is shown in Fig. 18.3. Some top loading machines have a horizontal drum.

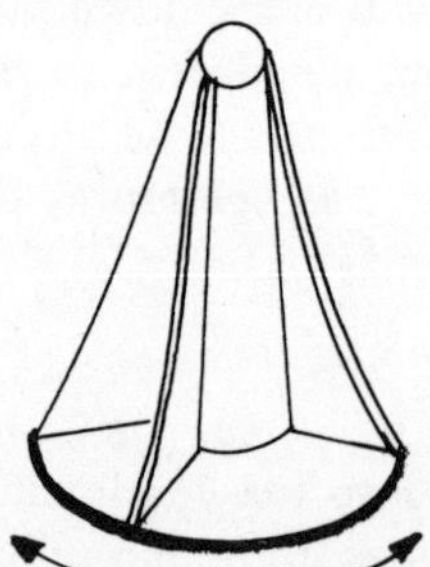

Fig. 18.3 – An agitator.

Front-loading washing machines have a horizontal perforated drum which may be made from stainless steel or enamelled steel. It has a number of baffles on the inside of the drum (Fig. 18.4) which lift the

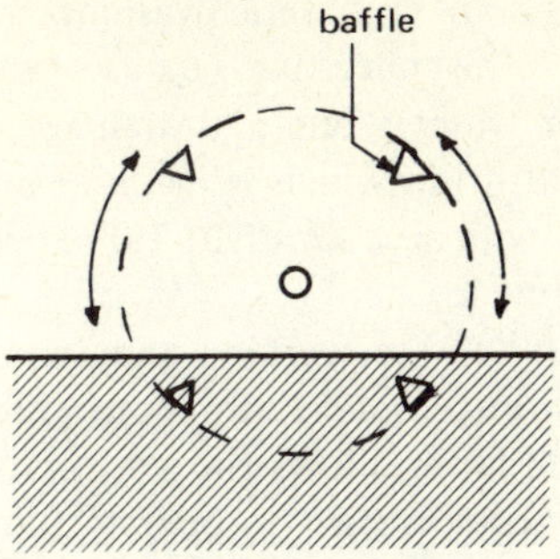

Fig. 18.4 – A horizontal perforated drum with baffles.

clothes out of the water as the drum rotates (at 50–60 rev/min). The clothes then tumble back into the water. On some machines the direction of rotation of the drum is reversed periodically.

(d) *Rinsing and emptying.* An automatic washing machine has a pump for drawing water out of the tub and discharging it, via a trap (section 6.7.2), to the drainage system. A number of rinses in hot or cold water (depending on the type of programme selected) follows.

(e) *Removal of water by spinning.* The drum rotates at high speed to remove water from the clothes by centrifugal action. Spin speeds vary from 400 to 1100 rev/min. High spin speeds result in much drier clothes than low spin speeds. This is important if the spinning is to be followed by drying in a tumble dryer (section 18.4). Clothes spun at 400 rev/min. take three times as long to dry as those spun at 1100 rev/min. An important factor during spinning, particularly with front-loading automatics, is the even distribution of clothes prior to spinning. Without this, severe vibration may result and the clothes will not dry evenly.

The washing sequence is controlled by a programmer. This gives control of time, temperature, water level, degree of agitation, number of rinses and spinning speed. The programmer may be an electrical clock which closes and opens electrical contacts as it rotates or it may be a solid-state controller. The latter has fewer moving parts and is, therefore, more reliable. Most machines offer a large number of programmes including those recommended by the HLCC and the International Wool Secretariat. Some machines use a microprocessor for control of the washing cycle. This allows a greater range of programmes and fault diagnosis.

Automatic washing machines may be free standing or built-in and may be front loading or top loading. Front-loading models are more convenient for built-in and under work-top applications and for stacking with a tumble drier. Top-loading models occupy a small floor area and are less prone to vibration during spinning. With both types, a load of washing weighs between 3 and 5 kg.

The outside cabinet of an automatic washing machine is made of steel panels coated with enamel. The inner tub is made of stainless steel or enamelled steel, with connections for water inlets and drainage. Inside this tub is a horizontal or vertical perforated drum which is made of stainless steel or enamelled steel. The tub has a door with a water-tight rubber seal. This door cannot be opened during the washing cycle.

A single motor is used for the washing and spinning action. Variation of speed is achieved either through split stator coils (Fig. 16.1) or by the use of electronic triac control. The latter allows the selection of a wide range of alternative speeds. A separate motor is used to operate the emptying pump. This is susceptible to blockage with lint, tissues etc. To prevent this some machines have a filter to protect the pump. A schematic diagram of a front loading washing machine is shown in Fig. 18.5.

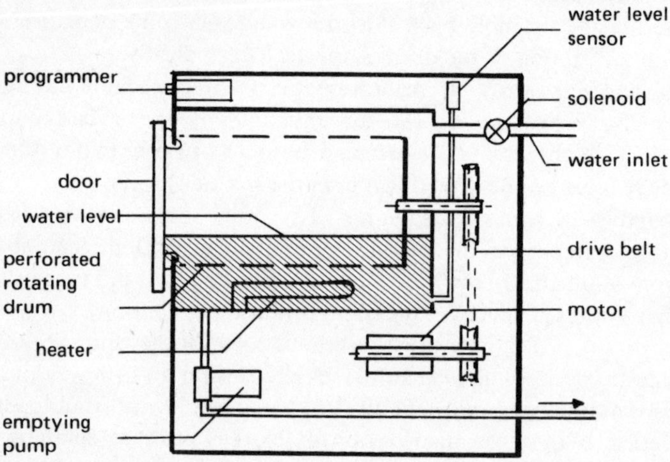

Fig. 18.5 – Schematic design of front-loading automatic washing machine.

The amount of water and energy used by an automatic washing machine is greater than that used by manual machines, although the actual amount used depends upon the programme selected and on the specific machine. Using programme 1, the Consumers Association (1980a) measured water consumption at between 50 and 200 litres and energy consumptions of between 1.3 to 3kW. Some machines now have low energy and low water consumption programmes.

Some automatic washing machines incorporate a tumble drier (section 18.4) in a single cabinet.

18.3.2 Non-automatic washing machines

A non-automatic washing machine has a number of advantages over an automatic one. It is both cheaper to buy and run and takes less time to wash a load of clothes. Against these factors must be balanced the disadvantage that the washing process must be controlled manually and more physical handling of the clothes is required.

A single tub washing machine has a single- or double-speed agitator (Fig. 18.3) and an electric heater of 2 to 3 kW. The machine has a pair of rubber covered rollers for squeezing water out of the clothes. These rollers may be rotated manually or by a reversible electric motor.

A twin-tub washing machine (Fig. 18.6) has one compartment for washing and a second compartment for spinning. The washing compartment has an agitator (Fig. 18.3) or a pulsator (a plastic disc with fan like blades rotating at 500 rev/min below the surface level of the water). Water is heated by a 2 to 3 kW heater with thermostatic control.

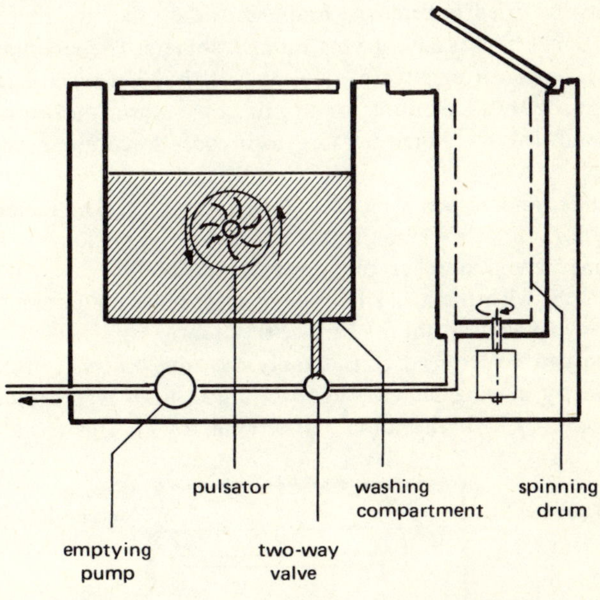

Fig. 18.6 – A twin-tub washing machine.

The second compartment has a horizontal perforated spinning drum which revolves at between 2500 and 3000 rev/min. Clothes can also be rinsed in this compartment. Water extracted from the clothes is removed from the compartment by a pump. (The same pump may also be used for emptying the washing compartment.) The lid of the spin dryer has a link to a microswitch, which switches off the spin motor when the lid is opened. The lid also operates a brake to stop the spin drier.

Independent spin driers are available which operate in the same way as the spin-drying compartment of a twin-tub. Models are available which hold between

2 and 3 kg of clothes. They have a pump for discharging the extracted water into a sink.

18.4 DRYING MACHINES

Clothes are dried by suspending them in a stream of dry warm air. The rate at which clothes dry depends on the following factors:

(a) The difference between the RH of the air and the ERH of the clothes (section 11.3). Since warm air can hold more water vapour than cold air, warm air is much more effective for drying than cold air.
(b) A moving stream of air accelerates the rate of evaporation from the surface of a fabric since it moves away the layer of moist air in contact with it.

Clothes may be dried outside the house providing that weather conditions are favourable. Some houses have drying rooms/cabinets. These contain a source of heat, such as a radiator or a warm-air vent and they have ventilation to outside the building. With all methods of drying, the warm moisture-laden air should be removed from the house because, as it cools down, there is a danger of condensation.

Small drying cabinets are available. These consist of an enamelled metal case with heaters (1 to 3 kW) in the base and ventilation louvres in the top. Inside the cabinet are a number of adjustable racks for hanging clothes to ensure good contact with the air. This type of drier relies on natural convection currents for removing water from the surface of a fabric.

Drying time can be reduced considerably by using forced convection (section 11.2.2) and by moving the clothes to produce more even drying. Both of these principles are used in the tumble drier (Fig. 18.7). This has a horizontal

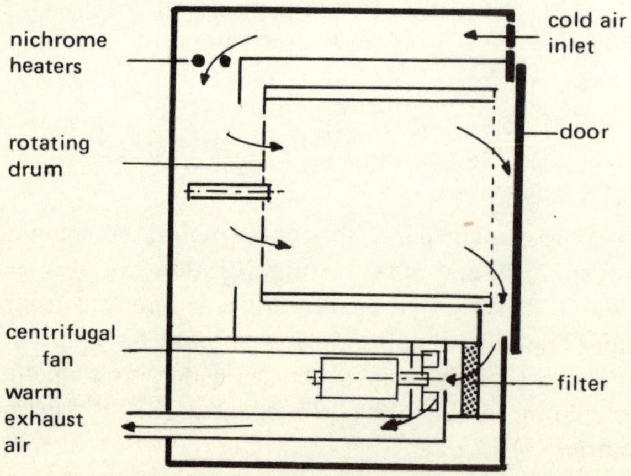

Fig. 18.7 – A schematic diagram of a tumble drier.

rotating drum with baffles. It is very similar to the drum of a front-loading automatic washing machine (Fig. 18.4). It rotates at 50 to 60 rev/min to produce a tumbling action in the clothes. On some machines the direction of rotation is changed periodically to increase the evenness of drying. The access door on the front of the machine has a microswitch which stops the rotation of the drum when the door is opened. In the end of the drum away from the door are holes through which hot air is blown. A single electrical motor is used to rotate the drum and to power a centrifugal fan. This fan draws cold inlet air over nichrome wire heating elements (2 to 3 kW) and then through the drum. Close to the heating elements is a bi-metallic strip thermostat to protect the elements from overheating should the fan fail to operate. A second thermostat in the drying compartment is used to control the air temperature. This may be a continuously variable thermostat or it may just have two heat settings (high and low). Once the air has passed through the tumbling clothes it is discharged, through a filter to trap lint, and then out of a vent. It is preferable to connect this vent to ducting which removes the warm moist air from the house.

In addition to the temperature control, a drying control is also provided. This may be:

(a) an electrical timer (relying on the experience of the operator to know how long drying will take);
(b) a temperature sensor in the exhaust air (this rises as the clothes dry);
(c) an electronic humidity sensor in the exhaust air.

The drying period is usually followed by a period of tumbling using cold air. This minimises the creasing of clothes.

The time and cost of drying depends on the size of load, the type of fabric and the water content of the wet clothes. A 5 kg cotton load which has been spun at 1100 rev/min takes one hour to dry and uses 2 to 3 kWh of electricity (Consumers Association 1980b). A load of synthetic fabrics will only consume half as much energy.

18.5 IRONS

There are three main types of iron: dry irons, steam irons and rotary irons.

18.5.1 Dry Irons

Dry irons have a thick metal sole plate which may be made of aluminium, stainless steel or chrome-plated steel. Some irons have a non-stick PTFE coating on the sole plate. On the upper side of the sole plate is clamped either a nichrome or a sheathed heating element (section 5.7.1). The sole plate also has a bi-metal type thermostat which is linked to a control dial on the top of the iron (Fig. 18.8). This control dial may be marked by temperature or with ITCL or HLCC

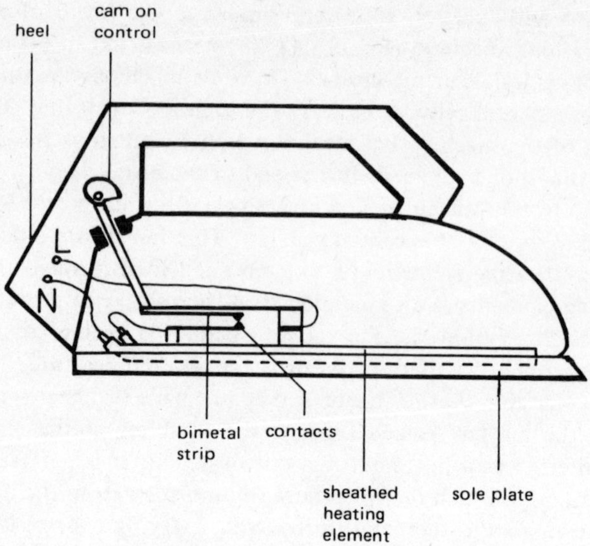

Fig. 18.8 – Dry iron.

settings. An indicator light is often used to show when the iron has reached the desired temperature. The heel of the iron is designed to act as a stand. The weight of the iron is 1 to 1.5 kg; light 'travel' irons weigh less than 1 kg.

18.5.2 Steam Irons

Steam irons generate steam at a temperature of 110°C which passes through holes in the sole plate. The steam condenses on the fabric. A reservoir is used to hold 200 to 300 g of distilled water. There is a switch on top of the iron which, when in the 'steam' position allows water to drip into a steam chamber. When the drops of water hit the sole plate they evaporate immediately (Fig. 18.9). Many steam irons also have a spray nozzle in the front of the iron. This is used to spray cold water onto a fabric which is too dry to iron.

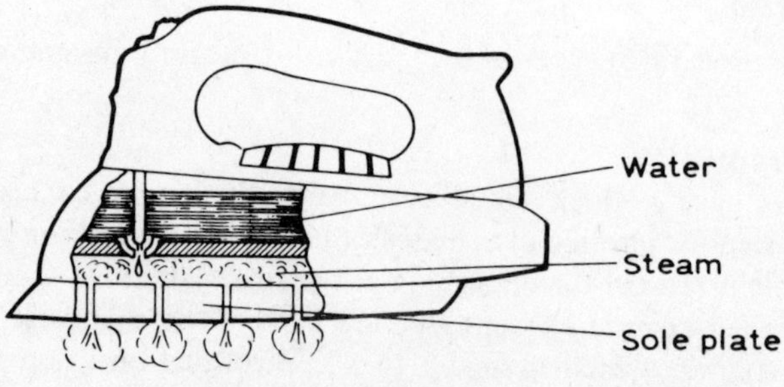

Fig. 18.9 – Steam iron.

18.5.3 Rotary Irons

Rotary irons consist of a curved polished chromed steel ironing plate. Fabrics to be ironed are held against this by a fabric covered roller driven by an electric motor. The heating element is rated at 2.5 to 3 kW and the fabric is drawn through the iron at 2.5 to 3 m/s. Rotary irons may fit on a work bench or they may have a floor stand.

REFERENCES

Collier, A. M. (1980), *A handbook of textiles,* Wheaton.

Consumers Association (1980a), Automatic Washing Machines, *Which,* **March,** p. 193-201.

Consumers Association (1980b), Tumble driers, *Which,* **Nov.,** p. 644-7.

Grandjean, E. (1973), *Ergonomics of the Home,* Taylor & Francis.

Mathews, W. (1978), *Watt Committee Report No. 3,* Appendix D.

BIBLIOGRAPHY

Consumers Association (1979), Steam irons and travel irons, *Which,* **Sept.,** p. 487-9.

Consumers Association (1980), Washing powders, *Which,* **Aug.,** p. 431-4.

Consumers Association (1981), Automatic Washing Machines, *Which,* **April,** p. 249-257.

Consumers Association (1981), Irons, *Which,* **June,** p.825-830.

Consumers Association (1981), Twin Tubs, *Which,* **Aug.,** 478-9.

Swaine, D. (1976), The cold wash, *Home Economics,* **22,** 20-21.

CHAPTER 19

Floor cleaning and polishing appliances

19.1 INTRODUCTION

Floors become dirty for a number of reasons. Dirt is carried into a house on shoes and clothing and is subsequently deposited on the floor. Any materials which are dropped or spilled fall to the floor. Even in a house which is not being lived in, dust particles from the air gradually fall and are deposited. Because of these reasons, floors require frequent cleaning. The best method and frequency of cleaning depends both on the rate of soiling and upon the type of floor covering:

(a) wood floors require sweeping (frequently), polishing (periodically) and re-sealing (occasionally);

(b) thermoplastic floors require sweeping (frequently) and washing (periodically);

(c) carpets require sweeping (frequently), stain removing (as necessary) and shampooing (periodically).

From this it can be seen that sweeping forms the most frequent of the floor cleaning activities. Sweeping is relatively easy to carry out manually on hard surfaces such as wood and thermoplastic. It is difficult to carry out manually on carpets where the dirt is held in the pile. Because of this mechanical carpet sweeping machines have been in use for many years, culminating in the modern suction (vacuum) cleaner.

19.2 PRINCIPLES OF CARPET CLEANING

Although there are many different types of carpet, there are two main methods of manufacture. The pile may be woven into the backing fabric at the same time as the latter is woven. Alternatively, the pile may be inserted into a previously woven backing fabric and held in place with latex. Whichever of these methods is used, all carpets have the same basic structure with a horizontal woven backing and vertical pile, as shown in Fig. 19.1. The pile, which may be of wool, nylon,

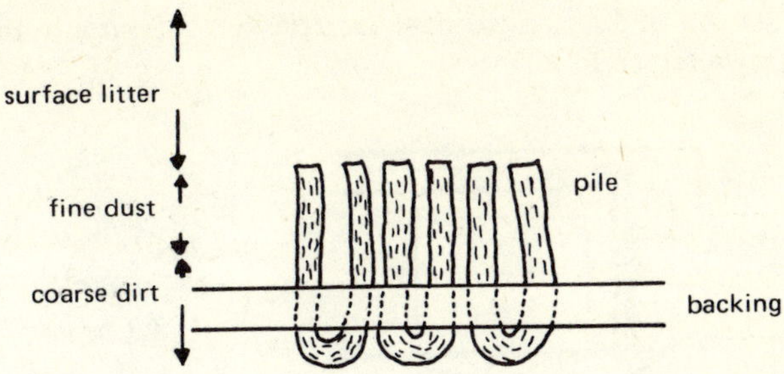

Fig. 19.1 – Basic structure of a carpet showing location of dirt.

acrylic or blends of these, varies in density and height (3 to 12 mm). The backing may be made of hessian, jute, cotton, rubber or expanded polyurethane. It is normal to have an underlay between the carpet and the floor which may be made of felt or expanded polyurethane. This acts as a soundproof layer and reduces wear of the carpet. In additon, it prevents dust from rising into a carpet through the boards of a supended floor. Latex backed carpets may have an underlay of thick paper to prevent them sticking to the floorboards.

Carpet dirt consists of a number of different types of material varying in size, density and nature (Fig. 19.1).

(a) *Surface litter* is usually material which is too large to sink into the pile. It remains on the surface (for example, thread, paper, pins etc.) from where it can be removed by suction and/or brushing.
(b) *Dry particulate dust* which varies in size and density. Fine dust particles collect in the upper layers of the pile. Dense material, such as grit, sand and clay sink to the base of the pile unless they are removed. They cause damage to the pile because their sharp edges cut the pile under pressure from user's feet. The pile of a carpet can hold a large volume of dry particulate dust. It can be removed by suction or by mechanical action (beating) or by a combination of the two.
(c) *Sticky materials* result from spillage. Dust particles stick to these materials, making the dust difficult to remove. If localised, sticky materials can be removed by sponging. More generalised soiling with materials must be removed by shampooing the carpet.
(d) *Staining* can be removed by localised treatment or by shampooing.

From the above discussion it can be seen that a mechanical carpet cleaner which is to remove dust from the pile can use either suction or mechanised agitation (beating) or a combination of both. Surface litter can be removed by suction or brushing or a combination of the two. Suction is generated by a centrifugal fan. This consists of a disc with curved vanes which rotates at high speed (Fig. 19.2). Air and entrained dust particles are sucked into the centre of the disc and thrown out to its edges.

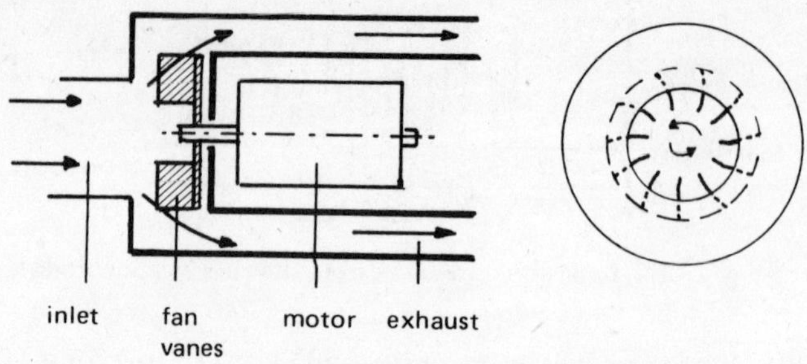

Fig. 19.2 – Centrifugal fan.

A mechanical brushing action can be produced by using brushes mounted on a roller which is in contact with the carpet. Agitation or beating can be incorporated by having metal bars on the roller in place of the brushes. On some cleaners there is a combination of brushes and beaters on a single roller as shown in Fig. 19.3.

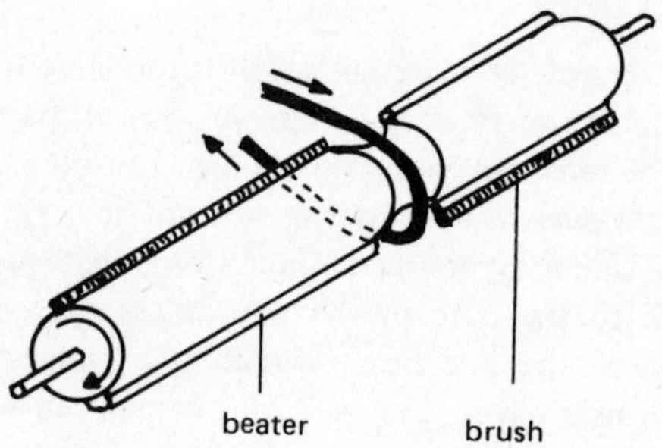

Fig. 19.3 – Combined beater and brush roller.

The beater is used in combination with suction from a fan. It loosens the dust and dirt from the pile and backing so that, once loosened, it may be drawn out of the carpet by suction.

The stages involved in cleaning a carpet are shown in Fig. 19.4. Dirt and dust are removed from the carpet using suction alone or suction combined with sweeping and beating. The dirt is carried along in the air through a filter which separates it from the air. In some cleaners the separation of dust/dirt and air takes place before the air passes through the fan. In others the air is passed through the fan before it is filtered. The filter may be made of fabric or it may be made of paper (in the form of a bag) which can be discarded once it is full.

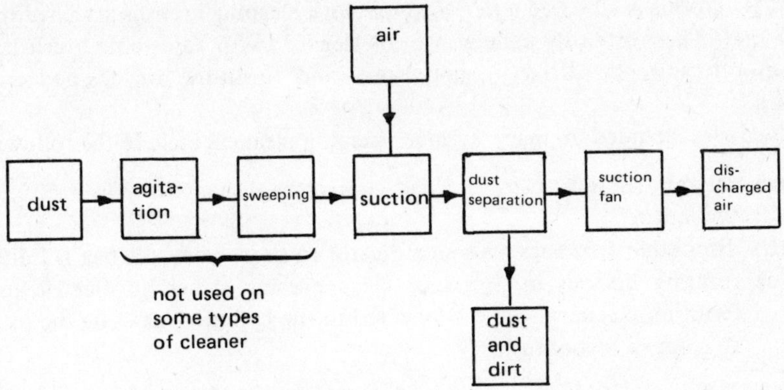

Fig. 19.4 – Stages involved in mechanical carpet cleaning.

19.3 TYPES OF SUCTION (VACUUM) CLEANERS

19.3.1 Cylinder and tank cleaners

Both of these cleaners have the same basic structure and method of cleaning. A cylinder cleaner is shown in Fig. 19.5. The body of the cleaner contains a powerful motor (600–1000 W), a dust container/air-filter, cable holder and

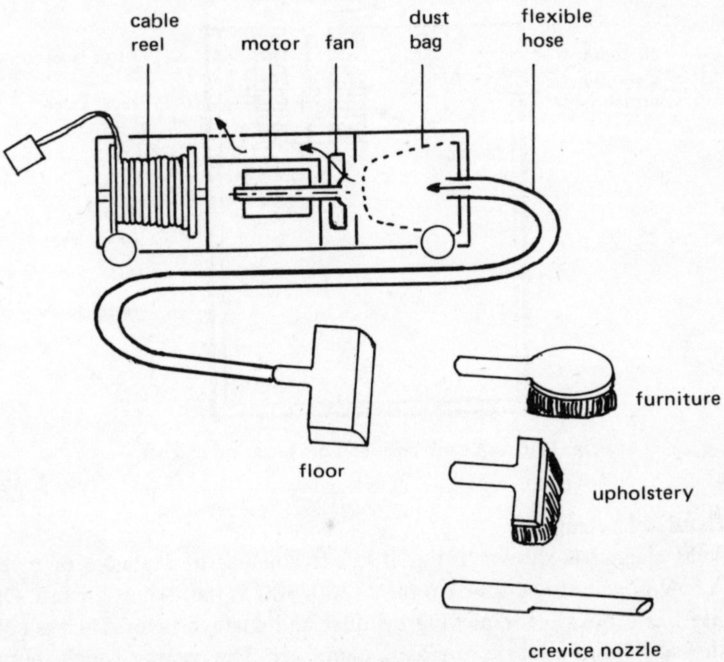

Fig. 19.5 – Cylinder suction cleaner.

controls. To this is attached a flexible hose with cleaning utensils attached to the other end. These utensils include a floor cleaner (with adjustable brush height for smooth and pile surfaces), upholstery and furniture brushes and crevice nozzles.

Features included on many cylinder vacuum cleaners include the following:

(a) variable suction, using a Triac control to adjust the power output of the motor;
(b) disposable dust bags with an indicator to show when the bag is full;
(c) rotating brushes in the floor cleaner using either an electric motor (with an electrical cable connected to the cylinder) or using the suction to rotate a turbobrush.

Tank cleaners (Fig. 19.6) are similar to cylinder cleaners except that the motor is mounted in the vertical position. Some tank cleaners have a receptacle which can hold water and other liquids. It can be used to suck water from floors, to unblock sincks etc. A float valve is used to switch off the motor when the tank fills with liquid. Models are available with tank capacities of 20 to 50 litres.

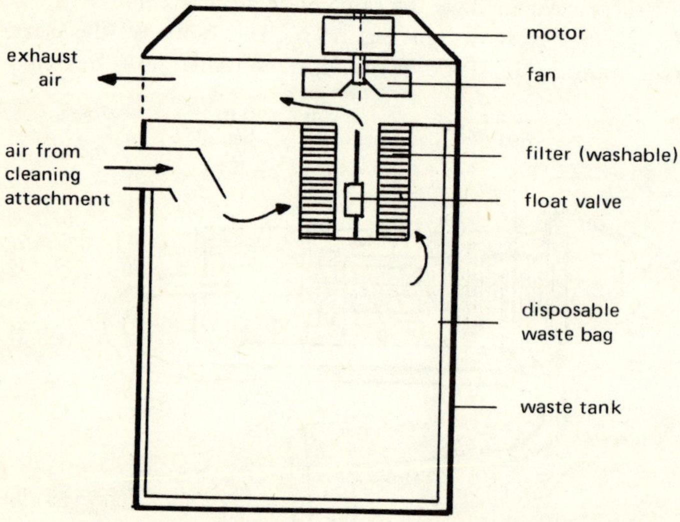

Fig. 19.6 – A tank (wet and dry) suction cleaner.

19.3.2 Hand-held cleaners

A hand-held cleaner is shown in Fig. 19.7. It consists of a plastic or metal case holding a low-power single-speed motor (200–300 W) and suction fan. One end of the case has a nozzle for picking up dust and surface litter. Dust is collected in a cloth bag held in a fabric or plastic container. The cleaner weighs between 2 and 3 kg. This type of cleaner is particularly useful for cleaning upholstery.

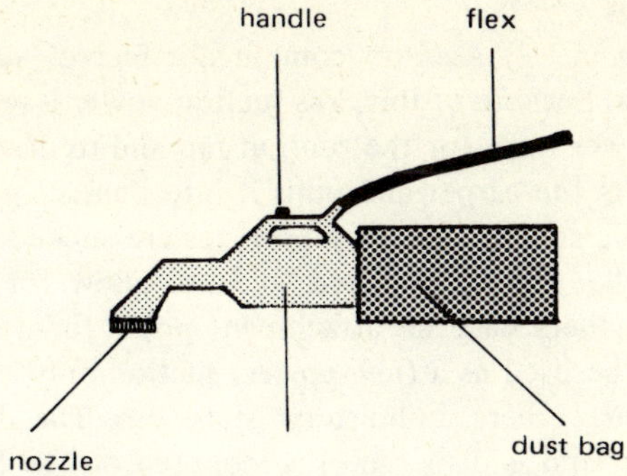

Fig. 19.7 – Hand-held suction cleaner.

19.3.3 Stick cleaner

The suction power of a stick cleaner is intermediate between that of a cylinder and a hand-held cleaner. The motor (300–600 W), fan and dust collector are mounted on a 1½ m long stick with a handle at one end and suction nozzle at the other (Fig. 19.8). Most models have a disposable dustbag. It is operated with two hands rather like a sweeping brush. The cleaning head has no wheels so that the weight (3 to 4 kg) must be supported by the arms. As with the cylinder cleaner, various cleaning nozzles are available.

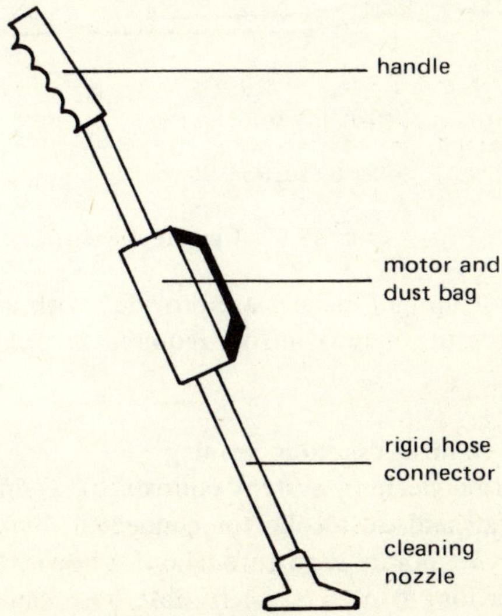

Fig. 19.8 – Stick suction cleaner.

19.3.4 Upright cleaner

Upright cleaners (Fig. 19.9) use the combined effect of suction, mechanical agitation and brushing. Because of this, less suction power is required. A 250–400 W motor provides power both for the suction fan and to drive the brush/beater roller. The suction lifts the carpet and pulls it into contact with the roller. The latter loosens the dust and dirt so that particles are sucked into the machine. The height of the roller can be adjusted so as to allow for varying depths of pile. Most upright cleaners have an attachment point for a suction tube. This allows the cleaner to be used as a (low-power) suction-only cleaner with appropriate attachments for corners, upholstery, stairs etc. The dust bag (which is usually made of paper and is disposable) is mounted on the handle in a plastic or fabric container. Many machines have a visible or audible warning to indicate when the dust bag is full.

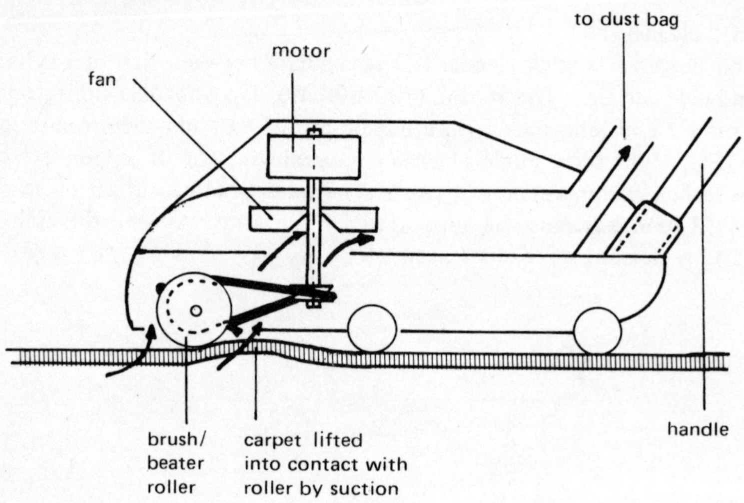

Fig. 19.9 – Upright cleaner.

Some models of upright cleaner are provided with a motorised drive to the wheels. This reduces the manual effort required in cleaning large expanses of carpet.

19.3.5 Centralised (built-in) suction cleaning

A centralised suction-cleaning system consists of a single suction unit with electrical motor, fan and dust collector connected, by a network of ducting, to a number of outlet points sited throughout a house. The outlet points have fittings to which a long (up to 8 m) flexible hose can be attached. This has cleaning fittings similar to those of a suction cleaner. Such a system avoids the necessity of carrying the whole vacuum cleaner from room to room.

19.4 FLOOR POLISHERS AND CARPET SHAMPOOERS

An electric floor polisher or carpet shampooer is similar in appearance to an upright suction cleaner. It consists of an electric motor (200–300 W) connected through a gear-box to two or three rotating polishing/shampoo heads. Various circular brushes can be attached to these heads for scrubbing, polishing and applying polish or shampoo. Some machines have a dispenser for liquid polish which is slowly released onto the polish-spreading heads. Shampooers have a tank on the handle which holds the detergent solution and slowly releases it to the cleaning brushes.

BIBLIOGRAPHY

Consumers Association (1978), Vacuum cleaners, *Which,* **Jan.**, p. 4–7.
Consumers Association (1978), Cylinder vacuum cleaners, *Which,* **June,** p. 361–3.
Consumers Association (1981), Vacuum cleaners, *Which,* **April,** p. 210–215.
Edwards, J. K. P. (1972), *Floors and their maintenance,* Butterworth.
Peet *et al.* (1975), *Household equipment,* Wiley.

Part V

SAFETY IN THE HOME

CHAPTER 20

Safety

20.1 ACCIDENTS IN THE HOME

Many accidents occur in the home. These vary from minor accidents, resulting in cuts and abrasions, to serious accidents, resulting in death. Because there is no formal system for reporting domestic accidents, accurate statistics are only available for accidents which result in medical treatment. The best statistics available are for accidents resulting in death. It has been estimated that for every accident leading to death, there are 150 to 200 accidents resulting in serious injury (Grandjean 1973).

Deaths from accidents in the home are of a similar magnitude to road accidents and much higher than from accidents at work (Table 20.1). Accidental deaths in the home are not evenly distributed amongst the population. Infants up to the age of four and elderly people are more at risk than the intermediate age group (Thompson 1980). Accidents to people over the age of 65 account for 62% of all accidental deaths in the home.

Table 20.1

Accidental deaths in the UK in 1978
(From Thompson 1980)

Site of accident	Percentage of accidental deaths
Road	53.6
Home	42.7
Work	3.5
Other	0.2

The most common causes of accidental death in the home are falls, fires, poisoning and suffocation. Figures for the UK for 1977 are given in Table 20.2. Equivalent data for other countries are given by Grandjean (1973). These figures show a similar pattern to those for the UK, with falls as the most common cause of death followed by fire and then poisoning.

Table 20.2
Causes of accidental death in the home (UK 1977)
(From Thompson 1980)

Cause	Percentage of deaths
Falls	65
Fires	11
Poisoning	14
Suffocation	10
Other	10

20.1.1 Falls

Falls are the most common cause of accidental deaths in the home and also result in a large number of cuts, bruises and fractures. The elderly are particularly susceptible to injury from falls. Many falls result from slipping on wet, greasy or polished floors and on irregular floors and steps. Less frequent are falls from a height such as from windows and from ladders. Precautions against this type of accident are outside the scope of this book.

20.1.2 Fires, burns and scalds

Many minor and serious accidents occur in the home involving burns and scalds and the kitchen is the most common site of these accidents. Children are frequently injured by pulling hot liquids onto themselves as a result of reaching up to hold pan handles and overhanging electrical flexes. Burns also result from people's clothing catching fire in front of an open radiant fire. Other causes of burns include the use of inflammable liquids, people falling asleep whilst smoking in bed and children playing with matches.

Serious fires resulting in the need to call out fire-fighting services are less common but result in considerable damage to property and a large number of deaths every year. In 1976 almost a third of all domestic fires in the UK were caused by cooking appliances (Thompson 1979). This is further illustrated by the fact that the commonest time of day for domestic fires coincides with mealtimes (11.00 a.m. to 2.00 p.m. and 4.00 p.m. to 7.00 p.m.). Space-heating appliances cause 10% of domestic fires.

Damage to property by fire may result from explosions caused by leaking gas, which is ignited by an electrical or mechanical spark. Short-circuits in electrical wiring and equipment may also cause fires. Spillage of paraffin and oil from liquid-fuel heaters or burning solid fuel from solid-fuel appliances can also cause home fires to start.

20.1.3 Poisoning with toxic gases

The number of deaths in the UK resulting from poisoning by gas fell with the introduction of natural gas. NG does not contain carbon monoxide (as did town gas) and is not toxic. In 1951 11% of all accidental deaths in the home were caused by gas poisoning but this fell to 4% in 1976. Death involving unignited natural gas is still possible but, since it results from suffocation, requires a much higher concentration of gas in the atmosphere.

A common cause of accidental death is inadequate ventilation (see section 11.2) or ineffective flues (section 2.7) for fossil-fuel burning appliances.

20.1.4 Poisoning with household chemicals and medicines

In the UK 600 deaths a year result from poisoning by household chemicals (Consumers Association 1979), medicine and garden insecticides and fungicides. Children are partickulularly susceptible to this type of accident. Accidents frequently occur because household and garden chemicals are stored in soft-drink bottles.

20.1.5 Suffocation and choking

In the UK 450 deaths a year result from suffocation and choking (Consumers Association 1979). There are many causes of these accidents and children are particularly susceptible.

20.1.6 Electric shocks

Every year in the UK 65 deaths result from electrocution (Consumers Association 1979). The most common causes of these accidents are ineffective earths and fuses, loose plugs and fittings and frayed flexes. Many accidents occur when repairs are attempted on electrical appliances without isolating the appliance from the mains.

20.1.7 Do-It-Yourself

The increase in DIY activities in the home, particularly in major projects such as painting externally, roof repairs, extensions and the like has increased the home accident rate due to inexperience with ladders, electrically operated power tools etc.

The cost of home maintenance has forced many people to resort to DIY without the necessary experience and without a proper awareness of safety precautions. The use of 'instant' adhesives has resulted in hospital treatment, such as the releasing of fingers which have been stuck together and the patching up of torn skin which has been damaged when trying to release an article which has become stuck to it. Another example of accidents caused by DIY is that of fire caused by the use of petroleum based adhesive. The fumes, which are heavier than air, may be ignited by static electricity, pilot lights on gas appliances or by cigarettes. A number of serious fires have been started in this way.

20.2 THE PREVENTION OF ACCIDENTS

Most accidents do not usually have a single cause but rely on a combination of errors. By eliminating one of these errors accidents can be prevented. Factors involved in accidents include:

(1) the design of the home;
(2) the maintenance of the home;
(3) the design of domestic appliances;
(4) the maintenance of domestic appliances;
(5) human error.

20.2.1 The design and maintenance of a house

The structure and design of a building can make a large contribution to safety. Structural aspects must be considered in the choice of a home for those families with young children or for the elderly since these two groups are particularly at risk. A discussion of this type of design feature is outside the scope of this book (see Grandjean 1973).

Electrical supplies should be designed and installed to meet national standards, many of which are concerned with safety. Particular attention should be given to the provision of correctly fused circuits (section 5.4.2) and earthing (section 5.4.3). Earth-leakage circuit breakers are an additional safety factor. Socket outlets, particularly those at ground level, should be shuttered safety sockets. These prevent infants inserting small objects into the socket. Safety factors are also important in the design of liquid fuel heating systems (section 3.7).

An important aspect of the design of a kitchen (section 17.4.4) is the location of the oven and the provision of work surfaces near the oven for placing hot dishes down immediately after removal from the oven.

The lighting levels, particularly for staircases, corridors and kitchens, can contribute to safety (section 12.3). Low levels of lighting contribute to falls and burns.

The maintenance of a house is of equal importance to safety. In old houses the electrical wiring often becomes dangerous because the insulation of the rubber perishes. This introduces the double hazard of electric shocks and fires. Old wiring should be checked and replaced before this situation can develop. As a guide wiring which is over 30 years old should be checked by a competent electrician. An additional problem with electrical wiring in old houses is that the number of sockets provided is insufficient for modern needs. This often leads to the use of adaptors which can take several plugs, leading to an overloading of wiring and circuits. Any overheating resulting from this can lead to a fire hazard. Where there is a problem additional circuits should be provided to give a greater number of sockets. Checks should be made so that each circuit has the correct rating of fuse. The practice of replacing fuses with higher rating of fuse or with wire is very dangerous because of the increased risk of fire and electric shock. In particular the operation of an earth depends upon the correct functioning of a fuse (section 5.4.3). The earthing itself should be checked since the replacement of metal pipes by plastic ones may have impaired its effectiveness.

20.2.2 The design and maintenance of domestic appliances

The safety of new domestic appliances are covered by national standards although in many cases testing is voluntary. The purchaser of an appliance should make certain that it conforms to the appropriate national standards and that it is supplied with the correct type of fuse.

Many safety features are incorporated into new appliances such as flame-failure devices (section 2.9), safety locks on washing machines (section 18.3), spin driers (section 18.4), microwave ovens (section 13.2.2.8) and self cleaning (pyrolytic) ovens (section 13.2.2.5). As a safety feature, portable paraffin room heaters (section 3.7) must be self extinguishing if they are knocked over. Ovens are designed so that the outside surface temperature presents no fire hazard when built into or stood next to wooden kitchen furniture.

From a safety point of view, maintanence of domestic appliances is very important. In the case of fossil-fuel burning appliances regular checks should be made on the ventilators and flues because of the danger of toxic fumes entering the room. Similarly electrical appliances should be serviced regularly. In particular damaged plugs and flexes should be replaced. Gas appliances and oil-fired boilers should be regularly serviced because of the risk of fires and toxic fumes. Gas appliances should be under maintenance contracts; if over 3 years old they should be checked annually.

20.2.3 Human error

Even in the best designed house with new appliances which have all the latest safety features, accidents can still occur because of human error. Many of these accidents occur because of a lack of understanding of the real hazards and dangers which exist in the home. Also many accidents are made worse because of the lack of training in first-aid and by the absence of first-aid kits, fire extinguishers etc. In general there is much more awareness of safety hazards at the place of work than there is in the home. This is in spite of the fact that many more accidents occur at home than at work. More safety education and training would lead to a reduction in home accidents.

REFERENCES

Consumers Association (1979), Home Safety, *Which,* **March,** p. 193–7.

Grandjean, E. (1973), *Ergonomics of the home,* Taylor & Francis.

Thomson, E. J. (Ed.) (1979), *Social Trends No. 9,* HMSO.

Thompson, E. J. (Ed.) (1980), *Social Trends No. 10,* HMSO.

Index